F.A.C.T.S.
Flexibility, Affordability, Conservation, Transition, Sustainability

F.A.C.T.S. Flexibility, Affordability, Conservation, Transition, Sustainability

Our Path to the Future of Energy and the Environment

M. Reza Fassihi
Beyond Carbon, LLC, Houston, TX, United States

Hamidreza Samouei
Texas A&M University, Brine Refining LLC,
College Station, TX, United States

ELSEVIER

Elsevier
Radarweg 29, PO Box 211, 1000 AE Amsterdam, Netherlands
50 Hampshire Street, 5th Floor, Cambridge, MA 02139, United States

ISBN: 978-0-443-52574-2

For Information on all Elsevier publications visit our website at https://www.elsevier.com/books-and-journals

Publisher: Priscilla Braglia
Acquisitions Editor: Fran Kennedy-Ellis
Editorial Project Manager: Somoshri Banerji
Production Project Manager: Paul Prasad Chandramohan
Cover Designer: Raman Kumar

Typeset by Aptara, New Delhi, India

Dedication

To Terry, Manny, Sammy, and Willow and
To the teachers who taught us to think critically, stay unbiased, and
serve future generations responsibly.

ابر و باد و مه و خورشید و فلک در کارند

تا تو نانی به کف آری و به غفلت نخوری

The clouds, the wind, the moon, the sun, and the heavens all labor
So that you may earn your bread and not eat it negligently.

***Saadi**, Persian poet, 13th century*

"Everyone is entitled to his own opinion, but not to his own facts."

***Daniel Patrick Moynihan**, the U.S. Senator from New York (1927–2003)*

Contents

Author biographies

Dr. M. Reza Fassihi is a BHP Distinguished Advisor Emeritus and Founder of Beyond Carbon, LLC, advising on reservoir management, carbon capture and storage, and low–carbon energy technologies. With more than 45 years of experience in the energy industry across major energy companies, he has led many technical and engineering teams with a focus on energy extraction, technology development, and field applications. His most recent focus areas have been in situ hydrogen generation, CO_2 sequestration, geothermal energy, and mineral extraction. He is widely recognized for his influential publications (more than 50 papers) and two books, and has received numerous honorary awards. Reza holds a BSc degree in General Engineering from Abadan Institute of Technology (AIT), Iran, and a PhD degree in Reservoir Engineering from Stanford University, CA, United States.

Dr. Hamidreza Samouei is a chemist and energy-water researcher at Texas A&M University, Texas, United States, and the founder of Brine Refining LLC. His research focuses on converting high-salinity waste streams, including oil and gas produced water, desalination reject brines, geothermal brines, and other industrial brines, into valuable resources by recovering critical minerals, producing reusable water, and sequestering CO_2 into stable carbonates. Samouei has contributed to advances in environmentally sensitive chemistry, separation and crystallization strategies, and carbon-enabled brine refining. He collaborates broadly with academic and industrial partners and translates research into practice through patents and technology development.

Foreword

The global energy system is experiencing a fundamental transformation. The imperatives of decarbonization and sustainability, enhancing fossil fuel-based energy to be a low-carbon and promoting other sources of energy such as nuclear and renewable alternatives, present one of the most difficult engineering and policy challenges of the 21st century.

This book provides a technical and systems-level overview of the energy transition, examining various pathways, technologies, and infrastructures shaping the future of energy. It addresses key components that influence investments and integration of various energy sources, grid modernization, energy storage, the added consumption of data centers, and policy requirements. It also considers lifecycle assessments, energy system modeling, and the role of demand-side management.

Examining the crossroads of opportunity and challenge, this book aims to illuminate the path forward for individuals, communities, and nations grappling with the complexities of sustainable energy sources and reshaping our world for future generations.

This book explores innovative energy solutions, highlights pioneering case studies, and provides a comprehensive analysis of the economic, social, and environmental implications of our choices. The scientific and engineering foundations are emphasized throughout the book. But it also acknowledges the interdisciplinary nature of energy transition. markets, regulatory frameworks, and social acceptance that play essential roles in shaping outcomes. Accordingly, technical discussions are contextualized within broader economic, environmental, and geopolitical frameworks. It advocates to policymakers and business leaders, as well as to everyday citizens, to foster a collective consciousness that emphasizes responsibility, collaboration, and creativity.

In summary, the book provides a rigorous yet accessible foundation for students, energy engineers, researchers, and professionals working in or entering the field of energy systems. By considering current knowledge and highlighting ongoing challenges and innovations, the book is intended to equip readers with tools to engage critically and constructively with all fields of energy systems.

Iraj Ershaghi, Ph.D., NAE, P.E.
Omar B. Milligan Chair, Petroleum Engineering and
Director, Petroleum Engineering Program,
The Mork Family Department of Chemical Engineering and
Materials Science
Executive Director, CiSoft and
Executive Director, E-CET
Viterbi School of Engineering,
University of Southern California
Los Angeles, CA, United States

Preface

M. Reza Fassihi

During my recent trip to Antarctica, I experienced a profound sense of awe (Fig. P.1). Witnessing the colossal icebergs adrift in pristine waters, breathing the crisp, clean air, and observing penguin colonies undisturbed by human presence, I was captivated by the tranquility and untouched beauty of this unique environment. The enforced zero-waste policy and the requirement to adhere to designated pathways served as a clear reminder of the fragility of this pristine landscape. I could not help but ponder why we cannot strive to replicate this level of environmental stewardship on the rest of our planet.

FIG. P.1 An iceberg in Antarctica.

Throughout my career in the energy industry, I have endeavored to educate the public about the complexities of energy production via my publications and speeches. Initially, my concerns about the environment lacked focus since that was not a priority in the companies where I was employed. However, as time went on, my employers allowed me to learn more about environmental sustainability through contact with the experts. As a result, I became increasingly more vocal about our responsibility to protect this planet for our future generations.

This concern culminated in my first book [1], a textbook for advanced petroleum engineering students, where I described strategies for reducing the carbon footprint within the petroleum industry. Then, I focused on carbon sequestration and renewable energy activities. Hence, my second book focused on carbon dioxide storage technologies, also written for engineers and geoscientists [2]. Upon retirement, I started my own company, *Beyond Carbon LLC,* to focus more on promoting energy and environmental sustainability. Through this journey, I have come to realize that the successful implementation of any greenhouse gas reduction initiative hinges on public understanding of energy production fundamentals and their associated environmental consequences. This realization ultimately led me to conceive this book and write it in a simple language for all readers who want accurate information about our energy options, environmental implications, and actions they can take.

Hamidreza Samouei

My research is centered in the energy sciences, with emphasis on CO_2 utilization, brine mining and refining, and water and wastewater treatment for resource recovery. Over the years, through conversations with the public, I have observed a persistent gap in understanding of energy and environmental sustainability. This gap can allow misinformation to shape decisions that are not always made with long-term consequences in mind.

As educators and practitioners, we share a responsibility to equip the next generation with clear, balanced knowledge about the benefits and limitations of all energy pathways, as well as practical approaches to reducing environmental impact. This conviction motivated me to contribute to this book. I aim to offer accessible, unbiased insights into energy, sustainability, and resource recovery, presenting both the challenges and the opportunities with clarity and care.

The key features of this book

- Comprehensive coverage: The book offers a holistic overview of the energy landscape, covering both traditional and renewable energy sources. It also presents specific case studies on different aspects of energy transition.
- Independence of chapters: Each chapter can be read independently, allowing readers to focus on specific areas of interest.
- Practical insights: The book provides practical insights into the challenges and opportunities associated with the energy transition.
- Action-oriented approach: The book empowers readers with knowledge and tools to become active participants in shaping a sustainable energy future.
- Clear and concise writing: The content is written to be accessible to a wide audience.

By providing a comprehensive overview of the energy landscape and offering actionable insights, this book equips readers with the knowledge and tools to navigate the complexities of the energy transition and contribute to a more sustainable future.

How to use this book

This book has been meticulously crafted to engage and inform a diverse audience, recognizing the unique perspectives and interests each group brings to the evolving energy landscape. For the academic community and the general public, it functions as a multifaceted educational resource. Educators, students, and researchers will find a comprehensive examination of the energy sector that fosters a profound understanding of contemporary energy issues. By presenting clear and accessible information, the book empowers the general public with crucial knowledge about various energy sources, their environmental impacts, and the complexities of transitioning to sustainable energy systems. This awareness equips individuals to make informed decisions about their personal energy consumption and to actively support policies that promote environmental responsibility. Additionally, the book catalyzes community engagement, encouraging collective participation in renewable energy projects, energy conservation efforts, and grassroots advocacy for a cleaner and more sustainable future.

For policymakers and decision-makers, this book is an essential tool for crafting well-informed energy policies. It offers a thorough analysis of different energy sources, highlighting their benefits, limitations, and environmental consequences, thus enabling policymakers to develop sound regulations, strategic investments, and effective policy frameworks. The insights provided also aid in long-term sustainable planning by addressing the challenges and opportunities inherent in the global energy transition. Industry leaders and business strategists will find value in the book's guidance on strategic decision-making related to energy investments, infrastructure development, and operational practices. Understanding the risks and potential rewards associated with various energy options allows businesses to implement robust risk management strategies that support long-term sustainability. Furthermore, this book shows new pathways for companies to reduce energy costs, enhance operational efficiency, and secure a competitive advantage in the marketplace, fostering innovation and resilience in a rapidly evolving energy sector.

Because this book draws on many technical and non-technical disciplines, we have supplemented the text with some simplified technical details that are shown within a "Tech Note" box to help those readers who want to dive a bit deeper into those topics. We have added many case studies to help the readers understand the significance of global energy usage and environmental pollution. In addition, the role of the public in adopting different energy sources or mitigating greenhouse gas emissions. Finally, unit conversion for different measurements is included, and a glossary of terms and acronyms is appended to this book.

References

[1] Fassihi and Kovscek, SPE monograph on "Low-energy processes for unconventional-oil recovery", 2017, https://onepetro.org/books/book/38/Low-Energy-Processes-for-Unconventional-Oil.

[2] Fassihi and Blangy, Integrated Aquifer Characterization and Modeling for Energy Sustainability, CRC Publishing, 2023, https://www.routledge.com/Integrated-Aquifer-Characterization-and-Modeling-for-Energy-Sustainability-Key-Lessons-from-the-Petroleum-Industry/Fassihi-Blangy/p/book/9781032224954

Acknowledgment

We are indebted to many subject-matter experts, especially Drs Behnam Askarian, Amir Mahmoodkhani, Bill O'brien, Kaveh Dehghani, Vahid Nourozi, Murali Venkatarama, and Bryan Williams. Shane Mohammadi, Gia Alexander, and Behzad Jazayeri, who reviewed different chapters and provided us with great suggestions to improve each chapter (of course, any errors in the book are our responsibility). Finally, I would like to thank Terry Fassihi, who spent many hours reviewing the final copy of this book before it went to the publisher.

Introduction

How are we talking about the climate these days? Let's review a few well-known quotes.

> *"Remember when they used to say global warming? They do not say that anymore. They say climate change because the planet's actually getting cooler,"*
>
> *President Trump [1].*

> *"Well, the old cliché "Denial ain't just a river in Egypt" should be joined by 'Despair ain't just a tire in the trunk.' Despair is just another form of climate denial. We do not have time for it. The stakes are so high",*
>
> *Vice President Al Gore [2].*

> *"It's time to take a different view of the climate change problem. The goal is not just to stop rising temperatures; it is to improve the human condition. That can be accomplished by doubling down on things that have the most impact and making tough choices about those that do not,"*
>
> *Bill Gates [3].*

Simply put, people are extremely divided about whether or not climate change is real, whether or not there are human causes, and whether or not we can or should do anything about it.

Mental health experts suggest that there are three types of stances we take that can get in the way of taking action:

- We can be frozen by fear – the climate phobic.
- We can dissociate from the problem or go into denial – the climate denier.
- We can look for reasons not to believe or trust the experts, the climate skeptics.

Mental health providers say their clients are often expressing worry about the weather catastrophes that surround us. Natural disasters precipitated by climate change, including hurricanes, heatwaves, wildfires, and floods, cause or exacerbate mental health issues such as depression, anxiety, post-traumatic stress, and substance abuse. The World Health Organization has officially called climate change the biggest threat to global health in the 21st century.

Climate anxiety (Ecoanxiety) is real. It is not an official diagnosis yet, but it is causing many people significant distress. It is defined as "the fear that the climate will collapse and there is no future for humanity." It involves feeling helpless and stressed. Symptoms associated with climate anxiety include panic attacks, insomnia, and obsessive thinking. It is particularly stressful for young people, who are more likely to be around for the worsening conditions, and who have no control over the choices adults are making. Their frustration over this is apparent. Ecotherapy is one treatment that explores how we interact with the earth [4].

In March 2019, an estimated 1.6 million school-aged protestors in 125 countries demanded that action be taken to combat climate change. The debates now are infused with political and religious ideology, but questions about climate change need to be addressed scientifically. Informed opinions about climate issues require us to learn the facts and understand science. This can be a daunting task, so most of us might be avoiding it.

So, where do we begin?

This book is written to provide accessible and factually accurate information about what is happening in the global climate and weather, how human industry and technology are involved, and a way to assess our options for protecting our environment and our way of life. More than one would expect, we really can have it both ways, but it will take a lot of effort, planning, and most importantly, collaboration.

During the past decade, growing concerns about the climate crisis led governments and investors to re-evaluate the world's reliance on greenhouse gas (GHG)-emitting fossil fuels. This led the global energy sector to start transitioning to renewables and other clean energy technologies, such as solar, wind, electric vehicles, and a range of different technologies, such as hydrogen and biomass.

This is great news, but despite this push, global reliance on oil and coal reached record highs in 2024, revealing that the shift is more of an "energy addition" where renewables supplement rather than replace conventional fuels. This is because as we add cleaner energy, we also add demand for more energy.

Also, the current US political environment is shifting away from clean energy and prioritizing traditional energy production. As one of the greatest global energy consumers, this US policy reversal will significantly slow down the results of global efforts to protect the environment.

The quest for a sustainable energy future is paved with good intentions yet riddled with the pitfalls of polarized viewpoints. On one hand, the fear of change and deep-seated ideological divides threaten to create gridlock, leading to stagnation and, ultimately, failure. On the other hand, the allure of quick fixes fuels all-or-nothing knee-jerk reactions, often overlooking critical social and environmental considerations. Navigating this minefield requires a nuanced understanding of these opposing forces and a commitment to finding common ground since we really all need the same things for our personal lives and the future of our planet.

At the heart of gridlock is the belief that we cannot transition out of traditional energy jobs, so many of us perceive the transition as a threat to workers. This fear can manifest in resistance to change, misinformation campaigns, and political maneuvering, all aimed at slowing down or even derailing progress.

On the other side of the spectrum lies the siren songs of a quick fix, which can lead to a push for all-or-nothing solutions with insistence on reaching net-zero carbon tomorrow. These often stem from a misguided belief in technological determinism, the notion that a single technological advancement can solve all our energy woes. This often leads to overemphasis on large-scale, utility-level projects, neglecting the potential of distributed, community-based solutions. Moreover, such approaches could fall prey to economic oversimplification, prioritizing cost-effectiveness at the expense of environmental and social impacts. The allure of quick fixes can blind us to potential downsides, such as ecosystem disruption, community displacement, and the creation of new dependencies on centralized systems.

Unfortunately, in such debates, energy conservation is almost forgotten. Many strategies for conservation are readily available and affordable and could have a huge potential impact, but they are scarcely used. Recently, an energy expert, when discussing the role of energy in human development, showed a map of the Earth. The northern hemisphere was significantly brighter at night with many lights compared with the southern hemisphere. His recommendation was to make energy more accessible to less developed countries. Maybe he should also have said something about making the lights in the north a bit dimmer by using less energy!

Today, at the macro level, the issue of climate change is deeply intertwined with politics and current global divisions. This divisiveness poses a significant barrier to meaningful discussion and action. However, as individuals, we can find ways to address the psychological distress caused by climate anxiety. We need to remember that our individual choices can make a difference. Each of us can find ways to make conscious choices in our daily lives that reduce our carbon footprint. In the public sphere, we can advocate for technologies that minimize emissions and promote cleaner air. Our main goal for preparing this book is to ensure public access to accurate, unbiased, factual information related to such matters that will eventually impact the future of our planet.

This book presents different views about key elements of energy transition and GHG reduction technologies. As with any other controversial issue, we are bombarded by narratives from many interest groups on a daily basis with ideas related to ways of combating global warming or reducing the carbon footprint. By presenting all aspects of energy transition, the authors give readers sufficient knowledge and tools in order to empower them to form their own opinions and take action for a better future.

This comprehensive guide to the energy landscape offers a holistic perspective on the challenges and opportunities associated with the transition to a sustainable energy future. Divided into five interconnected parts, the book provides a comprehensive look at both traditional and renewable energy sources, explores the environmental and social implications of energy production and consumption, and outlines the pathways to a cleaner and more sustainable future.

In Part I, we introduce the current energy landscape and its challenges. The fundamental concepts of energy and its crucial role in driving our modern world are reviewed in Chapter 1. Traditional nonrenewable energy sources are described in Part II, with a discussion of fossil fuels in Chapter 2 and nuclear energy in Chapter 3, where we highlight their contributions, limitations, and the environmental challenges associated with their use.

In Part III, we shift our focus to emerging clean energy technologies, where we explore the vast potential of renewable sources. Chapters 4, 5, and 6 explain the potential of solar power, wind power, and aqua energy. These chapters provide a detailed understanding of these technologies, their applications, and the challenges and opportunities associated with their widespread adoption.

In Part IV, we address the critical issue of climate change and the urgent need to transition toward a net-zero carbon future. Chapters 8 and 9 explore the interconnectedness of energy, air quality, water, and soil, highlighting the environmental impacts of energy production and consumption.

The final part of the book (Part V) offers a vision for a sustainable energy future, outlining the short-term and long-term solutions that can drive this transformation. Chapter 10 focuses on immediate actions, such as improving energy efficiency and embracing smart grid technologies. Chapter 11 delves into the long-term vision, emphasizing the need for policy reforms, technological innovation, and international cooperation. Government incentivization and carbon economy, including carbon offset and credit, are included in this chapter. Finally, in Chapter 12, we discuss the role of the public in shaping the energy transition, including ways of engaging them in the path forward.

This book provides scientific and economic information for readers who are not fully familiar with the intricacies of energy sources and environmental issues but want to have a deep understanding of the critical factors. We also have recommended additional readings for those who are interested in diving even deeper into these topics.

The traditional energy trilemma: affordability, sustainability, and security, is no longer sufficient to address the complexities of today's energy landscape. A more comprehensive framework is needed: the energy pentalemma, encompassing flexibility, affordability, conservation, transition, and sustainability. Notably, energy security is absent from this framework, and deliberately so. It is not an end goal in itself, but an outcome that naturally emerges when the other pillars are properly addressed. By building flexibility into our energy systems and embracing conservation to reduce demand and waste, nations can organically achieve the energy security they seek, without enshrining it as a standalone objective. The pentalemma thus offers a more dynamic, forward-looking approach to energy governance – better equipped to navigate the clean energy transition while keeping energy accessible and resilient for all.

References

[1] Presidential campaign rally; Uniondale, NY, 2024.
[2] D. Remnick, Al Gore does not say I told you so, The New Yorker 2023, https://www.newyorker.com/news/q-and-a/al-gore-doesnt-say-i-told-you-so (Accessed 20 Jan 2026).
[3] Gates Notes, Three tough truths about climate, 2025, https://www.gatesnotes.com/work/accelerate-energy-innovation/reader/three-tough-truths-about-climate?WT.mc_id=20251028100000_COP30_TGN-NYTFL (Accessed 20 Jan 2026).
[4] M. Castelloe, Coming to terms with ecoanxiety. Psychology Today 2018, https://www.psychologytoday.com/ca/blog/the-me-in-we/201801/coming-terms-ecoanxiety? (Accessed 20 Jan 2026).

Part I

Facing the crossroads—The energy landscape and its challenges

The flick of a light switch. The hum of the refrigerator. The wheels of a car crunch on a gravel road. These are the sounds of a hidden force: Energy. From the ancient flames of a campfire to the modern marvels of electric vehicles, humanity has sought to harness energy throughout history. Energy underpins all human activities, serving as a cornerstone for meeting basic needs such as access to clean water, cooking, housing, healthcare, agriculture, communication, and transportation.

But in today's world, our understanding of energy sources must go beyond mere utility. The choices we make about where our energy comes from hold profound consequences for our planet, our society, and our future.

A massive new demand for energy is on the horizon. Currently, huge additional demands for electricity are quickly increasing from: Crypto mining, data centers and AI, largely centered within China and other advanced economies. Financial investment in data center infrastructure is forecast to reach USD 580 billion in 2025, a figure that exceeds the USD 540 billion devoted to the global oil supply chain. Projections indicate that data center electricity consumption will triple by 2035.

Part I of this book serves as a compass, guiding us through the diverse landscape of energy sources. We will embark on a journey to dissect both renewable and nonrenewable resources, peeling back the layers to understand their fundamental principles, how they are harnessed, and the intricate dance between their benefits and challenges. A more detailed description of the energy resources and the emission reduction technologies will be discussed in Parts II–V.

In Understanding energy sources, we will establish a common language, defining key terms like primary and secondary energy and unveiling the distinctions between renewable and nonrenewable resources. We will then delve into the world of nonrenewable energy sources, starting with the fossil fuels—coal, oil, and natural gas—that have powered our world for centuries. We will explore their influence on economic development and the sobering reality of their environmental consequences.

Next, we will shift our focus to the rising stars of the energy arena: Renewable energy sources. From the sun's radiant power harnessed by solar panels to the wind's kinetic energy captured by turbines, we will unveil the captivating technologies that tap into natural, renewable resources. We will dive into the potential of hydropower, geothermal energy, and even the ocean's tides, uncovering both their exciting possibilities and the hurdles they still face.

Throughout this exploration, we will keep a critical eye on the challenges associated with each energy source. We will delve into the environmental impacts, from air and water pollution to greenhouse gas emissions. We will also address the social implications, including issues of energy security, access, and the future of jobs in the energy sector.

By the end of this part, we will have gained a good overview of the energy landscape and its role in our future energy mix. We will be equipped to ask informed questions, engage in meaningful discussions, and make conscious choices about the energy we consume and the future we desire. This knowledge is not simply academic; it is an empowering tool, enabling us to navigate the exciting yet critical era of energy transition. So, let's begin our journey, unlocking the secrets of the energy sources that fuel our world and secure a sustainable future for generations to come.

Climate change

The frequency and severity of natural disasters worldwide are escalating. The Horn of Africa is grappling with its fourth consecutive failed rainy season, leading to widespread famine, particularly among children. Central America, a bridge between rapidly warming oceans, is enduring a crippling drought, making agriculture nearly impossible. When rain does arrive, it often comes in the form of catastrophic storms. Wildfires have become more frequent, and heavy rains are causing major floods in many locations in the United States.

These crises are not the result of actions by those affected. The average Somalian emits less than 1% of the carbon produced by an average American, while Hondurans and Vietnamese contribute even smaller fractions. The United States, comprising just 4% of the global population, is responsible for about 15% of all atmospheric greenhouse gas emissions [1].

The famine in Somalia, the hurricanes in Honduras, and the flooding in Vietnam are direct consequences of the actions of industrialized nations. The so-called "invasion of immigrants" that has strained border security in many countries is driven by a complex interplay of socioeconomic factors. Climate change is a significant catalyst for these mass migrations [2].

Just transition

The energy sources need to be sustainable. i.e., meeting the energy needs of the present should not be at the expense of compromising the ability of future generations to meet their own needs. It is a comprehensive approach that moves beyond simply using renewable resources and requires balancing three core dimensions of environment, economics, and social sustainability.

How can we look at this dramatic development from a human rights perspective? The concept of human rights in the energy sector encompasses two fundamental aspects: Recognizing energy access as a human right in itself and applying human rights principles to the energy sector's operations and impacts [3]. Progress on global access to electricity has been remarkable in recent decades. Since 2000, the number of people without electricity has fallen by almost a billion, and 40 countries have reached near-universal access. However, the surge in energy production has brought about significant environmental degradation. Thus, there is an urgent need for a swift and equitable global energy transition to cater to the 730 million individuals still lacking electricity access and to offer clean cooking solutions to the 2 billion people reliant on the most polluting fuels for cooking [4].

Many publications advocate for a human rights-based approach to energy, aiming to facilitate inclusive, holistic transitions aligning with environmental imperatives across nations [1,5]. Potential human rights violations in the energy sector could encompass community displacement, land grabbing, and violations of indigenous rights. Additionally, working conditions in energy industries, particularly fossil fuel extraction, can be hazardous, resulting in labor exploitation and health risks. The connection between energy and human rights is evident within the context of the UN's Sustainable Development Goals (SDGs) [6]. The SDGs, also known as global goals, aim to go further to end all forms of poverty. These are displayed in Fig. P1.1. Although not legally binding, these 17 goals are closely linked to human rights law and explicitly aim to realize the rights of all individuals. A human rights approach to energy can ensure that policies are driven by the needs of citizens and communities rather than by elitists or corporate interests.

Energy justice advocates for equitable access to energy services and benefits, recognizing energy as a fundamental human right essential for poverty reduction, economic development, and improved quality of life. However, energy justice extends beyond access; it also encompasses the ethical and sustainable production and consumption of energy. This includes ensuring that energy development is environmentally responsible and does not perpetuate social inequalities.

The Intergovernmental Panel on Climate Change (IPCC), a United Nations body, established in 1988 by the World Meteorological Organization (WMO) and the United Nations Environment Program (UNEP), in their report [7], referred to a "just transition" as a set of principles, processes, and practices aimed at ensuring that no individuals, workers, communities, or regions are left behind

FIGURE P1.1 United Nations Sustainable Development Goals [6].

in the shift from a high-carbon to a low-carbon economy. It emphasizes respect and dignity for vulnerable groups, the creation of decent jobs, social protection, employment rights, fairness in energy access, and democratic consultation with relevant stakeholders.

To promote human rights and energy justice, collaboration among governments, corporations, and civil society organizations is essential. Governments should implement robust environmental and labor regulations, protect the rights of indigenous communities, and ensure that energy development benefits all citizens. Corporations must adopt responsible business practices, respect human rights throughout their supply chains, and engage meaningfully with affected communities. Civil society organizations can advocate for policy reforms, monitor human rights abuses, and support communities impacted by energy projects. By working together, we can build a future where energy serves as a force for positive change rather than a source of harm and inequality.

In the transition to achieving net-zero emissions, the International Energy Agency (IEA), in its report on Global Commission on People-Centered Clean Energy Transitions [8], emphasizes the importance of fairness and inclusivity. This principle highlights the need to provide developing economies with the financial support and technological expertise necessary to build sustainable energy systems that can meet the demands of growing populations and economies. Extending access to electricity for the millions currently deprived of it is not just a practical necessity but a moral imperative. Recognizing that not every worker in the fossil fuel industry can easily transition to clean energy jobs, governments must prioritize training and allocate resources to create new opportunities. Citizens should also be actively engaged in this transition, ensuring they feel like valued participants rather than passive subjects.

Public awareness

A major milk brand used to carry a sentence on their milk cartons, "Our goal is to become net carbon positive by 2030". This is an example of how misleading statements can be made about their carbon reduction aspirations, even by corporations. When addressing the complex challenges of climate change, particularly in understanding various mitigation options, their potential impacts, and their practicality, there exists a significant knowledge gap among the general population across different countries. Furthermore, the public expresses unease regarding the operation of emission reduction projects by major oil companies. Many perceive these companies as viewing CO_2 sequestration projects primarily as a way to prolong fossil fuel use rather than transitioning to alternative clean energy sources. This skepticism is compounded by a historical track record of environmental disasters linked to major oil companies, which raises concerns when faced with new greenhouse gas (GHG) mitigation proposals from these entities. Such distrust underscores the necessity for transparency, accountability, and robust regulatory frameworks to navigate the complexities of CO_2 sequestration and foster public confidence in climate mitigation efforts.

In the early 2000s, extremist groups within the environmental movement, such as the Earth Liberation Front, began setting fire to symbols of ecological destruction, including timber mills, SUV dealerships, and ski resorts. This group, which warned of an imminent ecological catastrophe, was widely demonized. Recently, some climate activists have started to openly consider the possibility—and necessity—of directly sabotaging the infrastructure of the carbon economy. Among them is academic Andreas Malm, whose recent book, How to blow up a pipeline, advocates for destroying the tools of fossil fuel extraction as a last-ditch effort to avert ecological collapse.

Public awareness of climate change in the United States can be categorized into six distinct levels [9]. The Alarmed (33%) are the most engaged, very concerned about global warming, and strongly support climate action. The Concerned (25%) recognize global warming as a significant threat but prioritize it less and are less likely to act. The Cautious (17%) are aware of climate change but uncertain about its causes and not particularly worried. The Disengaged (5%) are largely unaware of global warming, while the Doubtful (10%) question its existence or human causes, viewing it as a low risk. The Dismissive (9%) firmly reject the reality of human-caused global warming and oppose most climate change policies. Over the years, the Alarmed segment has grown, and in 2021, it outnumbered the Dismissive by more than three to one (33% vs. 9%).

Recent surveys indicate a growing awareness among company executives, including those in the energy and industrial sectors, about the importance of sustainability and climate action. However, many companies seem to prioritize easier, low-impact actions over more challenging, "needle-moving" initiatives. Obstacles to meaningful progress include difficulties in measuring environmental impact, insufficient supplies of sustainable inputs, and concerns about costs. Standardization of measurement, transparency, and access to proven lower-emission technologies are crucial factors in driving meaningful change to address climate challenges effectively.

Public perception can be transformed through proactive efforts by the private sector and the dissemination of accurate information about various technologies that mitigate climate change. Open and positive engagement in the environmental, social, and governance (ESG) space has proven critical in addressing the deeply held concerns of local stakeholders and can foster an atmosphere of trust. Similarly, engagement in the policy sphere is essential to ensure that local, state, and national groups, lawmakers, industry, and other policymakers shape the energy agenda. Once engagement in both the project and policy spheres is established and objectives align, the process of involving the public becomes more effective and widespread. However, even with a foundation of trust, cultivating supportive relationships may take years.

In the past, proposals made to the USA governmental bodies for funding such as Department of Energy (DOE) required a Diversity, Equity and Inclusion (DEI) section where the applicant had to provide a plan on how they planned to educate the public and get them behind their proposed technology, improve work environment in indigenous and lower income areas, and provide science, technology, engineering and mathematics (STEM) amongst many other things. The Trump administration's action on eliminating the DEI section requirement could profoundly impact the public engagement and trust in new energy initiatives.

International energy corporations, whether they are operating within a system with strong or weak laws, must proactively develop appropriate engagement processes, recognizing that the industry faces a material threat to its social license to operate and that the health and welfare of human lives and the environment depend on their actions. Investors have been increasingly pressing companies to disclose systematic policies and practices to identify and respond to key community concerns effectively. Careful assessment of community risk and consultation with impacted communities can build broad community support for new projects. Otherwise, they may face community and government opposition and associated costly delays to new projects or expansions of existing operations. Companies must openly and continuously address community concerns, especially because these issues directly affect local people.

Toward a cleaner future—Adopting a design standard for sustainable industrial plants (4Z index)

Industrial facilities are usually judged by what they produce. Yet, in practice, long-term performance is just as strongly determined by what the facility *discharges*—water, salts and solids, gases, and wasted energy. Today's sustainability playbooks tend to tackle these streams one at a time: A plant might pursue zero liquid discharge to comply with state or federal regulations, while still exporting waste solids elsewhere; another may cut emissions through flare reduction without closing the wastewater loop; a third may electrify and buy renewables while continuing to reject low-grade heat to the environment. The result is progress, but it is often fragmented—and sometimes it simply shifts burdens from one "pipe" to another.

To push beyond incremental optimization, we recommend a single, unifying design philosophy and evaluation metric, the *4Z Index*. The premise is straightforward: A truly sustainable industrial process should be constrained from the earliest design stage to avoid discharging (1) liquids, (2) minerals, (3) gases, and (4) energy. In other words, the process should be built to achieve zero liquid discharge (ZLD), zero mineral discharge (ZMD), zero gas discharge (ZGD), and zero energy discharge (ZED), not as four independent initiatives, but as one integrated engineering objective.

The first "Z," ZLD, is already a recognized target in industrial water management: It refers to systems designed to eliminate liquid effluent by recovering water for reuse and converting remaining constituents into solids for handling or valorization. Yet ZLD alone is not a guarantee of circularity—because the solids produced can still represent an offsite discharge of minerals, cost, and liability.

That gap motivates the second "Z," ZMD. ZMD extends the goal from *no liquid leaving the fence line* to *no dissolved or solid mineral value being discarded to the environment or exported as unmanaged waste*. In practice, this means designing separation and purification pathways so that salts and critical elements become controlled products (or benign, stable solids with a defined end-use), rather than a disposal problem.

The third "Z," ZGD, addresses the reality that many industrial sites still vent, flare, or otherwise release gases during normal operations. While "ZGD" is not a universal industrial standard, there are well-known, field-specific frameworks that align with the same intent—for example, the World Bank's zero routine flaring by 2030 initiative, which targets the elimination of nonsafety-related routine flaring. A 4Z-conformant plant treats gas streams as resources: Captured for reuse, converted to stable products, or permanently stored—while maintaining safety systems for true emergencies.

The fourth "Z," ZED, is the one most often overlooked—and arguably the most transformative. Even when a plant meets emissions and water goals, it may still "discharge" energy continuously through unrecovered heat, pressure let-down, mixing losses, and other forms of wasted exergy. The 4Z Index takes the next step: Not merely reducing purchased energy, but designing the system so usable energy is cascaded and recovered wherever practical before it is rejected.

Importantly, the 4Z Index is not a slogan; it is a call for a standardized method that prevents "partial victories." The same engineering move that closes one loop can open another: ZLD can increase energy intensity; mineral recovery can introduce gas-handling challenges; electrification can hide emissions upstream; exporting solids can relocate environmental impacts offsite. A credible 4Z standard, therefore, must define clear system boundaries, measurement and verification requirements, and rules that discourage burden-shifting.

Based on a scan of current practice and literature, mature examples exist that pursue individual "Z" targets (ZLD systems are well-established, and energy-neutrality and flare reduction programs are widely documented). However, a single, widely recognized and transparently audited industrial facility that publicly demonstrates all four objectives simultaneously—ZLD, ZMD, ZGD, and ZED under one unified framework—does not appear to be established as an industry standard today. That absence is precisely the opportunity: 4Z provides a practical north star for designing the next generation of industrial infrastructure, where "waste" is treated as a design failure and every output stream is intentionally converted into a product, a reusable utility, or a permanently managed form.

The 4Z Index can be developed as a rigorous, engineer-friendly evaluation tool—intended to guide concept selection, front-end engineering design, and performance benchmarking—so that "true sustainable process" stops being an aspiration and becomes a measurable design requirement.

References

[1] Statista, *Worldwide Annual Carbon Dioxide Emissions by Select Country*, https://www.statista.com/statistics/500524/worldwide-annual-carbon-dioxide-emissions-by-select-country/ (Accessed 18 Jan 2026).

[2] G. Vince, in: Nomad Century: How Climate Migration Will Reshape Our World, Flatiron Books, New York, 2022, p. 260.

[3] M. Wewerinke-Singh, A human rights approach to energy: realising the rights of billions within ecological limits, Rev. Eur. Comp. Int. Environ. Law 31 (1) (2022) 16–26. https://doi.org/10.1111/reel.12412.

[4] International Energy Agency (IEA), World Energy Outlook 2025, IEA, Paris, 2025, https://www.iea.org/reports/world-energy-outlook-2025 (Accessed 18 Jan 2026).

[5] K. Huhta, Conceptualising energy justice in the context of human rights law, Nord. J. Hum. Rights 41 (4) (2023) 378–392. https://doi.org/10.1080/18918131.2023.2210443.

[6] United Nations, Sustainable Development, https://www.un.org/sustainabledevelopment/ (Accessed 18 Jan 2026). b) V.M. Taghvaee, A. Assari Arani, M. Nodehi, J. Khodaparast Shirazi, L. Agheli, H.M. Neshat Ghojogh, N. Salehnia, A. Mirzaee, S. Taheri, R. Mohammadi Saber, H. Faramarzi, R. Alvandi, H. Ahmadi Rahbarian, Sustainable development goals: transportation, health and public policy. Rev. Econ. Polit. Sci. 8 (2) (2023) 134–161. https://doi.org/10.1108/REPS-12-2019-0168.

[7] P.R. Shukla, J. Skea, R. Slade, A. Al Khourdajie, R. van Diemen, D. McCollum, M. Pathak, S. Some, P. Vyas, R. Fradera, M. Belkacemi, A. Hasija, G. Lisboa, S. Luz, J. Malley, Intergovernmental Panel on Climate Change (IPCC), Climate Change 2022: Mitigation of Climate Change. Contribution of Working Group III to the Sixth Assessment Report of the Intergovernmental Panel on Climate Change, Cambridge University Press, Cambridge, U.K.; New York, NY, 2022 Eds., https://www.ipcc.ch/report/ar6/wg3/ (Accessed 18 Jan 2026).

[8] International Energy Agency (IEA), Recommendations of the Global Commission on People-Centred Clean Energy Transitions, IEA, Paris, 2021, https://www.iea.org/reports/recommendations-of-the-global-commission-on-people-centred-clean-energy-transitions (Accessed 18 Jan 2026).

[9] A. Leiserowitz, E. Maibach, S. Rosenthal, J. Kotcher, L. Neyens, J. Marlon, J. Carman, K. Lacroix, M. Goldberg, Global Warming's Six Americas, September 2021, Yale University and George Mason University, New Haven, CT, 2022, https://climatecommunication.yale.edu/publications/global-warmings-six-americas-september-2021/ (Accessed 18 Jan 2026).

Chapter 1

Understanding energy sources

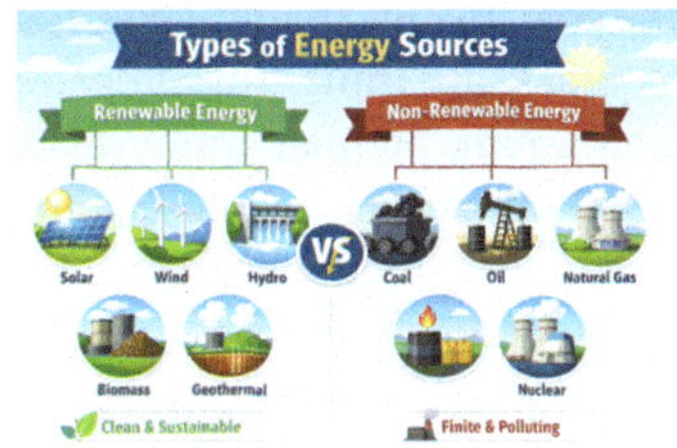

Synopsis

The journey toward a sustainable energy future demands a clear understanding of the fuel choices that power our world. This chapter provides an overview of the fundamental classifications that illuminate the unique characteristics and challenges associated with different energy sources, equipping us to navigate the complexities of this critical landscape. Each of the energy sources mentioned in this chapter will be discussed in detail in the upcoming chapters.

Definitions

Renewable versus nonrenewable: A matter of time and replenishment

Renewable energy sources, nature's perpetual gifts, are continuously replenished over human timescales. Sunlight bathes the Earth, wind whips across plains tirelessly, and tides rise and fall, offering perpetual sources of clean energy. Solar panels harness the sun's radiance, wind turbines capture the wind's kinetic energy, and hydropower converts the movement of water into electricity. These sources boast numerous benefits, including their environmental friendliness, contribution to energy independence by reducing reliance on imported fuels, and often lower operating costs. However, challenges remain. Intermittency, particularly for solar and wind, necessitates robust storage solutions and grid

F.A.C.T.S. Flexibility, Affordability, Conservation, Transition, Sustainability: Our Path to the Future of Energy and the Environment. DOI: https://doi.org/10.1016/B978-0-443-52574-2.00011-2

integration strategies. Additionally, their lower energy density requires more space or infrastructure compared to traditional sources.

Nonrenewable energy sources, on the other hand, tell a different story. Fossil fuels like coal, oil, and natural gas, formed over millions of years through geological processes, are finite treasures buried deep within the Earth. Nuclear energy utilizes uranium, another finite resource. While readily available and boasting established infrastructure, these sources come with significant baggage. Air and water pollution, greenhouse gas emissions contributing to climate change, and resource depletion leading to geopolitical tensions—these are just some of the environmental and social consequences they carry. Additionally, nuclear energy raises concerns about safety and waste disposal.

Primary versus secondary: From nature's bounty to usable power

Primary energy sources, untouched by human intervention, are directly extracted from nature. Fossil fuels slumbering underground, sunlight streaming down, and geothermal heat bubbling within the Earth's crust are all examples. Secondary sources, however, are derived from these primary sources through various conversion processes. Electricity generated from fossil fuels, gasoline refined from crude oil, and hydrogen produced from water electrolysis are all examples. Understanding this distinction is crucial, as the environmental impact of a secondary source hinges on its primary source. For instance, electricity generated from solar panels is undeniably cleaner than that generated from coal.

The nuances of sustainability: Going beyond simple labels

The classifications, however, are not always black and white. Biomass, for example, can be both a renewable and nonrenewable source, depending on management practices. Sustainably harvested wood can be a renewable fuel, while unsustainable practices can deplete forest resources, rendering it nonrenewable. Additionally, energy storage technologies play a vital role in managing the intermittency of renewable sources like solar and wind, enabling their effective integration into the grid. Furthermore, the energy landscape is constantly evolving. Technological advancements are pushing the boundaries, with new renewable sources like ocean energy emerging and conversion methods like biofuels undergoing continuous improvement. Staying informed about these advancements empowers us to make informed choices about the energy we consume and advocate for solutions that promote a sustainable future.

Renewable versus sustainable: A critical distinction

While renewable and sustainable energy are often used interchangeably, it is important to understand the subtle differences. A renewable energy source may

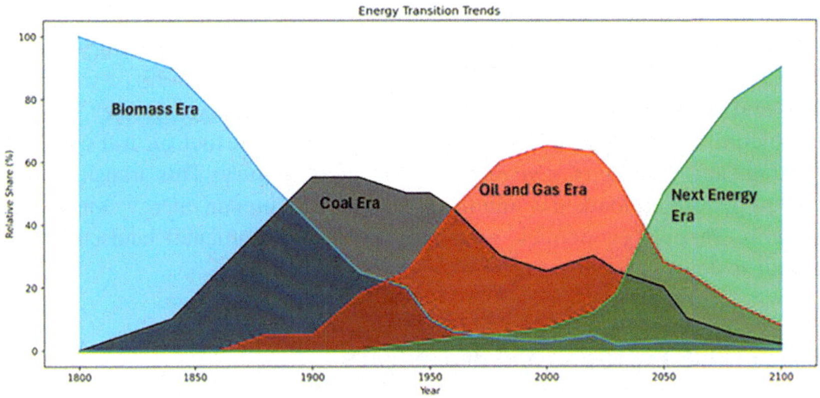

FIGURE 1.1 Historical energy landscape [1].

not be inherently sustainable if its extraction or utilization has negative environmental or social impacts. For example, unsustainable harvesting of biomass for biofuels can lead to deforestation and soil erosion. Conversely, some nonrenewable sources can be utilized sustainably through practices like carbon capture and sequestration. Therefore, true sustainability goes beyond the source of energy and encompasses the entire life cycle, from extraction and conversion to use and disposal. It demands responsible practices that minimize environmental impact, ensure social equity, and consider long-term resource availability.

A brief history of energy addition and transition

Having established the fundamental classifications of energy sources, we can now delve deeper into the global energy consumption landscape. This exploration will unveil current trends, emerging challenges, and exciting possibilities for the future. The energy landscape throughout history is schematically shown in Fig. 1.1.

As shown, energy addition and transition have been intertwined throughout history. Initially, societies relied on biomass, such as wood and animal dung, for heating, cooking, and lighting. However, as populations grew and industrialization expanded, the energy demand outstripped the capabilities of biomass. The Industrial Revolution drove the transition from biomass to coal. Its widespread availability and high energy density made it a cornerstone of economic development throughout the 18th and 19th centuries.

1900–1950: The rise of fossil fuels

The early 20th century saw the dominance of coal as the primary energy source, fueling industrialization and urbanization. As industrialization surged, so did the energy demand, leading to the subsequent rise of petroleum as a prominent energy source. Petroleum's ascendancy began in the late 19th century with the invention of the internal combustion engine, which revolutionized transportation and manufacturing. The discovery of vast oil reserves further solidified petroleum's role, propelling it to become the dominant fuel for transportation, heating, and electricity generation in the 20th century. This transition from coal to petroleum marked a significant evolution in human energy consumption patterns, shaping economies, geopolitics, and environmental landscapes on a global scale.

1950–1980: Expansion and dependency

In the post-World War II period, we witnessed a surge in energy consumption, with fossil fuels continuing to dominate. Geopolitical factors, technological advancements, and environmental considerations also influenced the transition from coal to oil and gas. Oil-rich regions like the Middle East gained strategic importance, shaping global politics and energy security. Additionally, innovations in drilling, refining, and distribution made oil and gas more accessible and cost-effective than coal in many contexts.

The advent of nuclear energy offered a promising alternative, although concerns about safety and proliferation emerged. Environmental awareness has grown, leading to the establishment of regulations aimed at mitigating pollution and protecting natural resources.

1980–2000: Shift toward renewable energy

Rising concerns over environmental degradation, coupled with oil crises and geopolitical tensions, spurred interest in renewable energy sources. Technological advancements, particularly in solar and wind energy, led to their increased deployment. Governments worldwide began implementing policies to incentivize renewable energy adoption and reduce greenhouse gas emissions.

2000–Present: Accelerated transition

In the 21st century, we have witnessed a rapid expansion of renewable energy, driven by falling costs, improved technology, and growing climate change awareness. Energy transition initiatives, such as the Paris Agreement, aimed to limit the rise in global temperature and promote sustainable development. Investments in clean energy surged, while coal-fired power plants faced increasing scrutiny and closures. Electric vehicles gained traction as alternatives to internal combustion engine vehicles, further reducing reliance on fossil fuels. The emergence of energy storage technologies and smart grids revolutionized the energy landscape, enabling greater integration of renewables and enhancing grid stability.

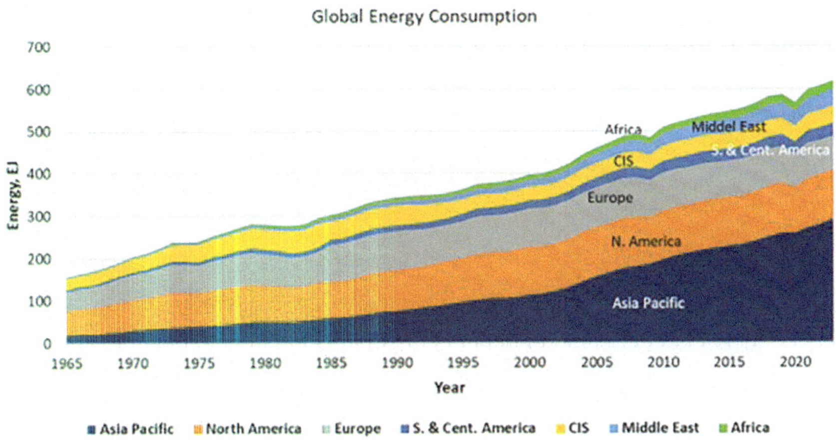

FIGURE 1.2 Global energy consumption between 1965 and 2023. (*Data from Ref. [2]*).

Toward decarbonization

The 2020s marked a pivotal decade in the energy transition, with a growing emphasis on decarbonization and achieving net-zero emissions. Governments, businesses, and communities worldwide have committed to ambitious climate targets and accelerated efforts to phase out fossil fuels. Renewable energy continued to outpace fossil fuels in terms of investment and capacity additions, signaling a fundamental shift in the global energy paradigm. Innovation and collaboration across sectors drove advancements in clean energy technologies, paving the way for a more sustainable and resilient energy future.

Navigating the global energy landscape

The fundamental objective of the energy transition has been to replace most of today's energy systems that rely on fossil fuels with a completely different system via renewable energy. Yet throughout history, no energy source, including traditional biomass of wood and waste, has declined globally in absolute terms over an extended period.

Global energy consumption by region between 1965 and 2023 is displayed in Fig. 1.2. As shown, Asia Pacific (specifically China), North America, and Europe have been the main energy consumers during this period. The energy consumption by energy type is presented in Fig. 1.3. [Notice that on this figure, oil, unless otherwise noted, includes crude (including shale oil and oil sands), natural gas liquids (NGLs), gas-to-liquids (GTLs), coal-to-liquids (CTLs), condensates, and refinery gains. Hydrogen-derived fuels are all fuels derived from low-carbon hydrogen, including ammonia, methanol, and other synthetic hydrocarbons. NG refers to natural gas. Renewables, unless otherwise noted, include wind, solar,

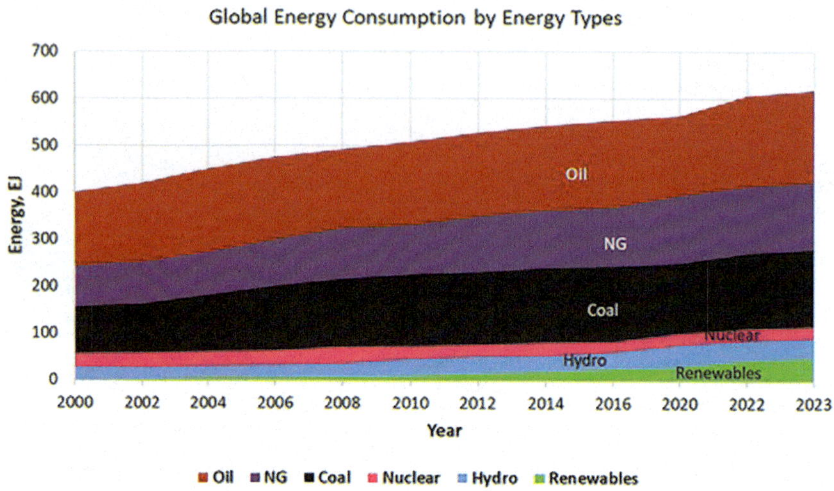

FIGURE 1.3 Global energy consumption by energy types between 1965 and 2023. (*Data from Ref. [3]*).

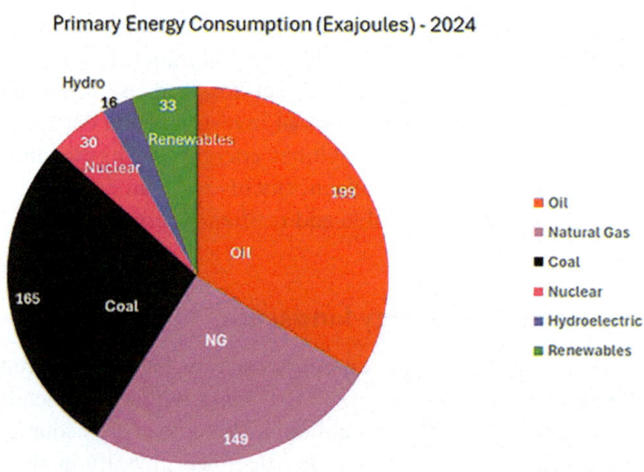

FIGURE 1.4 Global energy consumption (EJ) by energy types in 2024. (*Data from Ref. [4]*).

geothermal, biomass, biomethane, and biofuels and exclude large-scale hydro. Nonfossils include renewables, nuclear, and hydro. Traditional biomass refers to solid biomass (typically not traded) used with basic technologies, e.g., for cooking].

We have seen an increasing trend toward renewable energy resources in recent years. The amount of energy consumption in 2024 (in Exajoule) and the share of each energy type are shown in Fig. 1.4. In the past 60 years (almost), the

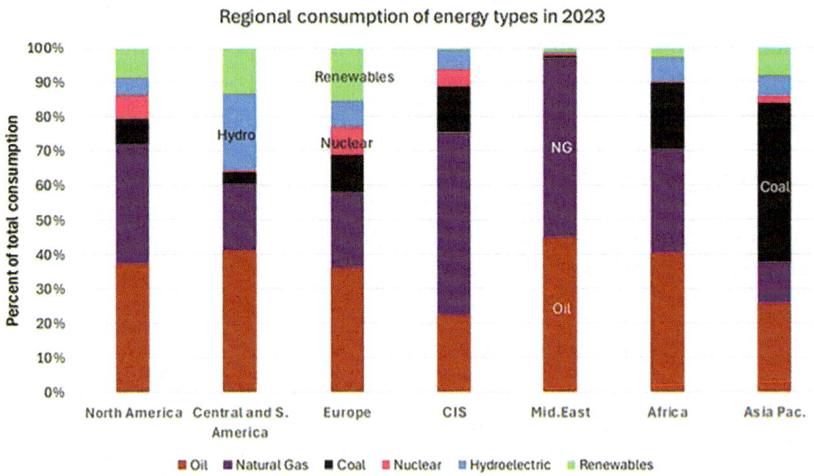

FIGURE 1.5 Regional energy consumption by energy types in 2023. (*Data from Ref. [3]*).

share of oil has fallen from around 38% to 32%, while the share of renewables has risen to about 14%. But oil, natural gas, and coal still play a dominant role in the energy field.

Despite recent advancements and heightened emphasis on sustainability, the relatively low percentage mentioned underscores our ongoing heavy dependence on nonrenewable resources. A significant reduction in this reliance necessitates a fundamental paradigm shift.

On a regional basis, the energy mix is dependent on the availability of energy resources. For example, as shown in Fig. 1.5, due to their abundant coal mines, the main energy resource in the Asia Pacific (specifically in China) was coal in 2023. Natural gas was the main fuel in the CIS (specifically Russia) and the Middle East.

Nonrenewable energy technologies: Powering progress with environmental cost

Navigating the complex landscape of energy demands a nuanced understanding of the technologies that convert natural resources into the power that fuels our lives. This section delves into the fundamental principles driving both renewable and nonrenewable energy generation technologies, equipping us with a professional-grade appreciation of their intricate inner workings.

Fossil fuels

Fossil fuels, namely coal, oil, and gas, are the key energy resources for many industries, such as transportation, power plants, chemicals, and agriculture. A brief introduction to different users of fossil fuels is provided in the following section. More in-depth discussions on fossil fuels are provided in Chapter 2.

Coal-fired power plants

Employ pulverized coal combustion within furnaces, generating heat that boils water to produce steam. This high-pressure steam drives turbines, converting thermal energy into electricity through a process known as the Rankine cycle. Switching from coal to natural gas fuel has been suggested for lower emissions.

Oil and gas engines and power plants

Utilize internal combustion engines or gas turbines to convert the chemical energy stored in these hydrocarbons into mechanical or rotational energy. Generators then convert this mechanical energy into electricity. Environmental impacts, resource depletion, use of significant water for cooling the power plants, energy security, and fluctuating fuel prices are the key challenges of fossil fuels.

Nuclear power

Leverage the controlled nuclear fission of uranium atoms, releasing immense heat. This heat boils water to generate steam for driving turbines, ultimately converting thermal energy into electricity through the Rankine cycle. More discussions on nuclear energy generation and utilization are provided in Chapter 3. The safety of nuclear power plants, nuclear waste disposal, proliferation risks, high upfront costs, public acceptance, decommissioning, and limited fuel supply are the key challenges of nuclear energy.

Renewable energy technologies

Renewables to deliver nearly 50% of power generation by 2030. In this section, we present a brief introduction to different types of renewable energy sources. A more detailed discussion about these technologies is available in the upcoming chapters.

Solar power

Solar power has undergone a lot of advancement in recent years and is readily available in many locations. A brief discussion of solar power types is provided below. More details are presented in Chapter 4.

Photovoltaic (PV) panels are the most common type of solar power. They employ semiconducting materials that convert sunlight directly into electricity through the photovoltaic effect. This phenomenon excites electrons within the semiconductor, generating a direct current that can be harnessed for various applications. Concentrated solar power utilizes mirrors or lenses to concentrate sunlight onto a receiver, heating a working fluid (molten salt or water) to high temperatures. The heat from the concentrated sunlight is then used to create steam, which drives a turbine to generate electricity. Alternatively, solar thermal energy (STE) collects heat from the sun to provide hot water or space heating. Solar hot water systems are commonly used in residential applications. Some of the new buildings use passive solar energy (PSE) design principles to capture, store, and distribute sunlight for heating and lighting buildings. This can be achieved through proper placement of windows and rooftops, using thermal mass materials, and building orientation.

Solar energy is sun-dependent and can be interrupted at night. Storing solar power with batteries can help, but battery technology still needs improvement to be perfectly reliable. The upfront installation cost is also high, although mass production in China has reduced its costs substantially.

Wind power

From ancient Roman times until now, people have been using this form of energy source. A brief introduction to different wind energy types is provided below. Please see Chapter 5 for more details.

The most common types of wind turbines are Horizontal-Axis Wind Turbines (HAWT). They resemble giant propellers with blades that spin vertically on a tall tower. The wind pushes against the blades, which rotate a shaft connected to a generator that produces electricity. Many wind farms, onshore or offshore, have hundreds of these turbines in a predesignated pattern. There have been changes in the design of wind turbines to make them more efficient. One of the newer designs uses a vertical-axis of rotation, with blades that spin horizontally [vertical-axis wind turbines (VAWT)]. These turbines can be omnidirectional, meaning they can capture wind from any direction, which can be an advantage in some locations. However, they are generally less efficient than HAWTs and are not as widely used.

Inconsistent wind could interrupt wind energy delivery. Also, wind farms need to be located in remote places that require long distances for transmitting electricity to populated areas.

Hydropower

This form of energy employs dams to capture the potential energy of falling water, which drives turbines to generate electricity. A brief introduction is provided below. Chapter 6 has more detailed discussions.

The run-of-river facilities utilize the natural flow of a river or stream without needing a dam. Water is diverted through a canal or penstock to spin a turbine that generates electricity. Dams are used for larger electricity generation. During peak electricity demand periods, water is released from the reservoir and flows through turbines to generate electricity. They offer greater control over power generation. Pumped storage hydropower is utilized like a giant battery for the electricity grid. They use surplus electricity during off-peak hours to pump water uphill into a reservoir. When there is high electricity demand, the stored water is released back down through turbines to generate power. This allows for energy storage and helps balance electricity supply and demand.

Dams can disrupt ecosystems, displace people and wildlife, and affect water quality. They also require high upfront costs. Finally, not all locations have suitable rivers or geography for efficient hydropower generation.

Geothermal energy

This technology harnesses heat from the Earth's core to generate electricity through various methods. In its simplest form, hot water from geothermal wells is extracted directly for space heating in buildings, greenhouses, aquaculture facilities, or industrial processes. Alternatively, one can use them as a heat pump, i.e., extracting heat in the winter for heating and rejecting heat in the summer for cooling buildings. They function similarly to traditional heat pumps but use the ground as a heat source or sink instead of relying solely on air. Please see below a brief introduction to different forms of geothermal energy. Chapter 6 delves deeper into this energy type.

The dry steam power plants utilize naturally occurring underground chambers of hot steam. Wells are drilled to access these chambers, and the steam travels directly to turbines to generate electricity. This is the most efficient type of geothermal power plant, but it is limited by the availability of these specific geothermal resources. There are other types of geothermal systems, namely, flash steam power plants and binary cycle power plants that work at smaller scale and are discussed in Chapter 6. These work with hot water reservoirs underground. As the hot water is brought to the surface through wells, the pressure decreases, causing some of the water to flash into steam. The steam then spins turbines for electricity generation. This method can utilize lower-temperature geothermal resources compared to dry steam plants.

Usable geothermal resources are not everywhere. Power plants need to be built near underground heat sources, which limit where you can get geothermal energy.

Ocean energy

The vast energy of the ocean can be harnessed in several ways, categorized broadly into capturing the mechanical energy of waves and tides or the thermal energy trapped within the ocean. A breakdown of the different types of ocean energy is provided below. Chapter 6 will provide more details.

Tidal energy technology utilizes the potential and kinetic energy generated by the rise and fall of tides. They are divided into two types. Tidal Barrage is a dam-like structure built across a bay or estuary that captures the difference in water height between high and low tides. As water flows through turbines placed in the barrage, electricity is generated. This method has a high energy output but a significant environmental impact. Tidal stream uses underwater turbines that are placed in areas with strong tidal currents. These are similar to wind turbines but operate underwater. These turbines capture the kinetic energy of moving water to generate electricity. This method has a lower environmental impact but may have a lower energy output compared to tidal barrages.

Installing and maintaining equipment in the harsh ocean environment is expensive. Also, many ocean energy technologies are still under development, and further research is needed to make them more efficient and cost-effective.

Biofuels

Biofuel is produced using organic mass such as plants, algae, or animal waste. Biofuels can be categorized into three generations based on the feedstock and production process. A breakdown of different biofuels is provided below. More discussions are provided in Chapter 7.

First-generation biofuels are the most common biofuels currently in use and consist of Ethanol production from crops like corn. It is blended with gasoline to reduce reliance on fossil fuels. Second-generation biofuels are made by using nonfood sources like cellulose content of plant materials like wood chips or animal fats. Third-generation biofuels are advanced biofuels under development, focusing on more efficient and sustainable feedstocks:

Growing crops for biofuels can compete with food production for land, potentially leading to deforestation. Also, current methods of converting biomass to fuel can be complex and require a lot of resources like water and fertilizer. Finally, not all biofuels are created equal. Some processing methods might not actually reduce greenhouse gas emissions compared to fossil fuels.

Hydrogen fuel

Hydrogen itself is not exactly a type of energy but rather an energy carrier. It can be produced from various sources and used in different ways. Hydrogen is called by its color depending on its source. The most environmentally friendly type, green hydrogen, is produced by electrolysis using renewable electricity sources like solar or wind power. When it is made from coal or lignite (brown coal) through gasification processes, it is called gray or black hydrogen. These clean-emitting technologies hold promise for transportation and stationary applications.

Both the fuel cells themselves and the hydrogen infrastructure (production, storage, and stations) are expensive. Hydrogen has a low energy density by volume, making it difficult to store enough on board a vehicle for a long driving range. Currently, a significant amount of hydrogen is produced from fossil fuels, which negates the environmental benefits. Large-scale, clean hydrogen production methods are still under development. Chapter 7 will describe more details on this technology.

Fusion energy

Fusion energy mimics the process that powers stars like our sun. It involves smashing light atomic nuclei together to form a heavier one, releasing a tremendous amount of energy in the process (thanks to $E=mc^2$). Fusion primarily uses isotopes of hydrogen, deuterium, and tritium. These are readily available, especially deuterium, which can be extracted from seawater. If achieved, fusion energy could be a game-changer. It offers a nearly limitless fuel source, with minimal waste and no greenhouse gas emissions. Chapter 3 will have more discussion.

Fusion requires incredibly high temperatures (over 100 million degrees Celsius) and pressure to initiate and sustain the reaction. Recreating these conditions and confining the hot plasma is a huge engineering feat. Also, it takes more energy to initiate and maintain fusion than we get out of the reaction. Achieving a sustained net energy gain, where more energy is produced than consumed, is a critical milestone. Finally, the environment inside a fusion reactor is harsh, with intense heat and radiation. Developing materials that can withstand these conditions for long periods is crucial.

The greenhouse effect and its role in climate change: A scientific perspective

Fig. 1.6 displays the ever-increasing global temperature over the last century. Recent data suggests that the atmosphere is warming at a rate of approximately 0.17°C or 0.306°F per decade [5]. According to the WMO 2024 was the first year in which global temperatures averaged more than 1.5°C or 2.7°F, above those the planet experienced at the start of the industrial age. For the past decade, the world has sought to avoid crossing this dangerous threshold.

This record-high temperature in 2024 was associated with the prolonged 40°C (104°F) heat waves in Bangladesh and India, warm ocean waters that supercharged hurricanes in the Gulf of Mexico and cyclones in the Philippines, and the roasting summer that caused the most destructive wildfires in California.

The phenomenon of climate change is intricately linked to the greenhouse effect, a well-established scientific principle. Fig. 1.7 displays the key features of this phenomenon. A breakdown of the mechanisms involved is shown as follows:

Solar energy and planetary response: Incoming solar radiation warms the Earth's surface. The planet then naturally emits a portion of this absorbed energy back toward space in the form of infrared radiation (heat).

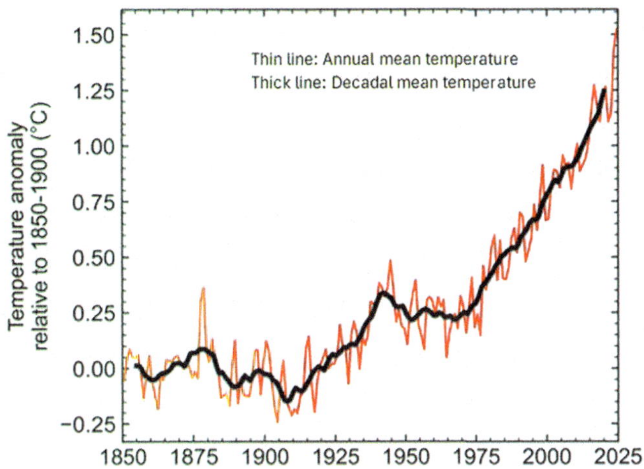

FIGURE 1.6 Global average temperatures compared with the late 19th century average temperature [6].

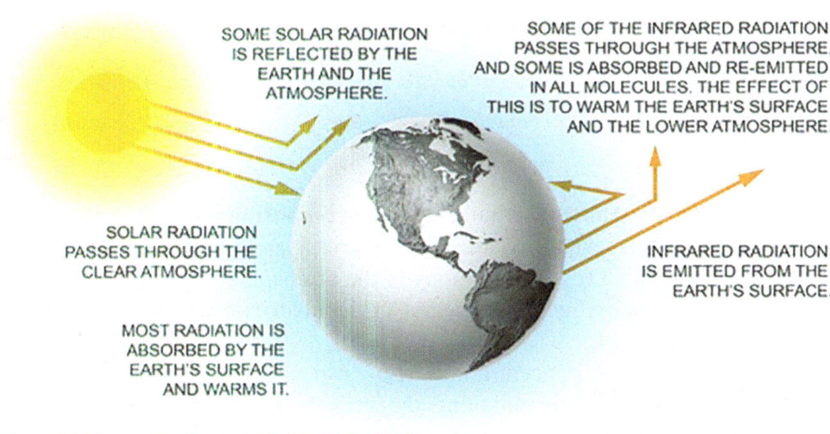

FIGURE 1.7 Greenhouse gas effect (EIA, Energy and the Environment Explained, 2019) [7].

The role of greenhouse gases: Specific atmospheric gases, including carbon dioxide (CO_2), water vapor, methane, and nitrous oxide, function as a radiative transfer agent. These gases absorb and re-emit infrared radiation within the atmosphere.

The amplifying effect: While some infrared radiation escapes back into space, a portion is trapped by greenhouse gases, causing a net warming effect within the atmosphere and at the Earth's surface. This phenomenon is known as the greenhouse effect.

Among these greenhouse gases, CO_2 plays a particularly significant role. Its molecular structure makes it highly efficient at absorbing and retaining infrared radiation, amplifying the greenhouse effect (Box 1.1).

BOX 1.1 Tech note—CO_2 emission

- On average, a person breathes out about 1 kg of CO_2 per day. This can vary based on factors like age, body size, and activity level. Generally, a human generates approximately 0.9 tons (900 kg) of CO_2 per year through respiration alone. However, lifestyle factors—like transportation, food consumption, and other activities—Significantly add to a person's overall CO_2 footprint. In developed countries, an average individual's total carbon footprint can range from 5 to 20 tons of CO_2 per year.
- An average gasoline-powered car emits about 4.6 metric tons of CO_2 per year. This estimate depends on a few key factors:
  - *Fuel efficiency*: A typical car with a fuel efficiency of around 22 miles per gallon (mpg) emits about 8887 grams of CO_2 per gallon of gasoline burned.
  - *Annual mileage*: If the average driver covers approximately 11,500 miles per year, this equates to around 4.6 metric tons of CO_2 annually.
- On average, a mature tree can sequester about 10 kg of CO_2 per year, although this amount varies based on species, age, location, and growth conditions. Over its lifetime, a single tree might absorb approximately 1 ton (1000 kg) of CO_2.
- Currently, nearly 40 billion tons of carbon dioxide are added to the atmosphere each year.

Human activity: Human activities like the combustion of fossil fuels for energy generation, deforestation, industrial processes, and agriculture significantly contribute to rising CO_2 concentrations in the atmosphere. Since the Industrial Revolution, these activities have driven CO_2 levels to unprecedented heights in the past 800,000 years.

As greenhouse gas concentrations rise—especially CO_2—the atmosphere holds onto more heat, strengthening the greenhouse effect in ways we can already observe. Average global temperatures continue to climb, ice caps and glaciers are retreating, and sea levels are rising as a result. At the same time, weather patterns are shifting, with extreme events such as heat waves, droughts, and floods becoming more frequent and, in many regions, more intense. These changes ripple through natural systems, stressing ecosystems and accelerating biodiversity loss.

The consequences are not limited to the environment. Food and water supplies become less reliable as rainfall patterns change and heat stress grows. Communities face greater exposure to floods, droughts, hurricanes, and other climate-driven hazards, while infrastructure—especially in coastal areas—must contend with higher seas and stronger storm impacts. It is also important to keep the bigger picture in mind: Greenhouse gases are not inherently harmful.

A natural level of them is essential for a livable planet; without this warming effect, Earth's average temperature would be about −18°C, far too cold for life as we know it. The problem is the rapid buildup of CO_2 from human activity, which has thickened this "blanket" beyond its stable range and pushed global temperatures upward.

A large body of research links rising greenhouse gas emissions to warming, but important questions remain about how long-term drivers interact and amplify one another over decades. We return to those findings later in Chapter 8. What is clear, however, is the scale of response needed: to keep warming near the critical 1.5°C threshold, the IPCC highlights the need to reach net-zero emissions by 2050.

Moving toward a low-carbon economy and net-zero by 2050

A growing group of organizations—including Equinor, BP, the IEA, Bloomberg NEF, and Rystad Energy—have developed detailed scenario studies to map how global oil demand could evolve through 2050. These exercises do not rely on a single forecast; instead, they test a range of futures built around a handful of key drivers. The policy environment is one of the most influential. As countries tighten climate regulations, expand carbon pricing, and set more ambitious emissions targets, the cost and acceptability of fossil fuels shift accordingly. Technology is another cornerstone, especially progress in renewables, battery storage, and electrification, which determines how quickly cleaner alternatives can replace oil in sectors like power generation and transportation. Economic conditions also matter: GDP growth, industrial expansion, and regional development patterns shape overall energy demand and the relative competitiveness of different fuels.

Despite the uncertainty these models are designed to capture, one message shows up again and again—fossil fuel consumption trends downward across essentially all scenarios. In several pathways, renewables scale rapidly, consistent with a world attempting to reach net-zero greenhouse gas emissions by mid-century. That faster transition requires coordinated changes: wider deployment of renewable electricity, accelerated adoption of electric vehicles, and major gains in energy efficiency.

Fig. 1.8 summarizes the IEA's outlook across fossil fuels, renewables, and other energy sources such as nuclear and biomass. The solid lines reflect energy demand under current policies, while the dashed lines show how demand could change if governments fully implement their stated policy commitments. Under this outlook, total global energy consumption is projected to reach about 838 exajoules by 2050, with roughly 30% supplied by renewables and about 60% still coming from fossil fuels. Within that fossil share, coal and petroleum use decline over time, while natural gas increases—highlighting a transition in fuel mix even in futures where fossil energy remains significant.

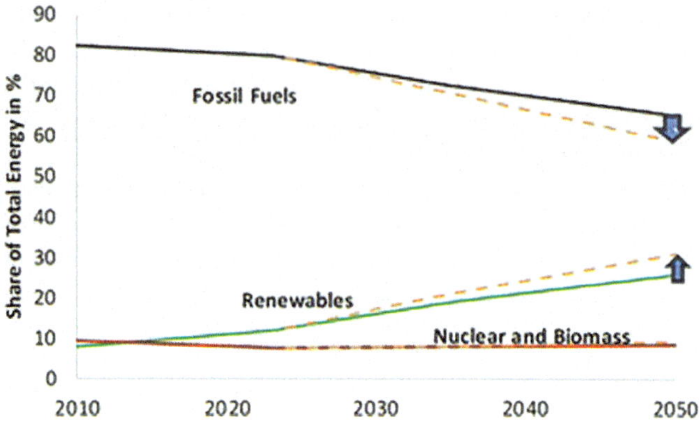

FIGURE 1.8 Projection of fossil and nonfossil fuel consumption until 2050. (*Data from Ref. [3]*).

The projected decline in fossil fuel consumption is largely driven by two forces. First, the rapid growth of electric vehicles is expected to reduce demand for gasoline and diesel as battery performance improves, costs fall, and charging infrastructure becomes more widespread. Second, tighter environmental regulations—along with stronger recycling systems and restrictions on certain plastic applications—are likely to curb fossil fuel demand in the petrochemical sector by reducing the need for virgin, fossil-based feedstocks.

The global commitment to net-zero

A growing number of countries are establishing ambitious net-zero emission targets and formulating comprehensive national strategies to achieve them. This growing international consensus reflects a paradigm shift toward clean energy, with a clear recognition of the urgent need to decarbonize the global economy. While current Nationally Determined Contributions (NDCs) may not be sufficient to keep warming below 2°C, recent revisions demonstrably showcase increased ambition, with significant emission reduction projected by 2030 [8].

The critical imperative of climate action necessitates acknowledging the inextricable link between global decarbonization efforts and the future of fossil fuels. The transition to clean energy will likely lead to a long-term decline in global demand for fossil fuels. This reality necessitates proactive planning and collaborative strategies between governments and the fossil fuel industry to manage this transition responsibly, minimizing economic disruptions and fostering a just and sustainable energy future.

Fostering innovation and shaping a sustainable future

Understanding the real tradeoffs in both renewable and nonrenewable energy systems helps people make better decisions and argue for solutions that are effective as well as responsible. That requires more than optimism about technology. Continued innovation is essential to improve the efficiency, affordability, and reliability of renewables and the infrastructure that supports them. At the same time, every energy project—whether it involves wind, solar, nuclear, natural gas, or oil—needs careful planning, credible environmental impact assessment, and practical mitigation measures to reduce ecological damage and community disruption. A sustainable transition also depends on demand-side action. Conservation and energy-efficient technologies can lower total energy use, easing pressure on supply systems and encouraging a culture of responsible consumption.

A credible pathway to a sustainable future, therefore, has to be multifaceted: better technology, responsible project development, supportive policies, and individual accountability working together. Choices made today will shape the energy landscape for decades, so decisions should be guided by evidence, long-term responsibility, and a commitment to passing on a cleaner and more reliable system to future generations.

Equity also has to be built into the transition. Many developing economies need financial and technological support to meet rising energy demand while protecting the environment. Universal access to electricity—especially across parts of Africa—remains both a practical necessity and a moral imperative, and it should be treated as a core objective rather than a side benefit. In parallel, governments need to prepare for workforce disruption by investing in retraining and transition support for workers and regions tied to fossil fuel production. The move toward net-zero emissions can create substantial employment opportunities, particularly in clean energy deployment and in energy efficiency—such as efficient appliances, electric vehicles, building retrofits, and construction. Yet those new jobs will not always appear in the same places, or require the same skills, as those lost in declining fossil fuel industries, and many displaced roles are well-compensated and geographically concentrated. That mismatch makes proactive planning essential: mapping community impacts early, directing investment toward regions likely to experience job losses, and offering targeted support such as skills training and, when needed, relocation assistance. A just transition is not automatic; it is a design choice.

Energy security remains another nonnegotiable. A stable energy supply underpins economic performance, public safety, and social stability. Shortages can trigger serious disruption, and heavy dependence on foreign energy sources can expose countries to price volatility and geopolitical risk. Even as nations pursue net-zero, they will continue to prioritize a level of energy reliability and resilience that protects households, industry, and critical services.

Finally, none of this works without public trust and participation. Successful transitions are built with people, not imposed on them. Citizens need to be engaged as informed stakeholders whose concerns and priorities shape decisions, not as passive observers. That principle aligns with the goals advanced by efforts such as the Global Commission on People-Centered Clean Energy Transitions, which emphasizes inclusion, fairness, and shared ownership of the path forward.

Summary

This introductory chapter was meant to provide an early and brief exploration of various energy sources, encompassing both renewable and nonrenewable options. Some of the challenges associated with each option were provided, from environmental concerns and resource depletion to economic considerations and social implications. More in-depth discussions are provided in the following chapters.

Key takeaways

- Understanding energy sources and their challenges is crucial for informed decision-making.
- Transitioning towards renewable energy sources is essential for mitigating environmental damage and ensuring long-term resource availability.
- Addressing societal challenges like energy security, access, and a just transition for the workforce is vital in achieving a sustainable energy future.

References

[1] Data from Rystad Energy. *Global Energy Scenarios 2025; Rystad Energy*: Oslo, 2025. https://www.rystadenergy.com/flagship-report-energy-scenarios-2025 (Accessed 18 Jan 2026).
[2] Energy InstituteStatistical Review of World Energy 2024 (73rd ed.), Energy Institute, London, 2024 https://www.energyinst.org/__data/assets/pdf_file/0006/1542714/684_EI_Stat_Review_V16_DIGITAL.pdf (Accessed 18 Jan 2026).
[3] International Energy Agency (IEA)World Energy Outlook 2025, IEA, Paris, 2025 https://www.iea.org/reports/world-energy-outlook-2025 (Accessed 18 Jan 2026).
[4] Vince, G. Nomad Century: How Climate Migration Will Reshape Our World; Flatiron Books: New York, 2022; 260.
[5] Z. Hausfather, Factcheck: what greenland ice cores say about past and present climate change, Carbon Brief (March 5) (2019). https://www.carbonbrief.org/factcheck-what-greenland-ice-cores-say-about-past-and-present-climate-change/. (Accessed 18 Jan 2026).
[6] P.M. Forster, C. Smith, T. Walsh, W.F. Lamb, R. Lamboll, C. Cassou, M. Hauser, et al., Indicators of global climate change 2024: Annual update of key indicators of the state of the climate system and human influence, Earth System Science Data Discussions (2025) 1–72.
[7] U.S. Energy Information Administration (EIA). Energy and the Environment Explained. https://www.eia.gov/energyexplained/energy-and-the-environment/ (Accessed 18 Jan 2026).
[8] M.G.J. den Elzen, I. Dafnomilis, N. Forsell, P. Fragkos, K. Fragkiadakis, N. Höhne, T. Kuramochi, L. Nascimento, M. Roelfsema, H. van Soest, F. Sperling, Updated nationally determined contributions collectively raise ambition levels but need strengthening further to keep paris goals within reach, Mitig. Adapt. Strateg. Glob. Change 27 (2022) 33. https://doi.org/10.1007/s11027-022-10008-7.

Part II

Non-renewable energy sources

Synopsis

On my visits to many oil fields around the world as an engineer in the 1990s, I frequently encountered scenes of horrific environmental degradation: Large ponds of produced wastewater, oil spills, and gas flares belching black smoke into the air, making breathing difficult (see Fig. P2.1). A few times, I saw streams of oil flowing like water around oil storage tanks and separators.

FIGURE P2.1 Environmental pollution around an oilfield.

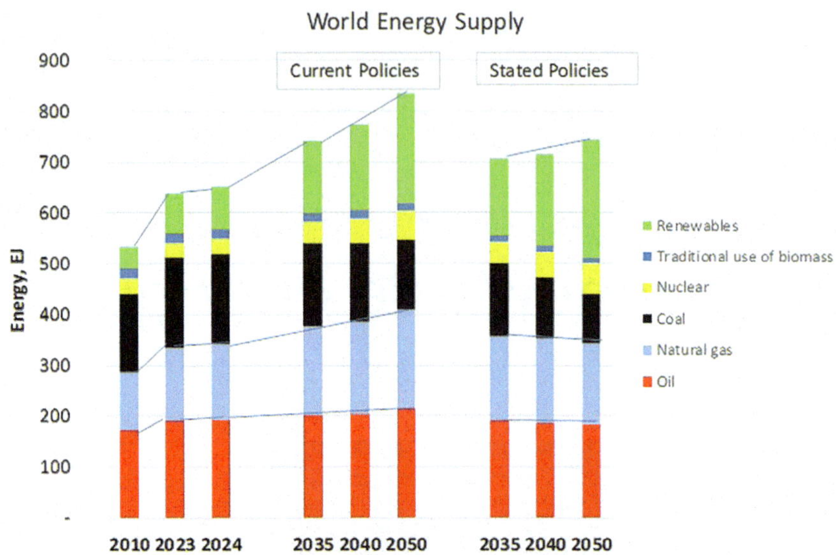

FIGURE P2.2 Historical world energy supply and future forecast [1].

While operational safety was emphasized by the operators, the environmental aspects were mostly ignored. Although, the environmental performance of the oil and gas industry has improved significantly, several disasters such as the Amoco Cadiz oil spill (1.6 million barrels in 1978 contaminating over 200 miles of coastline), Exxon Valdez oil spill (260,000 barrels in 1989 impacting 1300 miles of shoreline), and BP Macondo blowout (about 5 million barrels oil release into Gulf of Mexico in 2010), profoundly shaped public perception, casting a long shadow and reinforcing the image of the industry as an environmentally irresponsible entity.

The term "nonrenewable energy" applies to energy sources that are depleted upon usage and cannot be replaced in a reasonable time. These include fossil fuels and nuclear energy. In their 2025 world energy forecast report, the International Energy Agency (IEA, 2025) projected the future energy supply (in Exajoules, EJ, these terms are defined in the Unit Conversion attachment) for two different scenarios as shown in Fig. P2.2 (please see Box P2.1 for description of these scenarios). Under the current policy scenario (CPS), the fossil fuel supply would increase to about 65% of the global energy supply. However, under the stated policies scenario (STEPS) developed to protect the environment, there would be a reduction in total energy supply and fossil fuels as shown on Fig. P2.2. Notice that both scenarios show sharp increases (2.5 times) in renewable energy sources such as solar, hydro, wind, and biomass. If the STEPS scenario could be implemented, compared to today's 80%, the fossil fuel share of global energy supply is expected to drop to around 50% by 2050.

BOX P2.1 Tech note: IEA future scenarios.

- The current policies scenario (CPS) considers a snapshot of policies and regulations that are already in place and offers a generally cautious perspective on the speed at which new energy technologies can be deployed in the energy system.
- The stated policies scenario (STEPS) considers the application of a broader range of policies, including those that have been formally tabled but not yet adopted, as well as other official strategy documents that indicate the direction of travel.

Reference

[1] International Energy Agency (IEA), World Energy Outlook 2025—Free Dataset (Tables for Scenario Projections), IEA, 2025, https://www.iea.org/data-and-statistics/data-product/world-energy-outlook-2025-free-dataset#tables-for-scenario-projections (Accessed January 19, 2026).

Chapter 2

Fossil fuels

Synopsis

Fossil fuels have been the main energy source in most of the world for many generations, since the 1800's with the expanded use of coal. Petroleum, natural gas (NG), and coal are the primary categories of fossil fuels. There is no question about the important role of petroleum products in our daily life and how it has revolutionized modern times. Unfortunately, what comes with it is man-made climate change. Paradoxically, the same material that sustains our modern way of life has become a threat. In this chapter, we will become more familiar with the production and usage of these fuels. We will focus on both the pros and cons of fossil fuel usage in the world and its impact on societies and the environment. Then, we will shift our focus to nuclear energy.

From whale oil to petroleum

In the 18th and 19th centuries, whale oil was widely used for lighting homes and streets. It was a superior illuminant compared to earlier alternatives like candles and olive oil. The demand for whale oil fueled a booming industry that resulted in the killing of many whales. A great story is told by Herman Melville in his book, *Moby Dick,* about a maniacal quest for vengeance by Ahab, the captain of the whaling ship, Pequod, against Moby Dick, the giant white sperm whale. Such struggles were common in capturing whales because the whaling industry was financially lucrative.

Petroleum was once thought to have mystical properties. In Western Pennsylvania, the original center of American oil exploration, shamans from the Seneca tribe used this black substance in their ritualistic healing ceremonies. They also used it to seal their canoes. In the Middle East, NG leakage to the surface, once ignited, could be the location of fire temples.

In 1859, Edwin Drake drilled the first successful commercial oil well in Titusville, Pennsylvania. This marked the start of the modern petroleum industry. As petroleum production became more common, the lower cost and greater availability of petroleum led to a decline in the demand for whale

F.A.C.T.S. Flexibility, Affordability, Conservation, Transition, Sustainability: Our Path to the Future of Energy and the Environment. DOI: https://doi.org/10.1016/B978-0-443-52574-2.00012-4

oil. Thus, some might conclude that the emergence of petroleum "saved the whales!"

Petroleum resources

Some hundred million years ago, in a warmer atmosphere rich in carbon dioxide, ocean-dwelling microorganisms like plankton thrived [1]. Photosynthesis, a process that combines atmospheric carbon dioxide with water to form chlorophyll, created the building blocks of earth's carbon-based life. Released oxygen enriches the atmosphere. As these plankton died and sank to the ocean floor, and as more sediment accumulated, the pressure and temperature of the organic material increased and caused chemical changes to the organic matter, transforming it into a waxy substance known as *kerogen*. With further burial, usually at depths of several kilometers, kerogen underwent thermal cracking, breaking down into simpler hydrocarbon molecules like oil and gas. These newly formed hydrocarbons migrated upward through porous rocks, such as sandstone or limestone, until they became trapped by impermeable rocks, forming reservoirs where oil accumulated. This complex geological process, involving organic deposition, burial, heat, and pressure, ultimately resulted in the creation of underground oil and gas deposits.

On land, in swampy environments, massive accumulations of dead plants were buried and began their long transformation into coal deposits. The combustion of fossil fuels is the reverse of this long conversion process. Atmospheric oxygen is consumed to produce energy while CO_2 and water are released back into the atmosphere.

Underground reservoirs hold oil and gas within their porous structure. The larger the pore space within a rock, i.e., *porosity*, the more the location can hold hydrocarbons. The ease of production is related to how these pores are connected. This is referred to as *permeability*. Notice that in this book, oil, unless otherwise noted, includes crude (including shale oil and oil sands), natural gas liquids (NGLs), gas-to-liquids (GTLs), coal-to-liquids (CTLs), condensates, and refinery gains. Hydrogen-derived fuels are all fuels derived from low-carbon hydrogen, including ammonia, methanol, and other synthetic hydrocarbons. Renewables, unless otherwise noted, includes wind, solar, geothermal, biomass, biomethane, and biofuels and exclude large-scale hydro. Non-fossils include renewables, nuclear and hydro. Traditional biomass refers to solid biomass (typically not traded) used with basic technologies e.g. for cooking.

Oil and gas are found and produced through a series of activities that are carried out in three sectors of the oil and gas industry, namely, upstream, midstream, and downstream. The upstream sector of the oil and gas industry is responsible for exploration, drilling, and extraction activities. Midstream conducts transportation and shipping of gas, crude or unprocessed oil. The downstream sector is responsible for the conversion of gas and crude oil and distillation into other products, such as petrochemicals. Throughout these processes, environmental considerations, safety protocols, and regulatory compliance are paramount to ensure responsible extraction and minimize negative impacts on the environment and surrounding communities.

Upstream sector

The process begins with geological surveys and analysis to identify potential areas where oil and gas deposits might exist. This involves studying rock formations, seismic data, and other geological indicators.

Seismic imaging: Advanced seismic imaging techniques are used to create detailed underground maps of potential oil and gas reservoirs. This involves sending sound waves into the earth and monitoring the echoes to determine the structure and composition of underground formations.

Once a promising location is identified, drilling rigs are used to drill deep into the earth. This can involve vertical, horizontal, or directional drilling techniques.

After drilling, if hydrocarbon (oil and/or gas) is found, it is brought to the surface using a variety of methods. In the case of oil, it may flow naturally due to high pressure underground or require pumping. Gas may also flow naturally or need to be pumped to the surface. In many cases, only a fraction of the oil in a reservoir, e.g., between 5% and 40% of the original oil, can be extracted naturally or via *artificial lift* such as pumping. Enhanced oil recovery techniques, such as injection of water, gas (e.g., CO_2 or hydrocarbon gas), or steam, may be used to access more of the underground resources.

Once extracted, the oil and gas are processed to remove impurities such as sand or water to make them suitable for midstream transportation to refineries.

Midstream sector

The midstream sector of the petroleum industry encompasses the transportation, storage, and wholesale marketing of crude oil, NG, and refined petroleum products. This sector plays a vital role in facilitating the efficient and reliable movement of hydrocarbons from production fields to refineries, processing plants, distribution terminals, and ultimately to end consumers. Key infrastructure components of the midstream sector include pipelines, storage tanks, marine terminals, railcars, oil tankers, and trucks. Pipelines are the primary mode of transportation for crude oil and NG over long distances, while storage facilities ensure adequate supply during fluctuations in demand and provide strategic reserves for emergencies. Additionally, marine terminals handle the import and export of petroleum products, while oil tankers, rail, and truck transportation serve regions not accessible by pipelines. Overall, the midstream sector serves as the backbone of the petroleum industry, ensuring the smooth flow of hydrocarbons throughout the supply chain.

Downstream sector

The major components of the Downstream sector are refining and petrochemicals. The marketing, transportation, and distribution are not covered in this book.

Refining

Oil and gas undergo extensive processing at refineries to produce various valuable products. Here is a concise overview of the refinery processes:

Fractional distillation: Crude oil is heated in a distillation tower, where it separates into different components based on their boiling points. Lighter components like gasoline and jet fuel rise to the top, while heavier components like lubricating oil, diesel, and tar settle at the bottom.

Conversion: Some fractions undergo conversion processes termed *cracking* and *reforming* to break down large hydrocarbon molecules into smaller ones or rearrange their structures to produce more valuable fuel products like gasoline, diesel, and jet fuel.

Treatment: Various treatments, such as desulfurization and hydro treating, are carried out to remove impurities like sulfur, nitrogen, and metals from the refined products to meet environmental standards and improve their quality.

Blending: Different refined products are blended to meet specific requirements and achieve desired performance characteristics. For example, gasoline blending involves mixing various octane-rated components to meet regulated standards and consumer demand.

Product distribution: Once the refining process is complete, the various refined products are stored and then distributed via pipelines, trucks, ships, or railroads to end-users such as gasoline stations, industries including petrochemicals, and retail consumers of NG.

Petrochemicals

The petrochemical industry plays a crucial role in the global economy by converting hydrocarbons derived from processed crude oil, NG, and coal into a diverse range of chemical products. These products serve as building blocks for countless everyday items, ranging from plastics and synthetic fibers to pharmaceuticals and fertilizers. Petrochemical plants utilize various chemical processes such as cracking, reforming, and polymerization to transform raw hydrocarbons into valuable intermediate chemicals like ethylene, propylene, benzene, and xylene. These intermediates are then further processed to produce a vast array of end products that are integral to modern life.

Petroleum production

The historical oil production by region is shown in Fig. 2.1. In 2024, the top three oil producers were: the Middle East (31%), the USA (29%), and the Russian Federation (CIS) (14%). The trend indicates that there has not been any

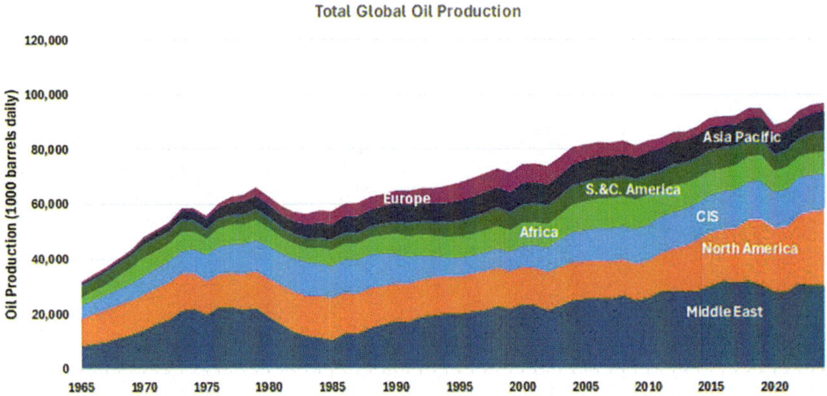

FIGURE 2.1 Historical oil production by region (CIS refers to the Russian Federation) [2].

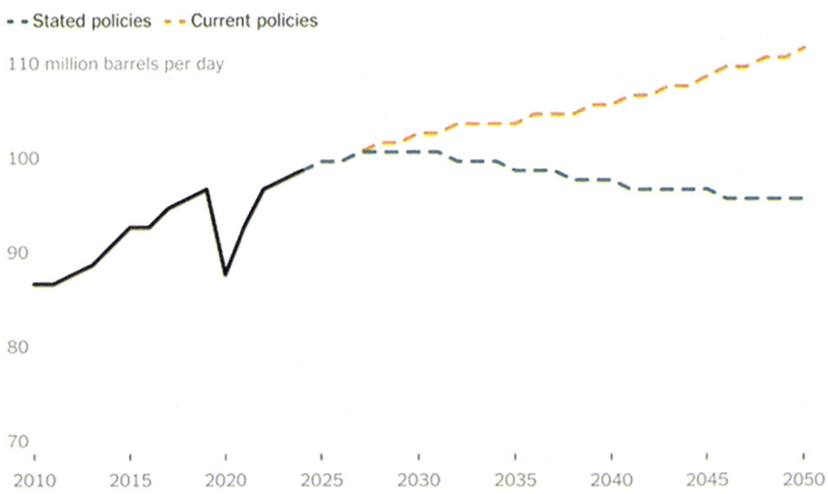

FIGURE 2.2 World projected oil (including petroleum products) demand [1].

slowdown in oil production. The dip in 2020 is due to the covid that resulted in low demand.

AS for the future, according to the IEA forecast, in their reference case, global oil consumption will increase, as shown in Fig. 2.2. This case represents business as usual, where net zero measures are not dominant. However, under the stated policies scenario, the oil demand will remain at the current level till 2035, and then, it will decline. The report's authors say that neither scenario should be considered a definitive prediction about the future. Instead, the scenarios are intended to allow politicians to see the consequences of different policy choices.

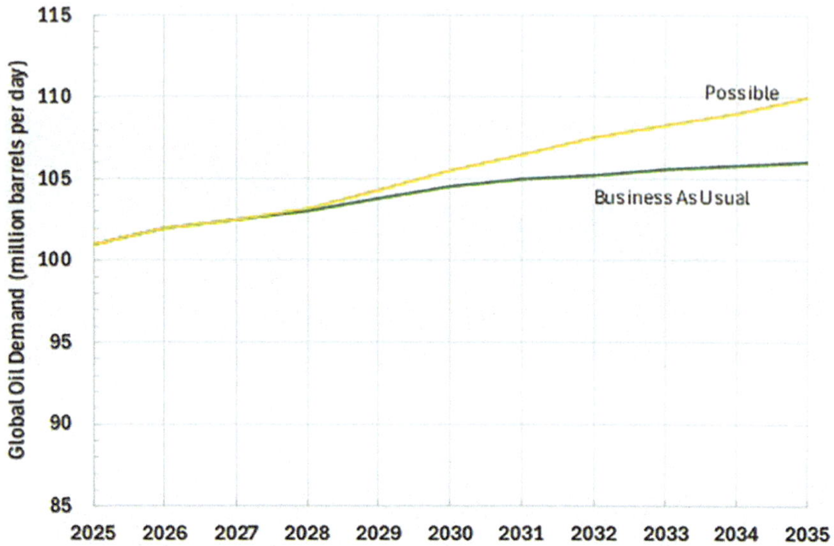

FIGURE 2.3 NCEA versus IEA global oil demand forecast [4]. *IEA*, International Energy Agency; *NCEA*, National Center for Energy Analytics.

Notice that road transport accounts for around 45% of global oil demand. In the stated policies scenario, the share of Electric Vehicles in total global car sales doubles by 2030 and rises above 50% by 2035. This puts the demand for road transport on a declining trajectory and leads to a peak in total oil demand around 2030.

Under the Stated Policy scenario, as climate policies are enacted and reinforced worldwide, new economic opportunities emerge, shaping the trajectory of future policy implementations. Additionally, shifts in geopolitical dynamics exert short-term deviation from this trajectory, underscoring the vulnerability of a global commodity like oil to fluctuations in price, demand, and politics.

Predicting future energy consumption scenarios is difficult and controversial. For example, the IEA oil demand scenarios have been challenged by other centers such as the National Center for Energy Analytics. They disputed the assumptions that the IEA had used to arrive at different future scenarios. They cite recent trends in countries not abiding by the Paris Agreement and a reduction in new renewable startups. Thus, they claim that in the case of oil, there is a high probability that demand will continue to grow in the foreseeable future and, possibly, quite significantly, as shown in Fig. 2.3[3]. It seems that their Obviously, their two scenarios (business as usual and possible) indicate that more diligence is required for combating greenhouse gas (GHG) emissions in the coming years.

Natural gas: A strategic bridge fuel in the fossil fuel era

Natural gas, primarily composed of methane (CH_4), is a hydrocarbon fossil fuel formed from the decomposed remains of organic matter over millions of years. Compared to coal and oil, it is lighter and exists in a gaseous state, requiring transport through pipelines. While classified as a fossil fuel and emitting greenhouse gases, NG occupies a unique and evolving position within the global energy landscape.

Natural gas has been by far the fastest-growing source of global energy between 2012 and 2024, providing 38% of the incremental energy added over that period. Global consumption reached 4.12 trillion cubic meters in 2024.

Liquefied natural gas market

Liquefied natural gas (LNG) is natural gas that has been cooled to about −162°C (−260°F), at which point it becomes a liquid. Liquefaction reduces the gas volume by about 600 times, making it economical to store and transport over long distances, especially where pipelines are not feasible. After transport, LNG is re-gasified and used like conventional NG. LNG produces ~50%–60% less CO_2 than coal and ~20%–30% less than oil (per unit of energy). One of the key advantages of LNG is about energy security. It diversifies energy supply sources and reduces dependence on single suppliers or pipeline routes.

Most European countries switched to LNG imports following the stoppage of NG imports from Russia. Global LNG trade was about 407 MT in 2024, showing continued growth in global markets. This accounted for about 60% of the global gas trade [5].

The ascendancy of natural gas

Projections indicate a significant rise in NG consumption, with estimates suggesting an increase exceeding 20% by 2050 (see Fig. 2). This trend can be attributed to several key factors:

NG offers a demonstrably cleaner burning profile compared to coal. It emits up to 50% less CO_2 during electricity generation. This environmental advantage is reflected in the ongoing conversion of coal-fired power plants to NG plants, as evidenced by data from the U.S. Energy Information Administration [6] showing over 100 such conversions between 2011 and 2019. Two different methods were used to switch coal-fired plants to NG. The first method was to retire the coal-fired plant and replace it with a new NG-fired combined cycle plant. The second method was to convert the boiler of a coal-fired steam plant to burn other types of fuel, such as NG. This trend not only highlights the environmental benefits of NG but also demonstrates the economic feasibility of modifications to existing energy infrastructure.

As the world transitions toward a future powered by renewable energy sources like solar and wind, NG serves a strategic role as a bridge fuel. while renewable infrastructure continues to develop and achieve cost competitiveness. Unlike solar and wind, which are weather-dependent, NG-fired power plants offer the advantage of being readily adjustable to meet fluctuating electricity demands, ensuring grid stability during the integration of renewables.

The versatility of NG extends beyond electricity generation. It serves as a vital fuel source for residential and commercial heating applications and plays a significant role in various industrial processes. NG is the main ingredient for producing nitrogen fertilizers. Thus, it contributes to global food production. This broad range of applications makes NG a crucial component of the current energy mix.

The future of natural gas: Opportunities and challenges

Despite its cleaner emissions compared to coal, NG still emits greenhouse gases when combusted to generate useful energy. Also, the methane leakage in the industrial sectors and power plants is a major concern since methane is a much more harmful pollutant than carbon dioxide. LNG's full cycle carbon footprint is 19.5 g CO_2 (equivalent) per Megajoules (MJ) of energy. This is much less than coal (90–110). Technological innovations, such as the production of methanol or hydrogen gas from NG with Carbon Capture, Utilization and Storage (CCUS), could offer a pathway for this resource (a topic that will be explored in chapter 8).

The long-term viability of NG depends on advancements in capturing carbon emissions and the competition from continued cost reduction of renewable energy sources. The development of new technologies and the geopolitical landscape will also play a role in shaping the future of NG Box 2.1.

Coal resources

Coal is formed underground through a process called coalification, which occurs over millions of years. It begins with the accumulation of organic material, primarily from plants, in swampy environments. Over time, as layers of sediment build up on top of organic matter, the increasing pressure and heat from the overlying sediment cause physical and chemical changes. This process, known as diagenesis, gradually transforms the plant material into peat, a soft and spongy substance with a high moisture content. With further burial and compaction, peat undergoes progressive stages of transformation, resulting in the formation of different types of coal. The degree of coalification depends on factors such as temperature, pressure, and the duration of burial. As a result of these geological processes, coal deposits are found in underground seams or layers, where they can be mined for energy production and other industrial purposes.

BOX 2.1 Tech note: Coal types

Coal was the main source of energy prior to the use of petroleum. Based on its carbon content, moisture content, and heating value coal has been classified mainly into five types: anthracite, metallurgical, bituminous, sub-bituminous, and lignite.

Anthracite: Anthracite is the highest grade of coal and has the highest carbon content and energy density. It is used in industrial processes and residential heating. It is typically hard, shiny, and black. Anthracite is primarily mined using deep mining methods such as underground mining.

Metallurgical: Also known as coking coal, it is used in steel production due to its ability to create a high-temperature, carbon-rich environment necessary for the production of steel.

Bituminous: Bituminous or thermal coal is the most abundant type of coal and is widely used for electricity generation and in industrial processes. It has a lower carbon content than anthracite but higher than sub-bituminous and lignite. Bituminous coal is usually found closer to the surface and can be extracted through both surface mining (open-pit mining) and underground mining methods.

Sub-bituminous: Sub-bituminous coal has a lower carbon content and heating value compared to bituminous coal. It is typically softer and contains more moisture. Sub-bituminous coal is often found at shallower depths than bituminous coal and is primarily extracted through surface mining methods due to its proximity to the surface.

Lignite: Lignite, also known as brown coal, is the lowest grade of coal with the lowest carbon content and heating value. It has a high moisture content and is relatively soft. Lignite is often found near the surface in relatively large deposits and is primarily extracted through surface mining techniques.

Coal extraction

The extraction of coal involves several methods depending on factors such as the depth of the coal seam, its location, and economic considerations:

Surface mining methods, including open-pit mining and strip mining, are used to extract coal deposits that are close to the surface. This involves removing layers of soil and rock to access the coal seam.

Underground mining techniques, such as room and pillar mining and long-wall mining, are employed to extract coal deposits found at greater depths. These methods involve excavating tunnels and chambers to access and extract coal seams.

In certain regions, particularly in the United States, mountaintop removal mining is used to extract coal deposits from mountains and hillsides. This method

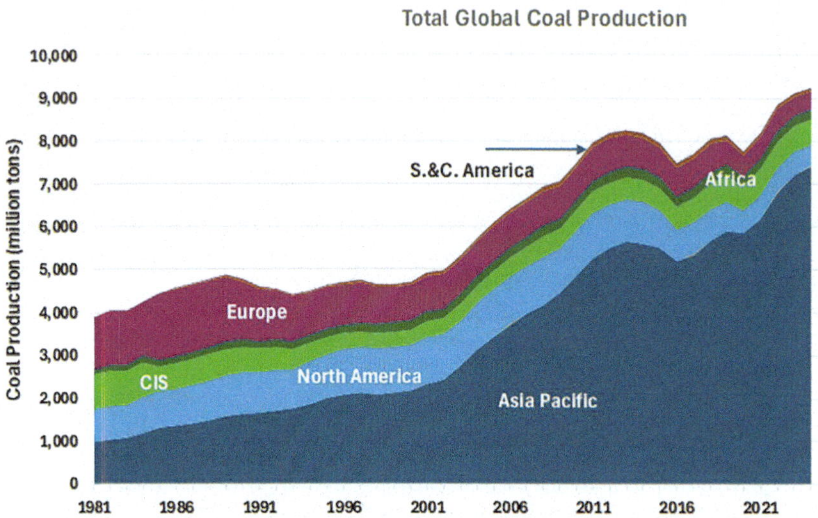

FIGURE 2.4 World total coal production in million metric tons [2].

involves removing the summit of a mountain to access coal seams beneath the surface.

Each extraction method has its own environmental and social impacts, and the choice of method depends on factors such as the geological characteristics of the coal deposit, accessibility, and regulatory requirements. Additionally, measures are often taken to mitigate the environmental impacts of coal mining, such as land reclamation and water management practices.

Coal production

The historical production of coal worldwide has steadily increased since the early days of its use as an affordable energy source (see Fig. 2.4). Notice that the data for European countries (EU27) are separated from the Rest of the World in 1990. In 2022, global coal consumption reached 8465 million metric tons (mt), with the top four consumers being China (55%), India (12.5%), Western Europe (7.5%), and the U.S. (6.7%).

The global coal demand forecast (in Million tons Coal equivalent, Mtce) for two scenarios, current policies and stated policies, is shown in Fig. 2.5. Most global coal demand, 55%, comes from electricity generation in emerging market and developing economies, largely China. According to the IEA forecast, the use of coal will be dominant in China through 2050 due to its huge supply. India's consumption will increase while the USA's consumption will drop, as shown in Fig. 2.5.

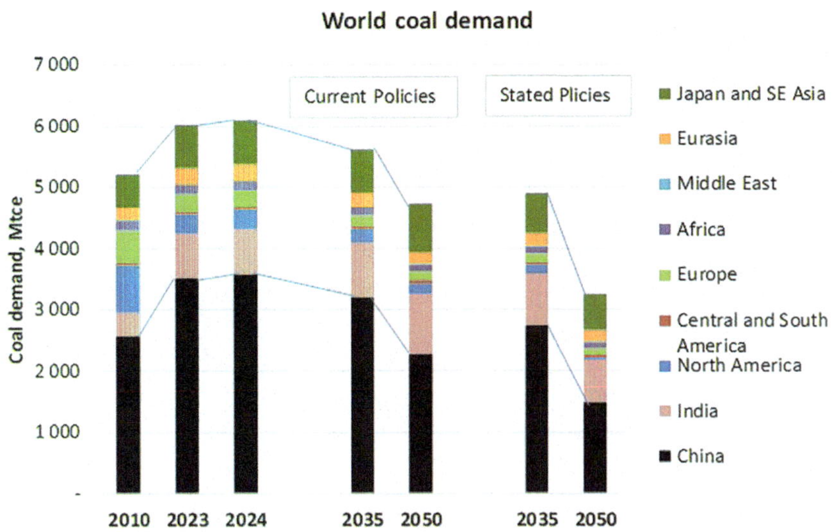

FIGURE 2.5 World coal demand by region [1].

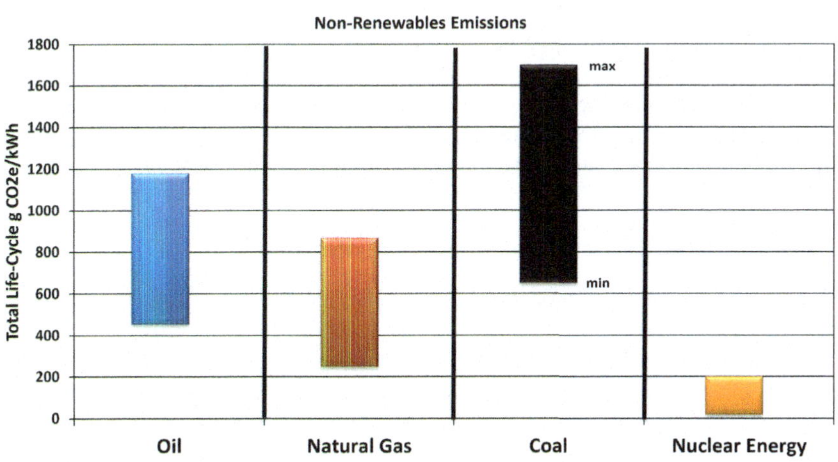

FIGURE 2.6 Life cycle greenhouse gas emission estimates [7].

Environmental impacts of fossil fuels

The primary human activity responsible for emitting CO_2 is the combustion of fossil fuels for energy and transportation purposes. Fig. 2.6 shows the ranges of CO_2 emission values (minimum, average, maximum) in g of CO_2 equivalent per kilowatt-hour for various nonrenewable energy sources. The term "life cycle" encompasses all emissions resulting from the production and consumption of a material. It is evident that simply transitioning from coal to NG in power plants

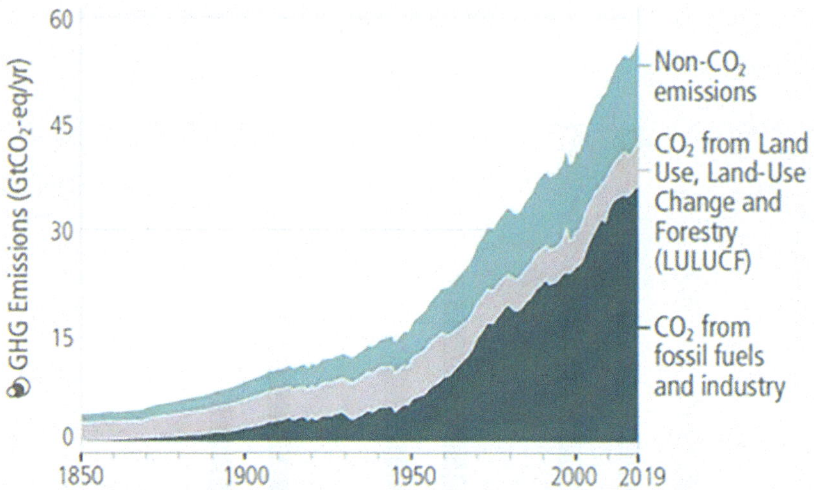

FIGURE 2.7 Main sources of greenhouse gas emissions resulting from human activities [8].

can halve GHG emissions. Obviously, the value for fossil fuel emissions would be less when combined with CCUS.

Additionally, various processes produce CO_2 emissions through chemical reactions unrelated to combustion. Examples include the production of mineral products like cement, metals such as iron and steel, and chemicals. A comparison is made between different sources of GHG emission in Fig. 2.6. It is apparent from this graph that fossil fuels constitute the dominant sources of emissions.

Fossil fuel industry and emission reduction

Historical sources of GHG emissions (in Gigatons CO_2 equivalent per year) resulting from human activities are presented in Fig. 2.7. generally, oil producers approach toward environmental concerns has been typically reactive rather than proactive, with attention to environmental aspects of oil and gas production often catalyzed by government regulations or the threat of penalties. Instances of environmental disasters such as oil blowouts, major spills, gas flaring, groundwater contamination, and seismic events have prompted public outcry, leading companies to implement follow-up measures to mitigate environmental impacts. By the same token, the initial response from some of the major oil companies to warnings about the impact of GHG emissions on global warming were often characterized by skepticism and attempts to discredit scientific reports. In the early 1990's a few companies started a handful of CO_2 sequestration projects (e.g., Equinor's Sleipner project in Norway) to demonstrate part of the solutions. It has taken years of continuous scientific studies, data analysis, and increased occurrences of natural disasters to raise public awareness about the severity of GHG impacts on the environment.

Subsequent public pressure and government regulations have compelled some oil companies to take corrective actions, with an increasing number of companies now showing an inclination toward incorporating green energy into their future portfolios. However, despite possessing the necessary technology and expertise, many oil companies cite costs and the need for government incentives as significant barriers to undertaking capital-intensive emission reduction projects. Some express concerns that unilateral action may raise expectations and result in broader regulatory mandates.

Some skeptic consumers perceive the current shift toward renewable energy by the oil companies as an act of "Greenwashing," a term used to describe misleading marketing practices where a company, product, or business is falsely or excessively promoted as environmentally friendly. This phenomenon undermines genuine efforts to reduce emissions and address the climate crisis. Through deceptive marketing and unfounded claims of sustainability, greenwashing misleads consumers, investors, and the public, eroding trust and hindering the ambitious action necessary to combat climate change and achieve a sustainable future. In subsequent chapters, we will explore how a proactive approach by the fossil fuel industry can restore public trust and garner confidence among environmentalists, fostering a constructive dialogue and facilitating meaningful progress toward sustainability. The key issues related to the impact of the fossil fuel industry on the environment are discussed in the following section.

Legacy wells, a key environmental issue

Once an oil well has gone through its useful life, the governmental regulations require the operator to plug and abandon that well. This is to prevent any future leakage of hydrocarbons into the atmosphere, protect groundwater and drinking water, avoid surface hazards, and reclaim the land to its original form. In 2018, the United States Environmental Protection Agency estimated that nationwide, there were 3.11 million abandoned (legacy) oil and gas wells. A total of 69% or 2.15 million of these wells are not even plugged [9]. Many were drilled prior to the existence of state regulatory programs, subsequently abandoned by their original owners or operators over a century ago and then left unplugged or poorly plugged. Also, there are at least 400,000 legacy wells in Canada, and tens of millions worldwide. Fig. 2.8 shows the legacy wells in the United States.

As the former owners of these abandoned wells cannot be traced or are unable to address the cleanup, the responsibility for plugging these wells typically falls on governments, who may require further guidance on managing these orphaned wells. Several studies have underscored the air, water, and human health hazards posed by orphaned oil and gas wells. State and federal regulators are in the process of plugging these wells, but the process is slow; many are still unplugged today.

In a recent study, researchers at McGill University found that over 4.6 million Americans, comprising approximately 13% of the nation's population, reside

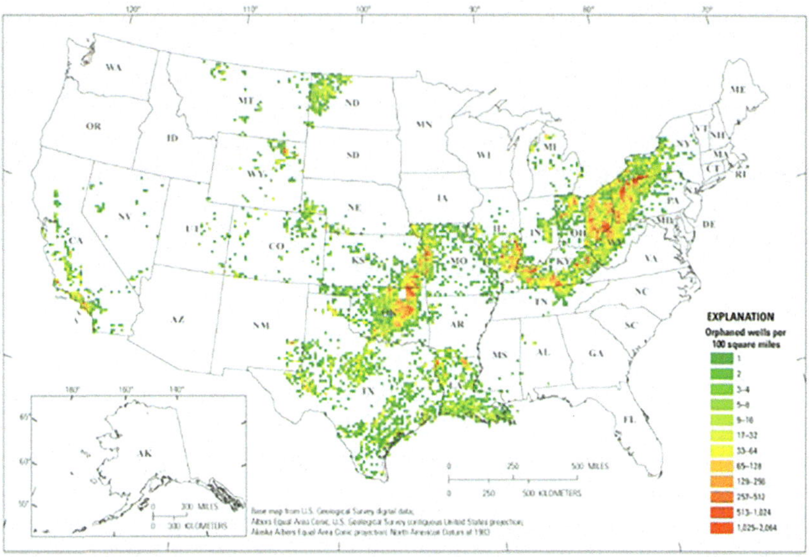

FIGURE 2.8 U.S. based legacy wells [10].

within 1 km of one of the more than 80,000 documented orphaned gas or oil wells in the US. Notably, there was a disproportionate representation of economically poorer populations among this demographic. Moreover, the study revealed that over one-third of these wells are located approximately 1 km from domestic groundwater wells, raising concerns about potential contamination risks [11].

Mitigating the impact of these legacy wells would come at a considerable cost. For instance, in California, the total cleanup bill for the onshore oil and gas industry is estimated to be around $21.5 billion. The oil industry has allocated only about $106 million for cleanup nationally, which regulators can access when a company ceases operations or otherwise neglects its responsibilities. However, this amount represents less than 1% of the estimated cleanup cost [12].

Nevertheless, there are alternative ways to utilize these legacy wells rather than reclaiming the land. The subsurface represents a valuable natural resource, and access to subsurface reservoirs unaffected by oil or gas leakage will be crucial for emerging energy resources. A significant portion of documented orphaned wells (91%) are situated in areas with subsurface storage potential for carbon dioxide, hydrogen, and NG, provided they meet containment standards. Rather than restoring the surface to its predeveloped state, the land could be repurposed for wind power generation, as nearly 75% of orphaned wells are in areas with top wind capacity. Additionally, approximately 33% of orphaned wells are situated in regions, such as North Dakota, that are moderately favorable for geothermal systems, while 1% are found in areas like Utah, Colorado, and California, which are deemed highly conducive to geothermal development.

Oil and gas pipelines

Just as there may be more cost-effective ways to deal with legacy wells, existing petroleum infrastructure can also be transitioned into uses for cleaner hydrocarbon or renewable energy sources. Existing gas pipeline infrastructure presents an opportunity for expedited deployment of CO_2 or hydrogen infrastructure through repurposing or reuse, thereby reducing lead times and mitigating the need for extensive new construction. Repurposing pipelines not only diminishes the environmental footprint of projects by curbing material demand but also allows infrastructure owners to optimize the economic longevity of their assets while deferring decommissioning expenses. With minimal adjustments, gas pipelines can accommodate hydrogen blends of up to 20% by volume (equivalent to 6% in energy terms), contingent on their material composition. However, full conversion to transport hydrogen is a more intricate process, necessitating threefold increases in volumetric flows to operate at their maximum capacity due to hydrogen's lower volumetric density compared to NG. Despite the requirement for more robust and costly compressors, repurposing pipelines could prove cost-competitive against constructing new hydrogen infrastructure, slashing investment expenditures by 50%–80% relative to new line construction [13]. However, practical experience is limited, and significant reconfiguration and adaptation will be necessary. Moreover, gas pipelines offer potential for CO_2 transport, with ongoing and forthcoming projects in the United Kingdom, the Netherlands, and the United States. Thorough assessments of pipeline strength, pressure rating, and integrity are imperative to ascertain their suitability for safe CO_2 or hydrogen transportation [14].

Conversion of power plants' fuel from coal to natural gas

Converting coal-fired power plants to NG is a practical and cost-effective way to reduce emissions, comply with regulations, and transition toward cleaner energy sources [15]. New coal plants generally cost about \$3–5 million per MWh while NG plants cost about \$1–1.5 million per MWh [16]. NG emits 50%–60% less CO_2 than coal per unit of electricity generated (from 1 ton per MWh for coal to 0.4 ton per MWh for gas). It also eliminates coal ash, sulfur dioxide (SO_2), and particulate matter pollution. They also require less water than coal boilers for cooling and steam generation.

While conversion offers economic and environmental benefits, challenges like infrastructure upgrades, fuel supply logistics, and upfront costs must be carefully managed. Ultimately, this transition serves as an intermediate step toward a fully renewable energy future. The conversion process can take different forms, depending on the plant's existing infrastructure and the desired level of efficiency. The simplest form of conversion involves replacing the coal burners with NG burners while keeping the existing boiler, steam turbine, and generators intact. A more efficient but costlier approach is the "combined cycle" method, in

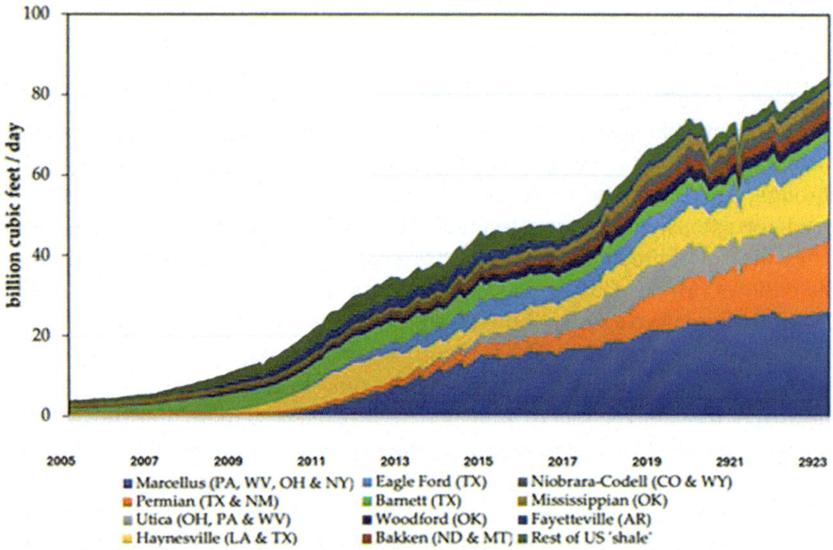

FIGURE 2.9 Shale gas production in the USA [19].

which an NG combustion turbine is used. For the coal plants that are too old, a "Simple cycle" method is used, where the coal plant is entirely decommissioned, and a new NG-fired turbine is installed. Retrofitting a coal plant to NG could cost between \$50–500 million, depending on the conversion from burner to full decommissioning [17].

Fracking – A savior or a demon

Many in the public sector believe that fracking is one of the most damaging processes in the oil industry, associating it with groundwater contamination, earthquakes, and pollution. This section explores both the pros and cons of this controversial practice.

Although fracking has been in use since the 1950s, its combination with horizontal drilling in the 1990s revolutionized the oil and gas industry, particularly in unlocking shale formations. Global proven NG reserves are estimated at 7257 trillion cubic feet (Tcf), nearly matching the 7299 Tcf of global shale gas resources [18]. These reserves could potentially meet the world's energy demands for over 60 years. In fact, the United States became the world's top NG producer largely due to shale gas extraction. Fig. 2.9 illustrates the contribution of various shale plays to total U.S. gas production.

Shale formations have extremely low permeability, meaning hydrocarbons cannot flow easily through them. Permeability measures a rock's ability to allow fluids to pass through. Due to the low permeability of shale, stimulation is necessary to extract oil and gas efficiently. Hydraulic fracturing (HF) (fracking) is a technique in which high pressure fluids are injected into rock formations to create fractures, enhancing hydrocarbon flow and increasing production rates. There are potential environmental and health risks with such technology, including:

Earthquakes: The injection of large volumes of water during HF and the disposal of produced water into underground wells can induce seismic activity. These injections increase pore pressure, potentially destabilizing existing faults and triggering earthquakes [20]. Addressing this issue requires identifying susceptible geological conditions, implementing operational controls, improving mitigation strategies, and enforcing science-based regulatory management [21].

Groundwater contamination: Fracking begins with seismic surveys that can disturb local ecosystems, followed by deep drilling operations that extend more than a mile underground. The process consumes millions of gallons of water and introduces chemicals deep into the earth. In some locations, this poses a considerable risk, as contaminants such as heavy metals and chemicals have been detected in nearby groundwater [22]. A balanced assessment of HF is necessary to weigh its economic benefits against its environmental impact.

Soil contamination: Shale gas extraction generates pollutants, including metals, salts, and organic compounds, which pose potential risks to soil quality [23]. To mitigate contamination, strict leak prevention measures must be enforced, such as securely storing flowback water in heavy-duty tanks to prevent leakage of fracturing fluids and pollutants into the soil [24].

Fresh water consumption: HF operations consume billions of gallons of water. In response to growing concerns, the industry is shifting toward wastewater reuse and recycling, a topic discussed further in Chapter 6.

Methane leaks: The gas collection phase presents challenges, particularly methane leaks, which significantly contribute to greenhouse gas emissions. The introduction of methane detection systems has begun to reveal the scale of the problem, and efforts are underway to reduce leakage.

To mitigate the risks associated with HF, continuous environmental monitoring, proactive safety measures, and regulations tailored to local geological and regional conditions are essential. This technology underpins the implementation of geothermal energy in tight rocks (as explained in Chapter 6).

Case study: Preserving Alaska

Alaska, often called "The Last Frontier," is a land of rugged beauty, vast wilderness, and untapped resources. While its breathtaking landscapes and extreme climate make it unique, its history has been shaped by two major economic booms—the Gold Rush of the late 19th century and the oil boom of the 20th century. These events not only transformed Alaska's economy but also

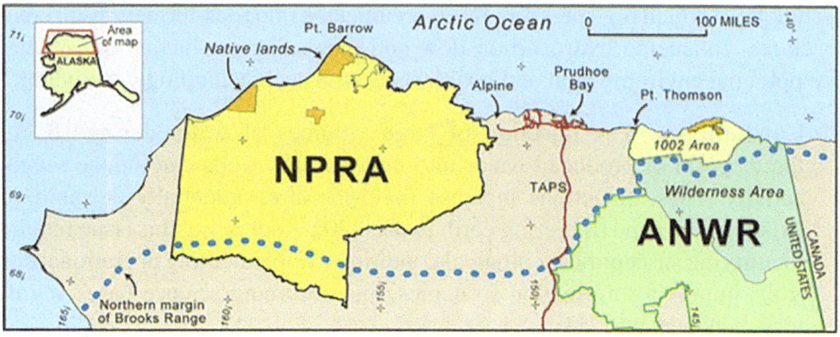

FIGURE 2.10 Map of northern Alaska showing the location of ANWR and NPRA [25]. *ANWR*, Arctic National Wildlife Refuge; *NPRA*, National Petroleum Reserves-Alaska

played a significant role in shaping its infrastructure, population, and geopolitical importance.

After the gold rush faded, Alaska remained rich in resources, but it was not until the 20th century that its most valuable asset—oil—was discovered. The search for petroleum began in the early 1900s, but the 1968 Prudhoe Bay oil discovery on the North Slope changed everything (see Fig. 2.10). To transport vast amounts of oil from Prudhoe Bay to the nearest ice-free port in Valdez, the Trans-Alaska pipeline system (TAPS) was constructed between 1975 and 1977. The 800-mile (1287 km) pipeline became an engineering marvel, crossing mountains, tundra, and rivers to efficiently deliver oil to tankers bound for refineries. TAPS traverses three mountain ranges and rests on nearly 80,000 elevated support structures designed to keep the permafrost from thawing.

To build the pipeline, the state was compelled to settle land claims with Alaska Natives, leading to the creation of the Alaska Native Corporations. These corporations were granted 44 million acres of land, much of it overlapping with oil, gas, and mineral resources. The legacy of oil and gas development is also reflected in the Alaska Permanent Fund, an $80 billion endowment that pays every Alaska resident, including children, an annual dividend of over a thousand dollars. The oil industry provided thousands of jobs, attracting workers from across the U.S. and fueling the growth of cities like Anchorage.

However, while the oil boom brought prosperity, it also raised concerns about environmental damage, pipeline spills, and the impact on indigenous communities. TAPS has suffered multiple spills over the years, including a 2010 leak of 5000 barrels due to corrosion. Offshore drilling in the Beaufort and Chukchi Seas carries the danger of spills in icy, remote waters, where cleanup efforts are extremely difficult. The 1989 Exxon Valdez spill devastated Prince William Sound, with its effects on marine life still lingering decades later. Oil infrastructure also affects wildlife, as seen with the Central Arctic Caribou herd, whose migration patterns are disrupted by drilling operations.

Debates over oil drilling in sensitive areas, such as the Arctic National Wildlife Refuge (ANWR), and the National Petroleum Reserves-Alaska (NPRA) continue today [26]. This 23-million-acre wilderness, home to caribou, polar bears, and over 200 bird species, potentially contains billions of barrels of oil. While proponents argue that drilling would boost the economy and enhance U.S. energy independence, opponents warn of irreversible environmental damage, habitat destruction, and threats to indigenous rights. The Inupiat people, who live near oil developments, remain divided—some benefit from oil-related jobs, while others worry about long-term environmental consequences. The Gwich'in people, whose livelihood depends on caribou herds migrating through ANWR, strongly oppose drilling that could endanger their way of life.

Alaska is warming twice as fast as the global average, largely due to climate change. Melting permafrost threatens pipelines, roads, and drilling sites, causing the ground to shift and sink. The Trans-Alaska Pipeline, for example, is at risk as its permafrost foundation becomes increasingly unstable. Oil extraction and the burning of fossil fuels contribute to greenhouse gas emissions, accelerating Arctic ice melting and rising sea levels. These warming temperatures disrupt wildlife migration patterns and threaten ecosystems that indigenous communities rely on for subsistence.

The cost and logistics of drilling in the Far North are notoriously prohibitive, which partly explains why much of the NPRA remains untouched. However, regulatory changes could open the door for new exploration activities, fundamentally reshaping the Western Arctic. Increased drilling could lock in decades of new fossil fuel extraction, with potentially irreversible environmental consequences. Oil companies are likely to move as quickly as possible to develop their remaining leases in these protected regions.

With growing concerns over climate change and a shift toward renewable energy, Alaska stands at a crossroads. Stricter environmental laws and policies are essential to ensure the sustainability of its pristine environment. Fortunately, Alaska possesses vast renewable energy potential, including wind, hydro, and geothermal power, which could help reduce dependence on fossil fuels and pave the way for a more sustainable future.

Summary

As we go forward, we need to recognize two primary stakeholders at play when it comes to fossil fuels and emission reduction. First, the fossil fuel industry itself faces an uncertain future. This is not surprising given that it is an industry characterized by fluctuating oil prices. Demand can go up or down depending on multifaceted causes that are difficult to predict. Second, Federal and provincial governments bear the responsibility, on behalf of the public, to ensure that the fossil fuel industry aligns its emissions with the country's climate commitments. Simultaneously, they must navigate the transition toward sustainable energy that will provide access to all with minimal interruption. Obviously, significant fossil fuel liabilities, including the cleanup of legacy wells and mine tailings, need to be addressed. These items will be discussed in more detail in upcoming chapters.

Key takeaways

- Many existing oil industry infrastructures, such as pipelines, offshore platforms, or legacy wells, can be repurposed for renewable energy utilization.
- LNG is a transportable, cleaner burning fossil fuel that enhances energy security, supports coal-to-gas transitions, and provides flexible, reliable power, making it a key transition fuel in the global energy system.
- Conversion of coal-fired power plants into NG-powered plants offers significant economic and environmental benefits.

References

[1] Oil, unless otherwise noted, includes crude (including shale oil and oil sands), natural gas liquids (NGLs), gas-to-liquids (GTLs), coal-to-liquids (CTLs), condensates, and refinery gains. Hydrogen-derived fuels are all fuels derived from low-carbon hydrogen, including ammonia, methanol, and other synthetic hydrocarbons. (2025).
[2] Data from Energy Institute. Statistical review of world energy 2025. (Accessed January 19, 2026). https://www.energyinst.org/statistical-review.
[3] National Center for Energy Analytics (NCEA). Energy delusions: peak oil forecasts—a critique of oil "Scenarios" in the IEA world energy outlook 2024 (PDF), 2025. (Accessed January 19, 2026) https://energyanalytics.org/wp-content/uploads/2025/01/2025-01-NCEA-Energy-Delusions-MillsAtkinson.pdf.
[4] M.P. Mills, N. Atkinson, Energy Delusions: Peak Oil Forecasts—A Critique of Oil "Scenarios" in the IEA World Energy Outlook 2024, National Center for Energy Analytics (NCEA, 2025 https://energyanalytics.org/wp-content/uploads/2025/01/2025-01-NCEA-Energy-Delusions-MillsAtkinson.pdf (Accessed January 19, 2026).
[5] CRU Group (BCI). Gas pricing and the LNG market. (2025). (Accessed January 19, 2026) https://www.bcinsight.crugroup.com/2025/05/03/gas-pricing-and-the-lng-market/.
[6] U.S. Energy Information Administration (EIA). More than 100 coal-fired plants have been replaced or converted to natural gas since 2011. Today in energy. (2020). (Accessed January 19, 2026) https://www.eia.gov/todayinenergy/detail.php?id=44636.
[7] Adapted from National Renewable Energy Laboratory (NREL). Life cycle greenhouse gas emissions from electricity generation: update; 2021.
[8] U.S. Energy Information Administration (EIA). International energy outlook 2023 (with projections to 2050). (2023). (Accessed January 19, 2026) https://www.eia.gov/outlooks/ieo/.
[9] U.S. Environmental Protection Agency (EPA). Greenhouse gas emissions from abandoned oil and gas wells (PDF), 2018. (Accessed January 19, 2026) https://www.epa.gov/sites/default/files/2018-04/documents/ghgemissions_abandoned_wells.pdf.
[10] M.D. Merrill, C.A. Grove, N.J. Gianoutsos, P.A. Freeman, in: Analysis of the United States documented unplugged orphaned oil and gas well dataset, U.S. Geological Survey, 2023, p. 1167. https://pubs.usgs.gov/publication/dr1167/full.
[11] McGill UniversityEnvironmental risks and opportunities of orphaned oil and gas wells, McGill Newsroom, 2023 (Accessed January 19, 2026) https://www.mcgill.ca/newsroom/channels/news/environmental-risks-and-opportunities-orphaned-oil-and-gas-wells-348742.

[12] ProPublica. Cost of California oil cleanup exceeds industry profits. (2022) (Accessed January 19, 2026) https://www.propublica.org/article/cost-of-california-oil-cleanup-exceeds-industry-profits.

[13] International Energy Agency (IEA). Global hydrogen review 2022: executive summary. (2022). (Accessed January 19, 2026) https://www.iea.org/reports/global-hydrogen-review-2022/executive-summary.

[14] International Energy Agency (IEA)The Oil and Gas Industry in Net Zero Transitions, IEA, Paris, 2023 (Accessed January 19, 2026) https://www.iea.org/reports/the-oil-and-gas-industry-in-net-zero-transitions .

[15] Siemens Energy. Coal-to-gas conversion (brownfield transformation). (2024). (Accessed January 19, 2026) https://www.siemens-energy.com/us/en/home/products-services/solutions-usecase/brownfield-transformation/coal-to-gas.html.

[16] U.S. Energy Information Administration (EIA)Capital cost and performance characteristics for utility-scale electric power generating technologies, U.S. Department of Energy, Washington, DC, 2024 (Accessed January 19, 2026) https://www.eia.gov/analysis/studies/powerplants/capitalcost/pdf/capital_cost_AEO2025.pdf.

[17] Woodway Energy. Coal-to-natural-gas conversions. (2023). (Accessed January 19, 2026) https://www.woodwayenergy.com/coal-to-natural-gas-conversions/.

[18] U.S. Energy Information Administration (EIA). International Energy Statistics. (n.d.) (Accessed January 19, 2026) https://www.eia.gov/international/.

[19] U.S. Energy Information Administration (EIA). Energy Information Administration (EIA) website (data source). (n.d.). (Accessed January 19, 2026) https://www.eia.gov/.

[20] M. Weingarten, S. Ge, J.W. Godt, B.A. Bekins, J.L. Rubinstein, High-rate injection is associated with the increase in U.S. mid-continent seismicity, Science 348 (2015) 1336–1340.

[21] R. Schultz, R.J. Skoumal, M.R. Brudzinski, D. Eaton, B. Baptie, W. Ellsworth, Hydraulic fracturing-induced seismicity, Rev. Geophys. 58 (2020) e2019RG000695.

[22] D.G. Abraham, H.K. Liberatore, M.T. Aziz, D.B. Burnett, L.H. Cizmas, S.D. Richardson, Impacts of hydraulic fracturing wastewater from oil and gas industries on drinking water: quantification of 69 disinfection by-products and calculated toxicity, Sci. Total Environ. 882 (2023) 163344.

[23] B.E. Fontenot, L.R. Hunt, Z.L. Hildenbrand, D.D. Carlton Jr., H. Oka, J.L. Walton, et al., An evaluation of water quality in private drinking water wells near natural gas extraction sites in the barnett shale formation, Environ. Sci. Technol. 47 (2013) 10032–10040.

[24] B. Hwang, Environmental implications of shale gas hydraulic fracturing, Water 15 (19) (2023) 3334. https://www.mdpi.com/2073-4441/15/19/3334.

[25] Courtesy of U.S. Geological Survey (USGS). U.S. Geological Survey 2002 Petroleum Resource Assessment of the National Petroleum Reserve in Alaska (NPRA). 2002. (Accessed January 19, 2026). https://pubs.usgs.gov/fs/2002/fs045-02/.

[26] **Sierra** (digital edition). Article (details not provided). https://digital.sierramagazine.org/publication/?i=837107&article_id=4901020&view=articleBrowser.

Further readings

– IEA, Global Energy Review 2025, www.iea.org.

– EI Statistical review of World Energy, 2025.

– Janetta McKenzie, Pembina Institute, The future of oil in the energy transition, December 2022.

– Bera and Kumar, Decarbonizing the Petroleum Industry, September 2025. Elsevier Publishing.

Chapter 3

Nuclear energy

Synopsis

The Fukushima nuclear disaster in 2011 caused catastrophic environmental, health, and economic damages in Japan, totaling nearly 200+ billion dollars and about 18,000 indirect deaths related to the earthquake and tsunami. Afterward, many countries downscaled their plans for installing new nuclear plants. Indeed, while nuclear energy is applauded for its capacity to generate significant amounts of baseload electricity with minimal GHG emissions, it concurrently evokes public anxieties surrounding safety, radioactive waste disposal, and potential proliferation risks.

In this chapter, first, we cover the fundamentals of nuclear energy generation. Then, we explore various reactor designs, highlighting their distinct characteristics and operational considerations. Through this evaluation, we will dissect the potential of nuclear energy as a clean energy champion and its challenges.

However, nuclear energy is as much a social and political issue as it is a scientific marvel. Therefore we will extend our analysis beyond the reactors themselves, engaging with critical questions that have shaped public discourse for decades: Can nuclear energy be a responsible partner in the collective pursuit of net zero emissions? What advancements are being made to mitigate historical safety concerns, and how effectively are these measures assuaging public apprehension? Finally, the critical issue of nuclear waste disposal demands exploration, examining proposed solutions and associated challenges.

F.A.C.T.S. Flexibility, Affordability, Conservation, Transition, Sustainability: Our Path to the Future of Energy and the Environment. DOI: https://doi.org/10.1016/B978-0-443-52574-2.00001-X

Nuclear energy and net zero emissions

The urgency of the climate crisis necessitates a multifaceted solution, and renewables alone may not suffice to meet the rapid decarbonization required. Nuclear energy offers a proven, large-scale, and carbon-free solution that can be implemented more rapidly than some renewable technologies, particularly in regions with well-established nuclear infrastructure. Furthermore, nuclear power plants have a high-capacity factor, meaning they operate consistently and can generate electricity for a significant portion of the year, unlike solar and wind, which is more variable. This consistency can provide a crucial backbone for integrating a growing share of renewables into the grid, mitigating the challenges associated with their intermittency.

With its capacity for large-scale and near-carbon-zero electricity generation, nuclear energy emerges as a potential ally for achieving energy sustainability. Unlike fossil fuel combustion, the fission process demonstrably generates no GHG emissions during electricity production. In contrast to renewable energy sources like solar and wind, which are intermittent and weather-dependent, nuclear reactors offer a reliable and dispatchable baseload power source. This consistent output is crucial for maintaining grid stability and ensuring a continuous electricity supply. Studies conducted by the International Energy Agency and the Intergovernmental Panel on Climate Change conclude that extensive deployment of nuclear energy alongside renewables is essential for achieving ambitious climate goals.

As of January 2024, the global operational nuclear power capacity was 371.5 GWh, provided by 413 reactors in 31 countries [1]. The top three producers of nuclear electricity were the United States (31%), China (16%), and France (13%). In November 2024, the United Kingdom and 30 other countries signed a global pledge to triple their nuclear capacity by 2050. The projected nuclear energy generation is shown in Fig. 3.1. China is planning at least 150 new reactors in the next 15 years and will have the highest growth in nuclear generation, followed by the United States.

Fundamental principles of nuclear fission

Nuclear energy, a cornerstone of modern energy production, harnesses the immense energy potential locked within the nucleus of atoms through a process called nuclear fission. This section delves into the core scientific principles underpinning nuclear fission, elucidating the intricate subatomic interactions that unlock this extraordinary energy source. Nuclear fission splits heavy atoms (like uranium) into smaller ones, releasing energy and neutrons in a chain reaction, used in current reactors but creates long-lived radioactive waste. Nuclear fusion combines light atoms (like hydrogen isotopes) to form heavier ones, releasing far more energy with helium as a byproduct and much less radioactive waste,

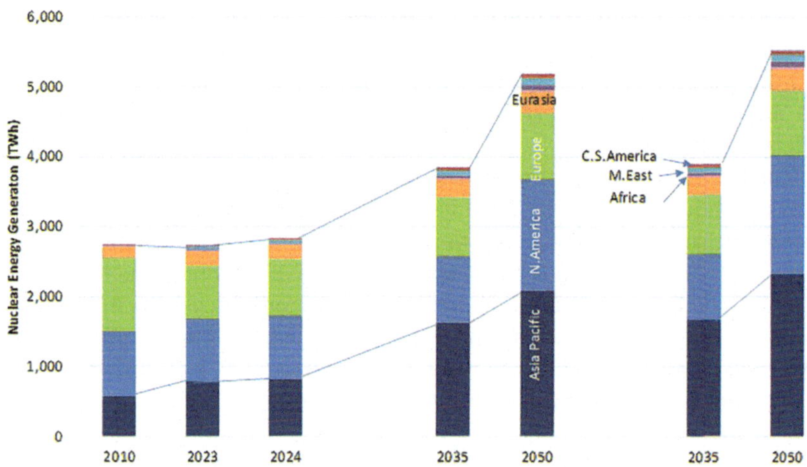

FIGURE 3.1 Projected global nuclear energy generation [2].

powering stars, but it is difficult to achieve on Earth. Nuclear fusion will be discussed later in this chapter.

The fission process: At the heart of every atom resides a dense core, the nucleus, comprised of protons and neutrons. Protons, positively charged particles, and electrically neutral neutrons bind together through the potent, strong nuclear force. This force, described by the theory of quantum chromodynamics, arises from interactions between even more minor constituents known as quarks, the fundamental building blocks of protons and neutrons. The strong nuclear force overcomes the inherent repulsion between positively charged protons, ensuring the stability of the nucleus. However, the strength of this force varies depending on the specific number of protons and neutrons within a nucleus.

The intricate dance of binding energy: The strong nuclear force acts like a cosmic glue, holding protons and neutrons within the nucleus. The energy required to overcome this attractive force and split the nucleus is known as "binding energy" (BE). BE is not a physical quantity stored within the nucleus, rather it represents the minimum amount of energy that must be supplied to separate the protons and neutrons and overcome the forces holding them together. Atoms with higher BE are more stable and require more energy input to undergo fission. Conversely, atoms with lower BE are less stable and readily release significant amounts of energy upon fission. Isotopes of heavier elements, particularly those with a high neutron-to-proton ratio, tend to exhibit lower BE, making them more susceptible to fission and thus more suitable as nuclear fuels.

Nuclear fission occurs when a heavy nucleus, typically an isotope of uranium (Uranium-235 or Uranium-238), is bombarded with smaller neutron particles. This neutron acts as a projectile and disrupts the delicate balance within the nucleus. The resulting unstable nucleus undergoes fission, splitting into two or more smaller nuclei, known as fission products, and releasing additional neutrons and a tremendous amount of energy. The specific fission products and the quantity of energy released vary depending on the particular isotope undergoing fission.

The neutrons released during fission are critical for sustaining the process. If they collide with other fissionable nuclei, these neutrons can induce further fission, releasing even more neutrons and perpetuating a self-sustaining chain reaction. This chain reaction is the fundamental principle behind a nuclear reactor. In a controlled chain reaction, the number of neutrons released is meticulously regulated to ensure safe and sustained energy production. Control rods, typically composed of materials with high neutron absorption cross-sections, are inserted or withdrawn from the reactor core to manipulate the neutron population and regulate the fission rate.

The fission rate within a reactor is directly linked to the number of neutrons released in each fission event and the probability of those neutrons causing further fissions. Several factors influence this probability, including the energy of the neutrons and the neutron absorption properties of the materials within the reactor core. The point at which the number of neutrons produced precisely balances the number lost through various processes, such as absorption by control rods, leakage from the reactor core, or unproductive capture by fuel nuclei, is known as criticality. A reactor operating at criticality maintains constant power output. Reactor control systems meticulously ensure the reactor operates within the critical range, preventing a runaway chain reaction that could lead to a nuclear accident. Sophisticated safety systems are also incorporated into reactor designs to automatically shut down the reactor in case of emergencies or abnormal operating conditions.

The immense energy released during nuclear fission originates from converting a small portion of the mass of the fissile nucleus into energy (following Einstein's famous equation, $E=mc^2$). A single fission event can release around 200 MeV (Megaelectronvolts) of energy, equivalent to burning millions of kilograms of coal. This immense energy density makes nuclear fission a powerful technology for electricity generation. Efficiently utilizing this energy requires converting it into heat, which can then be used to produce steam to drive turbines and generate electricity within a nuclear power plant.

The byproducts of nuclear power: The fission products resulting from nuclear fission (i.e., nuclear wastes) are typically radioactive isotopes of various elements. These radioactive isotopes decay over time, releasing additional energy through radiation. The specific types and half-lives of the fission products vary depending on the fissile isotope and the specific fission reaction. Managing these radioactive materials throughout the entire nuclear fuel cycle, from fuel fabrication to reactor operation and spent fuel storage, is crucial for ensuring nuclear energy's safety and environmental responsibility.

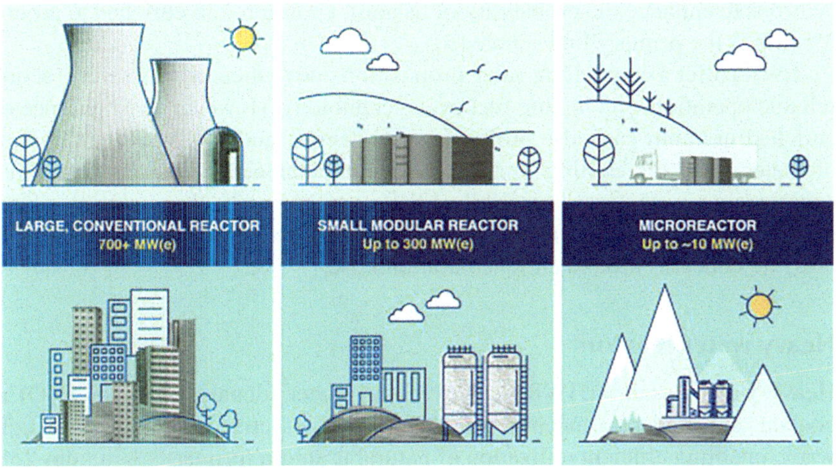

FIGURE 3.2 Different reactor types [3].

BOX. 3.1 Tech note: Uranium

Uranium-235, the most readily fissile isotope used in commercial reactors, constitutes only a tiny fraction of uranium ore (around 0.7%). The remaining uranium is primarily Uranium-238, which is not readily fissile with slow neutrons but can be transformed into fissionable plutonium (Pu) with fast neutrons in a process known as breeding. Additionally, research is ongoing into exploring alternative nuclear fuels, such as thorium, which has the potential to offer advantages in terms of fuel availability, waste minimization, and proliferation resistance (i.e., lower risk of being misused to develop weapons).

Review of reactor designs

Nuclear reactors, the workhorses of nuclear power generation, come in diverse configurations, each boasting distinct characteristics and operational considerations. A large nuclear reactor can generate about 24 million kWh per day (electricity for about 3.2 million homes). A smaller modular reactor's output is about 7.2 million kWh per day (9,70,000 homes). This section delves into several prominent reactor designs, highlighting their strengths, limitations, and suitability for various applications within the global energy landscape.

Light water reactors

Large commercial reactors are generally light water reactors (LWRs), accounting for over 90% of operational reactors (See Fig. 3.2). They leverage ordinary water (H_2O) for both coolant and moderator functions (a material that slows down neutrons to enhance the probability of fission). Uranium-235 enriched to around 3%–5% is the primary fuel source [3].

LWRs offer a compelling value proposition due to their proven safety record, reliable operation, and mature fuel cycle technology. However, their reliance on enriched uranium can raise proliferation concerns, and the significant volume of water used for cooling presents environmental considerations in regions experiencing water scarcity. While LWRs are currently the mainstay of the nuclear industry, advancements in fuel recycling technologies can potentially mitigate concerns around enriched uranium use.

Heavy water reactors

Heavy water reactors (HWRs) utilize heavy water (deuterium oxide, D_2O) as coolant and moderator. Heavy water is a more effective moderator than light water, enabling efficient utilization of natural uranium (primarily Uranium-238) without enrichment. It reduces proliferation concerns as natural uranium is less suitable for weapons development. Additionally, HWRs can efficiently burn plutonium, a byproduct of nuclear reactions in LWRs, further reducing waste generation.

HWRs offer the benefits of efficient fuel utilization, reduced proliferation risks, and the ability to manage plutonium waste from LWRs. However, heavy water production is energy-intensive, and these reactors may be more expensive to build and operate than LWRs. Additionally, heavy water necessitates robust safeguards due to its potential for use in weapons-grade plutonium production.

Gas-cooled reactors

Gas-cooled reactors (GCRs) utilize carbon dioxide (CO_2) gas as a coolant. It allows for higher operating temperatures than water-cooled reactors, potentially leading to greater efficiency and improved thermal applications beyond electricity generation, such as industrial process heat. Graphite serves as the moderator in these reactors. The Magnox and advanced gas-cooled reactor are prominent GCR designs. GCRs offer the advantages of high-temperature operation and the potential for cogeneration (simultaneous production of electricity and heat). It makes them suitable for applications beyond power generation, such as desalination or industrial heat requirements. However, CO_2 leaks can pose environmental concerns, and using graphite as a moderator raises safety considerations in case of fire. GCR development has also been less prevalent in recent decades than LWRs due to these risks.

Fast neutron reactors

Fast neutron reactors (FNRs) utilize fast neutrons (neutrons with high kinetic energy) to induce fission. It allows for using a more comprehensive range of fissile isotopes, including plutonium, and enables a process known as breeding. In breeding, fertile isotopes like Uranium-238 are converted into fissile plutonium, potentially extending fuel resources and reducing reliance on enriched uranium. Liquid sodium metal is often used as a coolant in FNRs due to its efficient heat transfer properties. Sodium-cooled fast reactors are a prominent FNR design.

FNRs offer fuel breeding and waste reduction potential, partially alleviating concerns regarding long-term nuclear fuel availability and waste management. However, they are a more complex technology with inherent safety challenges due to the use of liquid metal coolants and the potential for rapid power surges. Additionally, FNRs are still under development, and their commercial viability remains to be fully established.

Small modular reactors

Small modular reactors (SMRs) are a new generation of reactors (see Fig. 3.2) with significantly lower power capacities (300 MWh) than traditional ones (1000 MWh+) [3]. This modular design allows for factory fabrication and on-site assembly, potentially reducing construction costs and timelines. Additionally, SMRs can be deployed modularly, enabling a more flexible approach to grid management and potentially catering to smaller communities or remote locations.

SMRs promise to provide clean and reliable power in regions with limited grid infrastructure or niche applications. Their inherent safety features, such as passive cooling mechanisms, further enhance their attractiveness. However, technology is still evolving, and the economic viability of SMRs at scale requires further demonstration and cost-reduction efforts. Although many countries are considering SMRs, only Russia and China have operational plants.

Microreactors

Microreactors are a subset of SMRs and are designed to generate electrical power typically up to 10 MWh (see Fig. 3.2). Microreactors have smaller footprints than other SMRs and will be better suited for regions inaccessible to clean, reliable, and affordable energy. Furthermore, microreactors could serve as a backup power supply in emergency situations or replace power generators that are often fueled by diesel, for example, in rural communities or remote businesses.

Choosing the right reactor

The diverse landscape of reactor designs underscores the continuous innovation and adaptation within the nuclear industry. Each design boasts distinct advantages and considerations, offering tailored solutions for various energy needs and constraints. LWRs currently dominate the nuclear power landscape, and advancements in fuel recycling and waste management hold promise for their continued viability. HWRs offer efficient fuel utilization and waste reduction, while GCRs showcase potential for high-temperature applications.

FNRs, though complex, present possibilities for fuel breeding and extended fuel resources. Finally, SMRs introduce a paradigm shift toward decentralized power generation.

The optimal reactor design for a particular application hinge on carefully evaluating factors like energy demand, water availability, safety considerations, grid infrastructure, and economic feasibility. Many renewable energy sources, such as solar and wind, may need modularized nuclear reactors to overcome the intermittent energy generation. As we strive toward a sustainable energy future, a balanced approach encompassing a diverse portfolio of reactor designs and advancements in renewable energy technologies and energy efficiency measures presents the most promising path forward.

Safety aspects of nuclear energy

Several major nuclear accidents have occurred during the past 70 years, leading to significant environmental, health, and economic consequences. The five worst nuclear accidents [4] are summarized in Table 3.1.The apocalyptic explosion in September 1957 of a nuclear waste dump in the town of Kyshtym in the Soviet Urals was successfully covered up until the late 1980s when its news surfaced. The 1986 meltdown at the Chernobyl power plant in Ukraine had the worst impact on populations and is still releasing some contamination.

Life cycle perspective

A comprehensive life cycle assessment is crucial when a nuclear power plant operation produces no GHG emissions. Uranium's full-life cycle carbon footprint includes uranium mining, milling, conversion, enrichment, fuel rod fabrication, nuclear power facility construction/operation/decommissioning, and spent fuel management/reprocessing. These steps in the process of nuclear energy production emit significant GHGs, which range between 3.7 and 110 g of carbon dioxide equivalents per kilowatt-hour of electricity generated (CO_2e/kWh), depending on the uranium ore grade, electricity source, and location. Up to 30% of the nuclear life cycle GHG emissions is attributed to uranium mining and milling. Differences in these reported values are due in part to real differences in production methods (e.g., ore grade, mine type, and electricity source) and in part to study methodology (e.g., study completeness, methods, assumptions).

The following sections will discuss nuclear energy waste management.

Nuclear waste management

The long-lived radioactive nature of nuclear waste necessitates a safe and secure long-term disposal solution. Geological repositories, where spent nuclear fuel is stored deep underground in stable geologic formations, are widely considered the most viable option for long-term waste management. Several countries, such

TABLE 3.1 Major nuclear disasters.

Disaster	Date	Location	Cause	Consequences
Chernobyl [5]	1986	Ukraine	Flawed reactor design and operator errors led to an uncontrolled reaction at Reactor 4 of the Chernobyl Nuclear Power Plant.	Massive radioactive release 30 immediate deaths Evacuation of over 100,000 people Large exclusion zone still in place.
Fukushima [6]	2011	Japan	A 9.0-magnitude earthquake and tsunami disabled the cooling systems of three reactors at the Fukushima Daiichi plant, causing meltdowns.	Large radioactive contamination. Over 150,000 people evacuated. Long-term cleanup is still ongoing. Increased thyroid cancer cases reported.
3-mile Island [7]	1979	USA	A cooling system failure at Unit 2 of the Three Mile Island plant led to a partial meltdown.	No immediate deaths, but increased concerns over nuclear safety. Strengthened U.S. nuclear regulations. Partial release of radioactive gases.
Kyshtym	1957	Russia	A cooling system failure in a nuclear waste storage tank at the Mayak Production Association caused an explosion.	Significant radioactive contamination across a 300 km^2 area. An estimated 200+ deaths and long-term radiation-related illnesses. One of the least-known nuclear disasters due to Soviet secrecy.
Windscale Fire	1957	UK	A graphite-moderated reactor fire at the Windscale nuclear plant.	Release of radioactive materials, including iodine-131. Increased thyroid cancer cases. No immediate deaths, but estimated long-term fatalities.

as Finland and Sweden, actively develop and construct geological repositories, demonstrating a commitment to responsible waste management. Continuous research and development in waste reprocessing and recycling technologies offer further possibilities for minimizing waste volumes and maximizing resource utilization within the nuclear fuel cycle. This section explores the proposed solutions for nuclear waste disposal, examining the associated challenges and ongoing efforts to ensure responsible waste management.

Geological disposal of nuclear waste

The prevailing international consensus for nuclear waste disposal centers on geological repositories. These repositories involve encapsulating the waste in robust containers and burying it deep underground in stable geological formations, typically within granite or clay-rich rock layers. The multibarrier approach to geological disposal leverages several layers of protection:

The waste is first encased in corrosion-resistant containers designed to withstand extreme temperatures and pressures for extended periods. These canisters are typically fabricated from high-integrity materials such as stainless steel or specially designed ceramics.

The chosen geological formation offers a natural barrier with properties that minimize the migration of radioactive materials into the surrounding rock and groundwater. Ideal formations are typically hundreds of meters underground and exhibit minimal fractures or water movement. Granite and certain clay formations are particularly well-suited due to their low permeability and self-sealing properties.

Bentonite clay, a naturally occurring material with excellent swelling properties, surrounds the waste canisters. This clay layer further impedes the movement of radioactive materials by acting as a buffer and absorbing any water that may seep into the repository. Bentonite clay can self-heal fractures, further enhancing its isolation capabilities.

Challenges and considerations for geological disposal

While geological disposal is widely considered the most viable option for long-term nuclear waste management, several challenges remain:

Identifying suitable geological formations with the necessary stability and isolation characteristics is complex. Extensive geological surveys and characterization are essential to ensure the long-term safety of a repository site. Factors like rock stability, absence of groundwater movement, and minimal seismic activity are all meticulously evaluated. The selection process often takes years and involves rigorous scientific assessments to guarantee that the chosen site can safely contain radioactive waste for millennia.

The concept of a nuclear waste repository nearby can evoke public concerns. Open communication and public education can help to remedy the situation. This aspect will be discussed more in the following section.

Alternative disposal options

While geological disposal remains the leading approach, alternative disposal options (e.g., Partitioning and Transmutation, P&T, Deep Borehole Disposal, DBD) are also being explored, offering potential solutions for specific waste streams or complementing geological repositories:

P&T technology separates long-lived radioactive isotopes from spent nuclear fuel, allowing for the transmutation of some of these isotopes into shorter-lived or stable elements through irradiation in nuclear reactors. This process can significantly reduce the volume and radioactivity of the waste requiring disposal. Partitioning and transmutation are complex technologies still under development, but they promise to minimize the long-term burden of certain types of nuclear waste.

DBD concept involves placing waste canisters in deep boreholes drilled into stable rock formations, typically several kilometers deep. While still under development, this approach offers potential advantages like reduced surface footprint and potentially lower construction costs compared to traditional geological repositories. Deep borehole disposal may be suitable for certain types of waste, but further research is needed to evaluate its long-term safety and feasibility.

The major hurdles of the nuclear industry

Major hurdles that the nuclear industry needs to overcome in order to bring public trust are listed as follows:

Safety first: Maintaining the highest safety standards throughout the nuclear fuel cycle, from reactor operation to waste management, is paramount. Continuous safety improvements, robust regulatory frameworks, and emergency response plans are essential for ensuring public trust and confidence. The International Atomic Energy Agency (IAEA), on its website [3], states that they can assist Member States in the safe, peaceful, secure, and sustainable use of nuclear power.

Transparency and public engagement: Open communication regarding safety protocols, emergency response plans, incident reporting, and radioactive waste management practices fosters public trust and allows for informed decision-making. Proactive disclosure of information, even about minor incidents, helps to counter public perception of secrecy and fosters trust. Misinformation concerning nuclear energy can readily propagate in the digital age. Countering misinformation with accurate scientific information and fostering open dialogue with the public is critical for promoting a more balanced understanding of nuclear energy's risks and benefits.

Economic competitiveness: The cost of new nuclear plants is high, and this significantly constrains the growth of nuclear power under scenarios that assume "business as usual." Modest carbon emission constraints in those parts of the world where a carbon constraint is not a primary factor make fossil fuels, whether coal or natural gas, a lower-cost alternative for electricity generation. Nuclear energy must remain economically competitive with other clean energy sources, particularly as renewable energy technologies mature and their costs decline. Future research, development, and demonstration funding should prioritize reactor designs that are optimized to substantially lower capital costs, including construction costs.

Technological innovation: Continued research and development efforts should focus on enhancing reactor safety features, minimizing waste generation through advanced fuel cycles, and exploring novel reactor designs like SMRs for greater flexibility and adaptability.

Passive safety systems: These systems rely on natural physical phenomena like gravity or convection currents to cool the reactor core in emergencies. It eliminates dependence on operator intervention or functioning pumps, a critical weakness identified in Chernobyl and Fukushima.

Better containment: Modern containment structures are designed to withstand significant internal pressure and prevent the release of radioactive materials into the environment in case of an accident. Advancements include more robust containment materials like prestressed concrete and improved leak-tightness measures.

Redundancy: Modern reactors incorporate multiple, independent safety systems, ensuring a layered defense against accidents. This redundancy minimizes the risk of a single failure compromising overall safety. For instance, some reactor designs include core catcher systems designed to capture molten reactor core material in the unlikely event of a meltdown.

The advancements outlined above have demonstrably enhanced the safety of nuclear power plants. Modern reactors incorporating these features significantly reduce the likelihood of accidents and mitigate the potential consequences should an incident occur. For instance, third-generation reactor designs like the European Pressurized Reactor boast advanced safety features, including passive cooling systems and core catcher technology, significantly safer than older reactor designs.

Robust regulatory frameworks: National and international regulations are essential for ensuring the safe operation of nuclear facilities, the secure management of nuclear materials, and the responsible disposal of radioactive waste. Independent oversight bodies play a critical role in maintaining public confidence.

Regulatory bodies worldwide have implemented stricter safety standards, requiring rigorous stress testing, comprehensive safety assessments, and enhanced emergency preparedness plans for nuclear facilities. Independent oversight agencies play a crucial role in ensuring adherence to these regulations. The IAEA is vital in fostering international cooperation on safety standards and best practices.

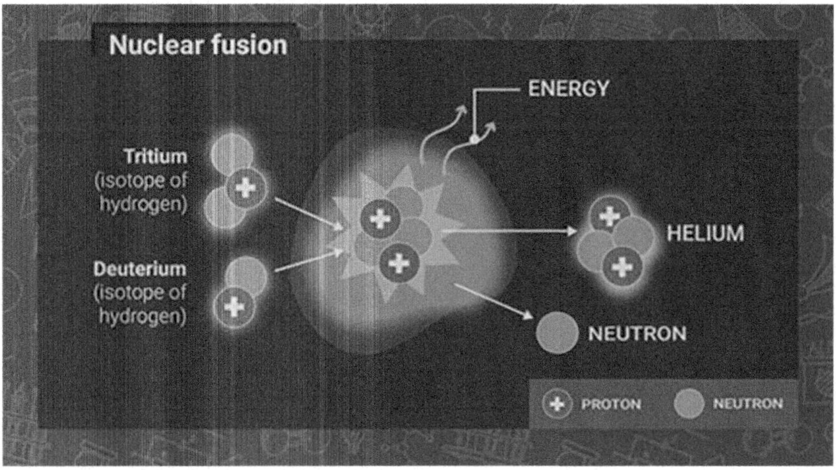

FIGURE 3.3 Nuclear fusion [8].

Global collaboration: International cooperation is crucial for knowledge sharing, technology transfer, and joint efforts in nuclear fuel cycle management and waste disposal solutions. Multilateral frameworks can foster responsible nuclear development and prevent proliferation risks.

Nuclear fusion

While nuclear fission, the splitting of atoms, is the basis of conventional nuclear reactors described in the earlier section, scientists have also been exploring the potential of nuclear fusion, the merging of subatomic particles, as a clean energy source, since the 1930s. Nuclear fusion occurs when two light atomic nuclei (such as hydrogen isotopes deuterium and tritium) combine to form a heavier nucleus, releasing a tremendous amount of energy (see Fig. 3.3). Fusion reactions take place in a state of matter called plasma; a hot, charged gas made of positive ions and free-moving electrons with unique properties distinct from solids, liquids, or gases. The collision of nuclei happens at extremely high temperatures, around a hundred million degrees Celsius. It has to be confined and controlled using the largest magnetic field ever produced. The high-temperature provides the nuclei with enough energy to overcome their mutual electrical repulsion. Once the nuclei come within a close range of each other, the attractive nuclear force between them will outweigh the electrical repulsion and allow them to fuse. For this to happen, the nuclei must be confined within a small space to increase the chances of collision.

It is worth noting that the Sun and other stars have been generating energy through hydrogen fusion since the beginning of time. Fusion is considered the "holy grail" of clean energy since it produces minimal radioactive waste and has abundant fuel sources such as hydrogen from water. In theory, with just a few grams of these reactants, it is possible to produce a terajoule of energy, which is approximately the energy one person in a developed country needs over 60 years. However, the need for extremely high temperatures and pressure makes this process very complex. Current researchers are aiming to make fusion viable for power generation. We are still in the early phases of harnessing this energy. Scientists and engineers from all over the world are collaborating to develop and test new materials and design new technologies to achieve net fusion energy.

Notice that the fusion process is inherently safe because it can only occur under strict operational conditions, outside of which (in the case of an accident or system failure, for example), the plasma will naturally terminate, lose its energy very quickly, and extinguish before any sustained damage is done to the reactor [8].

Summary

This chapter explored nuclear energy, delving into its scientific underpinnings and the broader social and political considerations shaping its future. This knowledge paved the way for a review of various reactor designs, highlighting their distinct characteristics and operational considerations. Through this evaluation, we found that new designs such as SMRs and microreactors are more flexible and more affordable.

Recognizing the limitations of a solely technical understanding, we extended our analysis beyond the reactors. We grappled with the critical question of whether nuclear energy is a responsible partner in the collective pursuit of net zero emissions. What advancements are being made to mitigate historical safety concerns, and how effectively are these measures assuaging public apprehension? We also explored the critical issue of nuclear waste disposal, examining proposed solutions and the associated challenges.

Key takeaways

- The central opportunity for nuclear energy over the next several decades is tied to its potential contribution to decarbonizing the power sector. It could also integrate with intermittent energy sources such as solar or wind.
- Aside from electricity generation, nuclear energy could be utilized for many other innovative technologies, such as water desalination or hydrogen production.
- A revival and expansion of nuclear power to achieve global goals will require robust safety guidelines, policy support, and tight cost controls.

References

[1] International Atomic Energy AgencyNuclear power reactors in the world; Reference Data Series No. 2, No. 44, IAEA, Vienna, 2024 https://www-pub.iaea.org/MTCD/Publications/PDF/p15748-RDS-2-44_web.pdf accessed Jan 19, 2026.

[2] International Energy Agency (IEA)World Energy Outlook 2025, IEA, Paris, 2025 https://www.iea.org/reports/world-energy-outlook-2025 accessed Jan 19, 2026.

[3] International Atomic Energy Agency. What Are Small Modular Reactors (SMRs)? https://www.iaea.org/newscenter/news/what-are-small-modular-reactors-smrs (accessed Jan 19, 2026).

[4] S. Plokhy, Atoms and Ashes: A Global History of Nuclear Disasters, W. W. Norton & Company, New York, NY, 2022.

[5] United Nations Scientific Committee on the Effects of Atomic Radiation (UNSCEAR), Health effects due to radiation from the Chernobyl accident; *Sources and Effects of Ionizing Radiation: UNSCEAR 2008 Report*, Scientific Annex D; United Nations, Vol. II, New York, 2011 Corrigendum May 18, 2016 https://www.unscear.org/unscear/uploads/documents/publications/UNSCEAR_2008_Annex-D-CORR.pdf accessed Jan 19, 2026.

[6] World Health Organization. Health Effects of the Chernobyl Accident and Special Health Care Programmes: Report of the UN Chernobyl Forum Expert Group "Health"; WHO: Geneva, 2006. https://www.who.int/publications/i/item/9241594179 (accessed Jan 19, 2026).

[7] U.S. Nuclear Regulatory Commission. NRC Library (Reading Room), 2025. https://www.nrc.gov/reading-rm (accessed Jan 19, 2026).

[8] International Atomic Energy Agency. What is Nuclear Fusion? IAEA News Center, 2023. https://www.iaea.org/newscenter/news/what-is-nuclear-fusion (accessed Jan 19, 2026).

Further readings

T. Murakami, A historical review and analysis on the selection of nuclear reactor types and implications to development programs for advanced reactors; A Japanese study, Energy Reports 7 (2021) 3428–3436.

S.A. Alameri, A.K. Alkaabi, Fundamentals of nuclear reactors, in: Nuclear Reactor Technology Development and Utilization, Woodhead Publishing, 2020, pp. 27–60.

A.S. Schneider, N. Yair, A review of safety analysis philosophies for nuclear reactors, Journal of Nuclear Engineering and Radiation Science 3 (3) (2017) 034501.

D.A. Arostegui, M. Holt, Advanced nuclear reactors: technology overview and current issues, Congressional Research Service Report for Congress (2019).

W. Younes, W.D. Loveland, An Introduction to Nuclear Fission, Springer, United States, 2021.

N. Schunck, D. Regnier, Theory of nuclear fission, Progress in Particle and Nuclear Physics 125 (2022) 103963.

A. Bulgac, J. Shi, I. Stetcu, Nuclear fission dynamics: past, present, needs, and future, Frontiers in Physics 8 (2020) 63.

I. Utami, M.A. Riski, D.R. Hartanto, Nuclear power plants technology to realize net zero emission 2060, International Journal of Business Management and Technology 6 (1) (2022) 158–162.

V. Nian, B. Mignacca, G. Locatelli, Policies toward net-zero: benchmarking the economic competitiveness of nuclear against wind and solar energy, Applied Energy 320 (2022) 119275.

A. Singh, S. Lal, N. Kumar, R. Yadav, S. Kumari, Role of nuclear energy in carbon mitigation to achieve United Nations net zero carbon emission: evidence from Fourier bootstrap Toda-Yamamoto, Environmental Science and Pollution Research 30 (16) (2023) 46185–46203.

T. Stone, The challenge of climate change—Complete energy systems transformation: No nuclear, no net zero, Nuclear Law (2022) 85.

S.M. Bragg-Sitton, The essential role of nuclear energy in achieving economy-wide net-zero solutions, Idaho National Lab. (INL) (2022).

J. Buongiorno, et al., Nuclear energy in a carbon-constrained world: Big challenges and big opportunities, IEEE Power and Energy Magazine 17 (2) (2019) 69–77.

M.V. Ramana, Technical and social problems of nuclear waste, Wiley Interdisciplinary Reviews: Energy and Environment 7 (4) (2018) e289.

S.S. Ho, S. Kristiansen, Environmental debates over nuclear energy: media, communication, and the public, Environmental Communication 13 (4) (2019) 431–439.

IAEA. What are small modular reactors? https://www.iaea.org/newscenter/news/what-are-small-modular-reactors-smrs?utm_source=chatgpt.com.

Part III

Renewable energy sources

Synopsis

Renewables offer a sustainable way to generate electricity, heat, and fuel. Harnessing these sources reduces carbon emissions and improves air quality. The flexibility of renewable energy sources can increase energy security and independence, making them a crucial part of a sustainable future.

In Part III, we describe the renewable energy sources including Solar (Chapter 4), Wind (Chapter 5), and Aqua (hydropower, geothermal and ocean). We also introduce emerging electrical energy sources (batteries, biomass and hydrogen).

Chapter 4

Solar power

Synopsis

Bill McKibben, one of the world's leading environmentalists, in his book, *Here Comes the Sun* [1], documents how solar and wind have unexpectedly become the cheapest forms of power on Earth. He argues that this is not just a technical upgrade, but a civilizational opportunity to move from "extractive" energy (mining and burning) to "ambient" energy (catching what falls from the sky).

In 2025, renewables surpassed coal as a source of electricity worldwide, and solar and wind power (2800 and 1300 GWh, respectively) grew fast enough to meet the entire increase in global electricity demand from January to June, according to energy think tank Ember [2]. In September, Chinese President Xi Jinping told the United Nations that China plans to cut its carbon emissions by up to 10% over the next decade, not by reducing energy use, but by accelerating the deployment of wind and solar. Meanwhile, solar panel imports have surged across Africa and South Asia, as households increasingly recognize that rooftop solar can provide affordable power for lighting, mobile phones, and fans. For many observers, the continued expansion of renewable energy now appears unstoppable. Solar energy is predicted to be the fastest-growing sector of renewable energy systems.

The International Energy Agency projected available solar energy in the world between 2024 and 2050 under two different scenarios is shown in Fig. 4.1. In 2050, solar energy is expected to provide 50% of the total renewable energy in the world.

However, the inherent variability of sunlight, known as intermittency, presents a key challenge for solar power. Energy storage solutions are crucial to ensure reliable electricity generation. This chapter will explore promising storage technologies like advanced batteries and pumped hydroelectric storage

F.A.C.T.S. Flexibility, Affordability, Conservation, Transition, Sustainability: Our Path to the Future of Energy and the Environment. DOI: https://doi.org/10.1016/B978-0-443-52574-2.00017-3

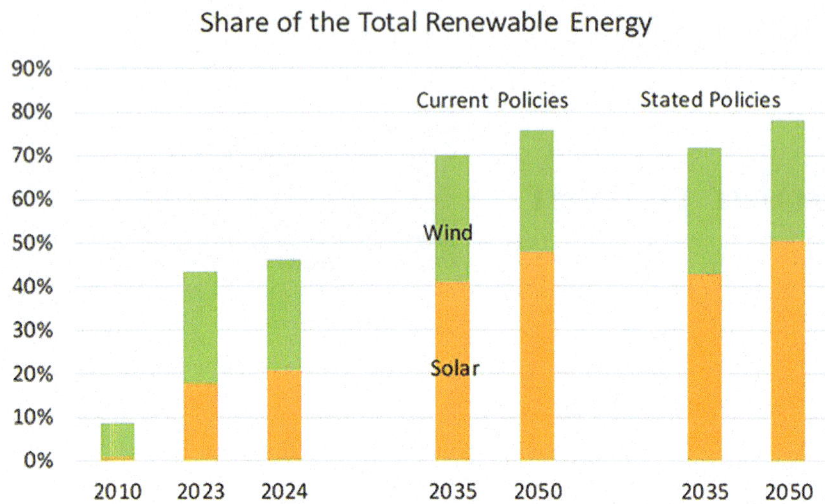

FIGURE 4.1 Projected world renewable energy generation [3].

(PHES), and their role in mitigating intermittency and maximizing solar energy utilization.

Effectively integrating solar power into the existing grid infrastructure requires careful consideration. The current grid is designed for dispatchable power plants, not the variable nature of solar energy. Understanding the complexities of grid management and the solutions required for seamless integration, such as smart grid technologies, is essential for a smooth transition to a solar-powered future.

The economic and environmental benefits of solar power are compelling. Declining solar power system costs, driven by advancements in manufacturing and economies of scale, are making solar power a financially attractive option across various scales. Furthermore, solar energy offers a clean and renewable source of energy, producing zero greenhouse gas emissions during operation. This translates to significant reductions in air pollution and combats climate change, paving the way for a more sustainable future.

This chapter will examine current trends and projections for solar energy's future growth, highlighting its transformative potential in the global energy mix. By understanding technologies, addressing challenges, and appreciating economic and environmental benefits, we can navigate the exciting path toward a solar-powered future. As we harness the sun's bounty, we can usher in a new era of clean, sustainable energy for generations to come.

Unveiling the power of the sun

The total amount of solar energy reaching the Earth's surface is 6–7 times greater than fossil fuel resources. In just a week, the Earth's surface receives such an amount of solar energy that it exceeds the energy of all the world's reserves of oil, gas, coal, and uranium. Using only 0.5% of this reserve could fully cover the world's energy demand in the long run.

Today, we harness the celestial power of the sun using two main categories of solar technology, photovoltaics (PV) and concentrated solar power (CSP). In this chapter, we will explore their underlying principles and their exciting potential for a sustainable future.

Dominating the solar sphere: Photovoltaic technology

Albert Einstein started scientific research in the field of solar energy. Many remember that he was awarded the Nobel Prize in 1921. However, few people know that he received it not for the theory of relativity but for explaining the laws of the external photoelectric effect. In 1905, he published a work based on Planck's hypothesis that the energy of radiation emitted or absorbed is proportional to its frequency. Einstein applied this idea to the photoelectric effect, arguing that light itself behaves as if it consists of localized energy quanta, a radical step that laid the foundation for quantum theory. He described how and in what quantities light quanta "knock out" electrons from metal. Using Einstein's research, Soviet physicists managed to obtain an electric current using the photoelectric effect in the 1930s. For the first time, they harnessed the sun's energy to generate electricity!

PV technology, the current leader in solar energy utilization, leverages the photovoltaic effects of the ability of certain materials to convert sunlight directly into electricity. Imagine sunlight cascading onto a solar panel. Within the panel, layers of specially treated silicon, the dominant material in most PV cells, reside. When sunlight strikes these layers, it disrupts electrons, setting them in motion and generating an electric current (Box 4.1).

BOX 4.1 Tech note— Silicon [4]

Silicon occupies the second place in mass in the Earth's crust after oxygen. Currently, 1 kg of silicon in a solar cell generates about 300 MWh of electricity over 30 years. Considering the heat of combustion of oil 43.7 MJ/kg, it gives 25 tons of oil per 1 kg of silicon. A thermal power plant using fuel oil has about 33% efficiency. Thus, in terms of generated electricity, 1 kg of silicon is equivalent to about 75 tons of oil.

(continued)

BOX 4.1 Tech note— Silicon [4]—*cont'd*

Tech note—Silicon PV module

A typical silicon (Si) PV module (Fig. 4.2) consists of a glass sheet for mechanical support and protection, laminated encapsulation layers of ethylene vinyl acetate for ultraviolet (UV) and moisture protection; 60 to 96 individual 6-inch-square (15-cm-square) solar cells, each capable of producing 4–5 watts under peak illumination (Wp); a homopolymer backseat for further environmental protection; and an aluminum frame for mounting. Common module dimensions are 1 m by 1.5 m by 4 cm, and peak power ratings range from 260 to 320 W.

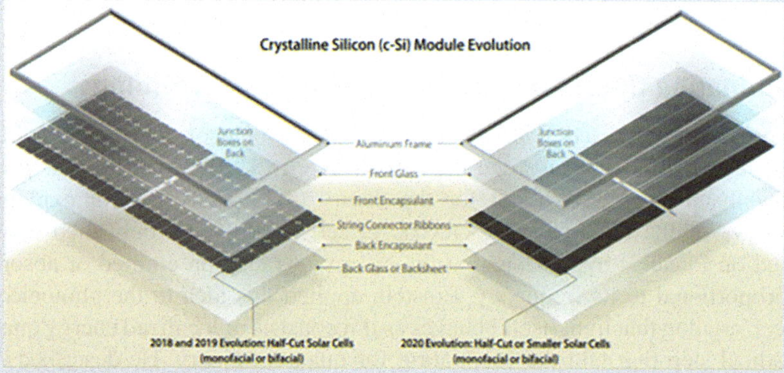

FIGURE 4.2 Principal components of a c-Si solar module [5].

During operation, the front surface of the PV module is illuminated by sunlight. Solar photons are transmitted into each cell, and those photons with sufficiently high energy (i.e., higher than the material-dependent energy bandgap) are absorbed. An absorbed photon transfers its energy to an electron and its positively charged counterpart (a hole). An internal electric field pulls electrons toward one electrode and holes toward the other, resulting in a direct electric current (DC).

There are two primary types of PV technologies:

Crystalline silicon solar cells (c-Si):

These highly efficient cells form the backbone of the PV market, offering a proven track record of reliability and performance. However, the manufacturing process for crystalline silicon can be energy-intensive.

Thin-film solar cells:

These lightweight, flexible cells represent a promising alternative. Fabricated from materials like cadmium telluride (CdTe) or copper indium gallium selenide, they boast advantages such as lower production costs and enhanced performance in low-light conditions. However, thin-film cells typically exhibit lower efficiency compared to crystalline silicon cells. Fig. 4.3 shows an overview of the general thin-film cell architecture.

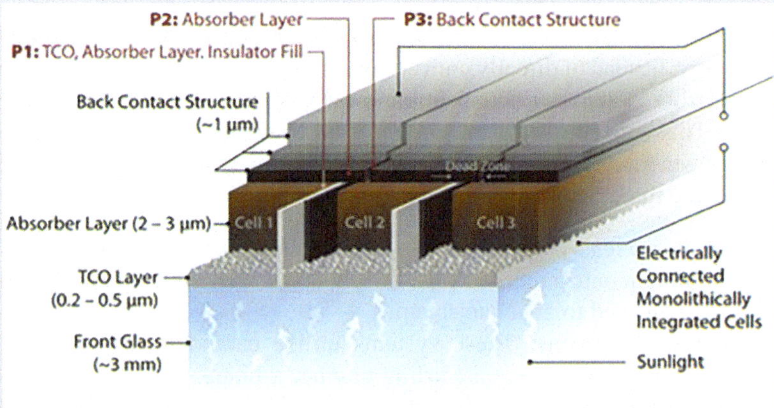

FIGURE 4.3 General architecture of a thin-film cell [6].

FIGURE 4.4 Large-scale and small-scale deployment of solar panels (Photos by Reegan Moen, DOE (*left*) and Dennis Schroeder, NREL 45218 (*right*) [13]).

The potential of PV technology is vast. Advancements in material science are constantly pushing the boundaries of efficiency. Researchers are developing next-generation solar cells capable of capturing a wider range of the solar spectrum, potentially leading to significant performance improvements. Additionally, the declining costs of PV systems and increasing government incentives are making solar power a more attractive option for residences, businesses, and utility-scale power plants. Furthermore, the modular nature of PV systems makes them highly versatile. Solar panels can be installed on rooftops, integrated seamlessly into building facades, or even deployed in portable configurations, enabling a wide range of applications. This adaptability is a key advantage of PV technology, allowing for effortless integration into existing infrastructure. Fig. 4.4 depicts the scalability of solar energy generation.

Concentrated solar power

CSP technology takes a fundamentally different approach to harnessing the sun's energy. Instead of directly converting sunlight into electricity, CSP systems utilize mirrors or lenses to concentrate a large area of sunlight onto a central receiver. This concentrated heat is then used to create steam, which drives a turbine to generate electricity, similar to a conventional power plant.

There are three main types of CSP systems:

Parabolic trough systems: These systems employ trough-shaped mirrors to focus sunlight onto a receiver pipe filled with a heat transfer fluid. The heated fluid is then used to generate steam.

Dish stirling systems: These systems utilize parabolic dish reflectors to concentrate sunlight onto a Stirling engine, a unique external combustion engine that converts heat into electricity.

Solar power tower systems: These large-scale systems employ an array of mirrors (heliostats) to concentrate sunlight onto a receiver tower. The heat from the receiver is used to generate steam and produce electricity.

CSP technology offers several advantages. Its ability to store thermal energy using molten salts or other heat storage media allows for electricity generation even during periods of low sunshine. This capability is particularly valuable in regions with unreliable sunshine patterns, providing a degree of dispatchability that photovoltaic systems lack. CSP plants can also achieve higher energy conversion efficiencies than traditional PV systems, potentially reducing the land area required for large-scale electricity generation.

However, the upfront costs of CSP plants can be higher than those of PV systems. The complex infrastructure required for concentrating sunlight and the need for specialized heat storage materials contribute to these higher costs. Additionally, CSP plants are typically large-scale facilities that require significant upfront investment, making them less suitable for distributed generation applications.

Selecting optimal technology

The choice between PV and CSP technologies hinges on various factors, including location, budget, and desired application. PV systems are generally more versatile and cost-effective for smaller-scale applications, such as powering homes or businesses. They are well-suited for both on-grid and off-grid applications, and their modular nature allows for easy integration into existing infrastructure. CSP plants, on the other hand, are better suited for large-scale electricity generation, particularly in regions with abundant sunshine and a demand for dispatchable renewable energy.

The road ahead: A solar-powered future

Both PV and CSP technologies offer immense potential for a sustainable energy future. As research and development efforts continue, these technologies will become even more efficient and cost-competitive. PV systems are likely to remain the dominant force for distributed generation due to their ease of installation and scalability. CSP plants, however, hold the promise of becoming a game-changer for large-scale, reliable renewable energy production. Hybrid systems, combining PV and CSP technologies, are also under development, offering the potential to leverage the strengths of both approaches for enhanced efficiency and dispatchability.

The most robust renewable energy systems of the future will likely be a blend of various renewable technologies working in harmony. Solar can be effectively integrated with wind energy, for instance, to create a more reliable and resilient power grid. When the sun is not shining, wind can take over, and vice versa. Energy storage solutions are also crucial for maximizing the potential of solar energy. By storing excess solar energy during peak production periods, it can be used to meet electricity demands when sunshine is limited.

Navigating the sun's variability: Challenges and solutions for solar power integration

While solar energy offers a transformative pathway toward a sustainable future, its full potential can only be realized by addressing key challenges. Here, we delve into three significant hurdles–energy storage, intermittency, and grid integration and explore promising solutions to ensure reliable and widespread adoption of solar power.

Unlike traditional power plants fueled by fossil fuels, solar energy production fluctuates throughout the day, peaking during sunshine hours and dropping significantly at night. This variability can disrupt grid stability and make it difficult to ensure a reliable supply of electricity to meet fluctuating demand. Imagine a scenario where a sunny afternoon leads to a surge in solar energy production, only to have that production plummet as the sun sets, creating a sudden shortfall that the grid must scramble to fill. Without proper mitigation strategies, intermittency can hinder the large-scale deployment of solar power.

Energy storage: Capturing the sun's bounty for continuous power

Energy storage solutions are crucial for mitigating the intermittency challenge and maximizing the utilization of solar energy. These technologies allow us to essentially capture the sun's bounty during peak production periods and release it back onto the grid when needed, such as during evening hours when demand is high, but sunlight is limited. Think of it as creating a virtual reservoir of solar energy that can be tapped into whenever necessary.

Several promising energy storage technologies are on the horizon, each offering unique advantages and limitations:

Advanced batteries: Lithium-ion batteries offer a versatile and scalable solution for storing solar energy. Advancements in battery technology are constantly improving their capacity, lifespan, and affordability. However, the current capacity of lithium-ion batteries may not be sufficient for large-scale, grid-level energy storage applications.

Pumped hydroelectric storage: This mature technology utilizes surplus electricity, often from solar sources, to pump water uphill to a reservoir. During periods of peak demand, the stored water is released through turbines to generate electricity. PHES offers large-scale storage capacity and mature infrastructure, making it a valuable asset for grid stability. However, geographical limitations restrict the feasibility of PHES plants in certain regions.

Emerging technologies: Novel storage solutions like compressed air energy storage (CAES) and flywheel energy storage are under development. CAES utilizes surplus electricity to compress air into underground caverns for later use in generating electricity. Flywheel energy storage captures and stores kinetic energy in a rotating flywheel, which can be released back into the grid when needed. These technologies have the potential to offer additional options for storing solar energy in the future, but their commercial viability and large-scale deployment potential are still being evaluated.

Integrating with the grid: Ensuring seamless and reliable power delivery

Solar panels generate DC power. In order to use this power for household appliances or to send it to the grid, an inverter is needed to convert it to grid-compatible AC power. These components are usually the higher expense in any solar system cost between 500–3000 USD. Fig. 4.5 Displays the required components for both utility grid and off-grid connections. For an average household with a weekly consumption of 30 kWh per day, we need to install between 15 and 22 panels. In 2025, the cost of installation is about 1800–30,000 USD. The federal tax credit usually reduces this cost by about 30%. In addition, storage batteries are needed when sunlight is not available. Depending on their capacity, Battery cost can go as high as 10,000 USD [7]. Based on the savings associated with using solar energy, the payback for such expense is roughly 8–10 years. It is expected that the cost of these components will decrease significantly in the future.

Traditional grids were designed for dispatchable power plants, where electricity generation can be adjusted on demand to meet fluctuating needs. Imagine a power plant burning coal or natural gas–the operator can control the amount of

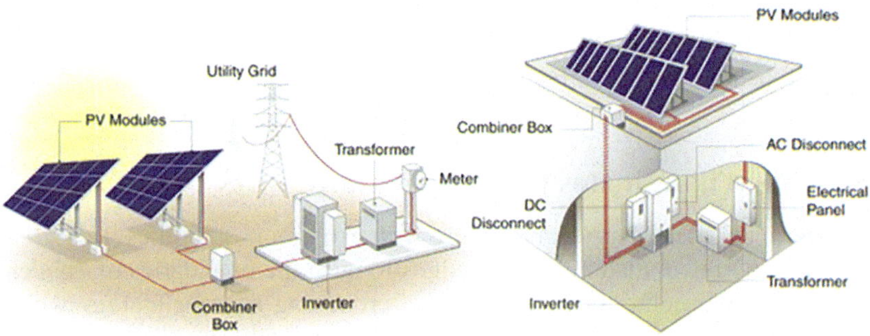

FIGURE 4.5 Principal components of photovoltaic systems [8]. *PV*, Photovoltaic.

fuel being burned to regulate the amount of electricity produced. Solar energy, on the other hand, is variable and cannot be readily controlled. The sun shines when it shines, and solar panels can only capture the available sunlight. This mismatch between the controllable nature of traditional power plants and the variable nature of solar energy can create challenges in maintaining grid stability and ensuring reliable power delivery.

Solutions for effective grid integration involve a multi-pronged approach:

Smart grid technologies: These advancements involve intelligent upgrades to the grid infrastructure, utilizing advanced communication and control systems. Smart grids can be likened to an intelligent traffic management system for the electricity grid. They can optimize energy flow in real-time, enhance grid flexibility, and accommodate the variable nature of renewable energy sources like solar power. By analyzing data on electricity consumption and production from various sources, smart grids can reroute power flows, activate backup generators when needed, and integrate stored solar energy to maintain grid stability.

Demand-side management: Encouraging consumers to shift their electricity usage patterns to align with peak solar production periods can ease the strain on the grid during high-demand times. This can be achieved through incentivized programs or time-of-use pricing structures. For instance, consumers could be offered lower electricity rates during periods of high solar production and encouraged to use energy-intensive appliances like dishwashers or washing machines during these times. By shifting some electricity demand to coincide with solar availability, especially for industrial activities during off-peak time, the pressure on traditional power plants during peak evening hours can be reduced.

Geographic diversity: Spreading solar installations across a wider geographic area can help mitigate the impact of local weather fluctuations on solar energy production. Imagine a cloudy day in one region; while solar production dips in that area, other regions with sunshine can continue generating power. By strategically distributing solar farms across a wider geographic area, the overall variability of solar energy production can be smoothed out, contributing to grid stability.

Innovation illuminates the path forward

The future of solar energy is brimming with exciting possibilities. Researchers are exploring novel materials and designs to push the boundaries of efficiency. Perovskite solar cells, for instance, are a promising new technology with the potential to surpass the efficiency limits of traditional silicon cells. The operation principle is similar to other semiconductor solar cells, but with strong light absorption and long carrier diffusion lengths. Additionally, advancements in nanotechnology are leading to the development of next-generation solar cells that can capture a wider range of the solar spectrum, including infrared and UV light.

Space-based solar power

Caltech scientists are developing a new method of capturing solar energy [9] using space satellites. This technology, first described by Isaac Asimov in 1941 in his short story "Reason," collects solar power in space and transmits the energy wirelessly to Earth through microwaves. This technology is eight times more powerful than Earth-based solar power and enables terrestrial power availability unaffected by weather or time of day. Solar power could be continuously available anywhere on Earth. Additional work is required to scale up this technology and make it available worldwide.

The challenges of energy storage, intermittency, and grid integration are not insurmountable. Through ongoing research and development in energy storage technologies, the implementation of smart grid infrastructure, and the adoption of innovative grid management strategies, we can pave the way for a future where solar energy becomes a reliable and ubiquitous source of clean power. Furthermore, fostering collaboration between scientists, engineers, policymakers, and the energy industry is crucial for developing and implementing effective solutions. By addressing these challenges head-on, we can unlock the full potential of solar energy and create a sustainable energy future powered by the sun's abundant and inexhaustible light.

The economic and environmental imperative of solar power

Solar energy is rapidly transitioning from a nascent technology to a pivotal force in the global energy landscape. This part delves into the compelling economic and environmental benefits of solar power, solidifying its position as a cornerstone of a sustainable future.

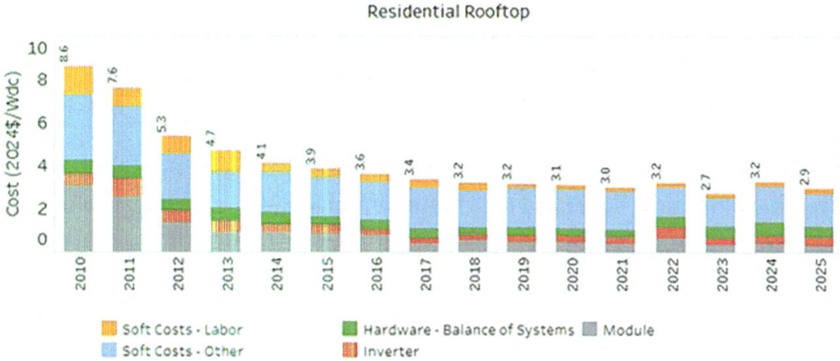

FIGURE 4.6 Historical prices for residential photovoltaic systems [10].

The economic rationale for solar energy is becoming increasingly robust. The cost of solar panels and installation has experienced a dramatic decline in recent years, making solar power a financially attractive option for a wider range of stakeholders. Homes, businesses, and even utility companies are recognizing the economic viability of solar investments. Fig. 4.6 shows historical system costs (in USD per Watt generated in DC current) for residential rooftop PV systems (before tax credit). The decreasing trend in cost is expected to continue, with some studies predicting costs falling to 0.165–0.481 USD/Watt in 2050.

Government incentives and tax credits further enhance the economic attractiveness of solar power, promoting widespread adoption. Beyond upfront cost reductions, solar power systems offer significant long-term operational savings. By generating clean electricity that offsets reliance on the traditional grid, users can expect a tangible decrease in their energy expenditures over time.

The burgeoning solar energy sector acts as a powerful economic engine, stimulating activity and job creation across various industries. From engineers designing cutting-edge solar cells to technicians installing rooftop panels, the solar industry fosters economic growth and empowers communities by creating new job markets that are independent of fossil fuels. This shift toward a solar-powered future holds the potential to create a more resilient and sustainable economy, less susceptible to the price fluctuations and environmental burdens associated with traditional fossil fuel dependence.

Environmental issues

Solar energy stands as a beacon of hope in the fight against climate change. Unlike traditional fossil fuel-based power generation, solar energy is a clean and renewable resource. Solar panels produce zero greenhouse gas emissions during operation, significantly reducing our carbon footprint and mitigating the effects of climate change. But still, they have a carbon footprint during panel and accessories production. By embracing solar energy, we can lessen dependence on fossil fuels. This transition to clean energy sources translates to cleaner air, improved public health, and a healthier environment for future generations. Reduced reliance on fossil fuels also minimizes air and water pollution associated with their extraction, transportation, and combustion.

Solar farms need a lot more land per unit of energy. We need about 1–2 hectares (2.5–5 acres) for generating 1 MWh. While they are projected to take up a fraction of the area dedicated to agriculture, solar panels cover the rooftops and lands occupied by cities, towns, roads, and all kinds of industries. As solar farms sprout across countries, loss of habitat is a major concern. The boom in solar, set to be the fastest-growing energy source, is predicted to fence off millions of acres across the nation, blanketing them in rows of glassy squares. Developers also need to be able to move electricity through farming land. This could be a major source of friction between farmers and developers. Proper local and state regulations are needed to govern large-scale solar farms. Also, additional research should be conducted on how animals interact with these farms. Site design for wildlife-friendly fencing needs to take high priority in solar farm development.

Installing solar panels in a desert environment apparently has some positive impact on the soil conditions. A team of researchers from Xi'an University of Technology studied the Gonghe Photovoltaic Park in China's Qinghai Province, a one-gigawatt solar farm covering vast stretches of desert [11]. Their study indicated that solar panels do not just harness the sun's power; they alter soil conditions by providing constant shade, which encourages vegetation growth and reshapes the local climate.

A standard solar panel usually has a lifespan of 25–30 years. A key environmental issue is how to recycle the used solar panels. The focus has been on building this renewable capacity; not much consideration has been put on the end-of-life of these technologies. Although 80% of a typical photovoltaic panel is made of recyclable materials, disassembling them and recovering the glass, silver, and silicon is extremely difficult. The National Renewable Energy Laboratory estimated that it costs roughly \$20 to \$30 to recycle a panel versus \$1 to \$2 to send it to a landfill. Regulations need to be in place to encourage producers to support their products through responsible end-of-life disposal. Also, manufacturers can come up with a standard design that makes panels easier and cheaper to recycle.

Energy justice: Making energy affordable worldwide

The inequitable adoption of rooftop PV systems has increasingly become a recognized energy justice issue. Households in under-resourced communities are significantly less likely to adopt rooftop PV and benefit from electricity bill savings and the satisfaction of participating in the clean energy transition. Although the adoption of rooftop solar is becoming more equitable as costs decline, a significant disparity remains. Consequently, under-resourced communities receive disproportionately small shares of the private benefits associated with rooftop PV adoption. Additionally, the inequitable adoption of rooftop PV may lead to cost shifting from high-income to low-income households through publicly funded incentives and utility rate structures. These adoption patterns mirror those seen with other emerging technologies. Fortunately, research indicates that various interventions—including financial support, community engagement, strategic siting, policy reforms, regulatory changes, and resilience measures—can improve equity in rooftop PV adoption. The grid transition will also entail an unequal distribution of economic opportunities. The transition will generate hundreds of thousands of jobs, but it will also displace jobs in fossil fuel industries. Solar is only one of several factors driving job displacement, and public retraining programs could help the rapidly growing solar industry absorb displaced fossil fuel workers.

A growing role in the future energy mix

The future trajectory of the global energy landscape is undeniably intertwined with the continued growth of solar power. Technological advancements are leading to increased conversion efficiencies and cost reductions, making solar power a more competitive player in the energy market. This, coupled with growing public awareness of the environmental and economic benefits of solar energy, is driving demand for clean energy solutions. Governments around the world are setting ambitious renewable energy targets, with solar power leading the charge. These ambitious goals, along with ongoing research and development efforts, will further propel the advancement of solar technologies and their widespread adoption.

As energy storage technologies improve and grid integration strategies evolve, the challenges associated with intermittency will be addressed. This will pave the way for even greater adoption of solar power, enabling a future powered by the sun's limitless energy. Imagine a world where cities, businesses, and even entire nations are powered by the sun, a future where clean energy is not just a dream but a reality–this is the promise of solar power, and it is a future worth pursuing with unwavering commitment.

Scientists have proposed creating a global energy system of solar stations evenly spaced in the Earth's equatorial belt so that some of the stations will always be on the daytime side of the Earth. All power plants will be connected with efficient power lines. Such a global energy system could generate 17,300 TWh per year of electrical energy continuously throughout the year. Such a project could increase the share of solar energy in the world's electricity consumption by 60%-70% and reduce greenhouse gas emissions by 10 times [12].

A collaborative path forward: Ushering in the solar age

The transition to a solar-powered future requires a collaborative effort on multiple fronts, as shown here:

Government support: A challenge has been unpredictable, shifting government policies toward solar power, discouraging investment in long-term projects. Implementing long-term, stable policies that incentivize solar adoption is crucial. This includes tax credits, feed-in tariffs that offer financial rewards for feeding excess solar energy back into the grid, and streamlined permitting processes for solar installations. Predictable policy frameworks will encourage long-term investments in solar technologies and foster market stability.

Research and development: Continued investment in research and development is essential for pushing the boundaries of solar technology. This includes exploring novel materials for more efficient solar cells, developing advanced energy storage solutions, and optimizing grid integration strategies. By fostering innovation, we can unlock the full potential of solar power and ensure its long-term viability.

Raising awareness: Educating the public about the economic and environmental benefits of solar power is critical for driving widespread adoption. Public awareness campaigns and educational initiatives can play a vital role in fostering understanding and encouraging individuals and communities to embrace solar energy solutions.

Community solar projects: Encouraging community-based solar initiatives allows households and businesses without suitable rooftops to participate in the solar revolution. Community solar farms, where individuals can purchase shares and benefit from the electricity generated, can democratize access to clean energy and empower communities to become active participants in a sustainable energy future. (See Chapter 12 for exciting examples of successful local initiatives).

Summary

Solar energy will play a key role in the future of energy generation. As technology advances, efficiency will increase, costs will continue to decline, and environmental concerns will be dealt with properly. Solar power is poised to take center stage in the global energy mix. By embracing a collaborative approach that fosters innovation, incentivizes adoption, and educates the public, we can harness the sun's boundless energy to create a cleaner, more sustainable future for generations to come.

Key takeaways

- There is currently an inequitable adoption of rooftop PV systems among communities, with low adoption in under-resourced communities that could benefit from electricity bill savings.
- Technological advancements are leading to increased conversion efficiencies and cost reductions, making solar power a more competitive player in the energy market.

- The intermittency of solar power requires the utilization of other renewables to provide continuity in power supply to the consumers.

References

[1] B. McKibben, Here Comes the Sun: A Last Chance for the Climate and a Fresh Chance for Civilization, W. W. Norton & Company, New York, 2025.

[2] T. Appenzeller, Good morning, sunshine, Science 390 (6779) (2025) 1208–1209. https://doi.org/10.1126/science.aee8000.

[3] Data from International Energy Agency (IEA)World Energy Outlook 2025, IEA, Paris, 2025 https://www.iea.org/reports/world-energy-outlook-2025 (Accessed Jan. 19, 2026).

[4] Massachusetts Institute of TechnologyThe Future of Solar Energy: An Interdisciplinary MIT Study, MIT Energy Initiative, Cambridge, MA, 2015 https://energy.mit.edu/research/future-solar-energy/ (Accessed Jan. 19, 2026).

[5] P. Basore, D. Feldman, Solar Photovoltaics: Supply Chain Deep Dive Assessment, U.S. Department of Energy, Washington, DC, 2022 https://doi.org/10.2172/1871588.

[6] M. Woodhouse, J. Cordell, E. Warren, D. Feldman, J. Zuboy, R. Margolis, Solar photovoltaic and storage supply chains and technology and market opportunities, in: Society of Vacuum Coaters (SVC) Conference, Washington, DC, National Renewable Energy Laboratory (NREL): Golden, CO, 2023, p. 14. https://docs.nrel.gov/docs/fy24osti/86128.pdf.

[7] National Renewable Energy Laboratory. Solar technology cost analysis. 2026 https://www.nrel.gov/solar/market-research-analysis/solar-cost-analysis (Accessed Jan. 19, 2026).

[8] M. Woodhouse, D. Feldman, V. Ramasamy, B. Smith, T. Silverman, T. Barnes, J. Zuboy, R. Margolis, Research and Development Priorities to Advance Solar Photovoltaic Lifecycle Costs and Performance, National Renewable Energy Laboratory, Golden, CO, 2021 https://www.nrel.gov/docs/fy22osti/80505.pdf.

[9] B. Abiri, M. Arya, F. Bohn, A. Fikes, M. Gal-Katziri, E. Gdoutos, A. Goel, P.E. Gonzalez, M. Kelzenberg, N. Lee, et al., A lightweight space-based solar power generation and transmission satellite, arXiv (2022). https://doi.org/10.48550/arXiv.2206.08373.

[10] National Renewable Energy Laboratory. Solar Installed System Cost Analysis. https://www.nrel.gov/solar/market-research-analysis/solar-installed-system-cost (Accessed Jan. 19, 2026).

[11] W. Wu, H. Chen, C. Li, G. Lu, D. Ye, C. Ma, L. Rei, G. Li, Assessment of the ecological and environmental effects of large-scale photovoltaic development in desert areas, Sci. Rep. 14 (2024) 22456. https://doi.org/10.1038/s41598-024-72860-8.

[12] E. Kabir, P. Kumar, S. Kumar, A.A. Adelodun, K.-H. Kim, Solar energy: Potential and future prospects, Renew. Sustain. Energy Rev. 82 (2018) 894–900. https://doi.org/10.1016/j.rser.2017.09.094.

[13] Solar futures study report, US DOE, Sep 2021. https://www.energy.gov/sites/default/files/2021-09/Solar%20Futures%20Study.pdf.

Further readings

E.W. Dickinson, Solar Energy Technology Handbook, CRC Press, Boca Raton, 2018.

M.A. Halabi, A. Al-Qattan, A. Al-Otaibi, Application of solar energy in the oil industry—current status and future prospects, Renew. Sustain. Energy Rev. 43 (2015) 296–314.

S.A. Kalogirou, Solar Energy Engineering: Processes and Systems, Elsevier, 2023.
T. Kuramochi, W. Turkenburg, A. Faaij, Competitiveness of CO_2 capture from an industrial solid oxide fuel cell combined heat and power system in the early stage of market introduction, Fuel 90 (3) (2011) 958–973.
T. Li, Regional Life Cycle Sustainability Assessment on Decentralised Electricity Generation Technologies in the Northeast Region of England, Newcastle University, 2019.
B. McKibben, Here Comes the Sun, W.W. Norton & Company Publishing, 2025.
M.E. Mackay, Solar Energy: An Introduction, Oxford University Press, USA, 2015.
J. Pinkse, D. Van den Buuse, The development and commercialization of solar PV technology in the oil industry, Energy Policy 40 (2012) 11–20.
G.N. Tiwari, A. Tiwari, Handbook of Solar Energy, 498, Springer, Singapore, 2016.
H. Tyagi, A.-K. Agarwal, P.R. Chakraborty, S. Powar, et al. (Eds.), Applications of Solar Energy, Springer, Singapore, 2018.

Chapter 5

Wind power

Synopsis

Windmills have been in existence for nearly 2000 years. The earliest known windmills were in Persia around 500–900 A.D. During my childhood in Iran, I often encountered operational vertical wooden windmills in villages near my hometown of Birjand (see Fig. 5.1). These 200+ year-old windmills were strategically installed in regions with strong winds that blew throughout most of the year. The tall walls surrounding the windmills not only supported the turbines but also funneled the airflow effectively. Unlike conventional windmills, these structures operated on drag rather than lift. The vertical arrangement of blades transmitted energy down the mast to the grindstone, eliminating the need for gears typically found in horizontal axis windmills. Their primary function was to turn heavy grindstones for milling wheat into flour.

Wind energy, a force long recognized for its power and potential, is rapidly evolving into a critical pillar of a sustainable energy future. In 2024, wind power provided about 10% of the electricity generated in the United States, making it the nation's largest source of renewable energy [2]. This chapter delves into the diverse wind turbine technologies that capture this readily available

F.A.C.T.S. Flexibility, Affordability, Conservation, Transition, Sustainability: Our Path to the Future of Energy and the Environment. DOI: https://doi.org/10.1016/B978-0-443-52574-2.00006-9

FIGURE 5.1 Ancient windmills in Iran. (*Courtesy M. Taghi Ref. [1]*).

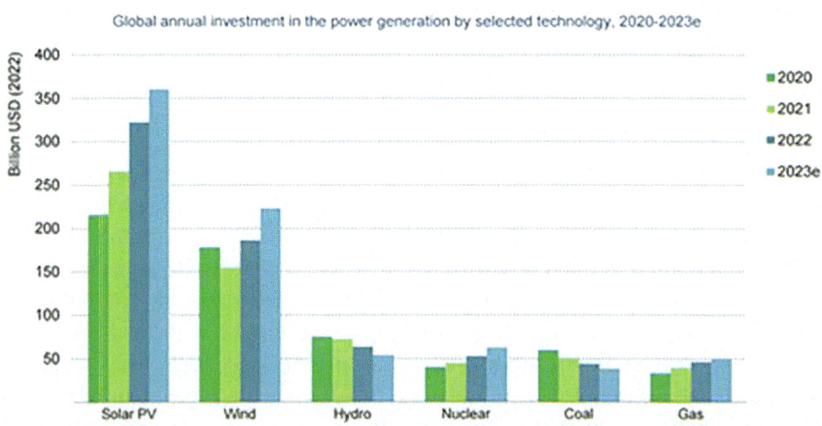

FIGURE 5.2 Global annual investment in power generation by selected technology, 2020–2023 [3].

resource and convert it into usable energy. We embark on a detailed exploration of the advantages and challenges associated with various turbine types, from the established land-based giants to innovative vertical axis designs that represent the future of wind energy capture.

Solar and wind retain a strong competitive advantage over other renewables. Fig. 5.2 shows global annual investment in different power technologies in recent years.

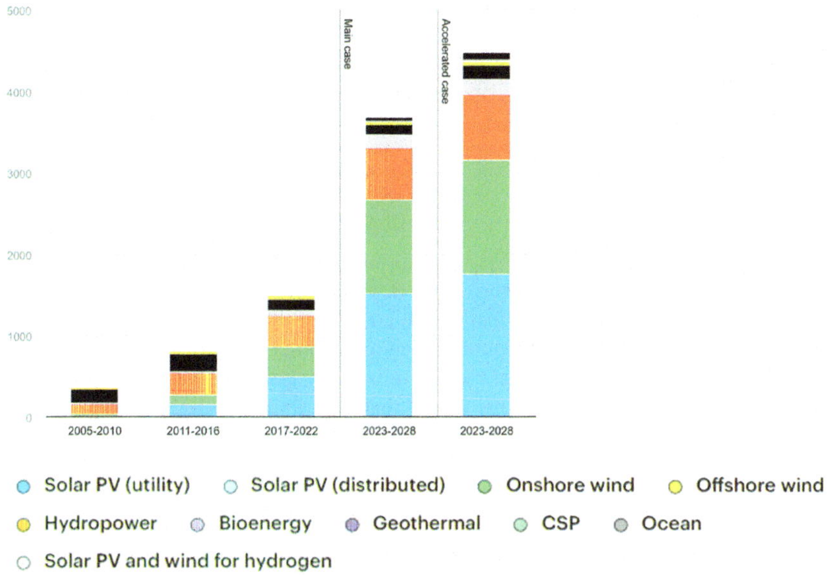

FIGURE 5.3 Renewable energy growth between 2005–2028 [4].

In this chapter, our examination extends beyond the technical aspects of wind power generation. We address the environmental considerations of wind farm siting, including visual impact and potential risks to wildlife, particularly bird mortality. By fostering a balanced approach that prioritizes responsible development alongside clean energy generation, we can harness the wind's power without compromising environmental integrity.

The potential of wind power extends far beyond its role in traditional electricity generation. It can also be used for transportation, fueling electric vehicles, etc. The historical consumption and the future forecast of renewable capacity growth (in gigawatts (GW)) by technology, for two scenarios, main (current practice) and accelerated (stated policies), are shown in Fig. 5.3. The global wind industry installed a record 120 GW of new capacity in 2024.

This chapter explores the exciting possibility of wind power as a key driver of decarbonization in the transportation sector. We envision a future where wind not only powers our homes and businesses but also propels electric vehicles and fuels sustainable marine transportation. By integrating wind energy with electric mobility solutions, we can revolutionize the way we travel, severing our dependence on fossil fuels and charting a course toward a cleaner transportation landscape.

Wind power can provide significant financial benefit, can create new job opportunities, and stimulate local economies. It reduces reliance on foreign energy sources, and bolsters national energy security.

Wind turbine technologies

Wind turbines, the modern-day windmills, stand as testaments to human ingenuity in harnessing the power of the wind through a series of steps (see also Box 5.1):

BOX 5.1 Tech note: Electricity generation with wind turbines

The amount of electricity a large wind turbine can generate depends on several key factors. Turbine capacity, measured in MW, represents the maximum power a turbine can produce under ideal conditions. Modern offshore turbines typically range from 8 to 12 MW, with some models achieving even higher capacities.

Wind speed plays a crucial role, as electricity generation is proportional to the cube of the wind speed. This means that doubling the wind speed results in eight times the power output. Offshore wind farms benefit from stronger and more consistent wind patterns compared to onshore locations, enhancing their energy production potential.

Air density is another important factor, as denser air allows turbine blades to capture more energy. Offshore turbines often operate in cooler ocean air with higher pressure, providing better conditions than those found at higher altitudes or warmer temperatures.

The characteristics of the site also influence energy capture. The specific offshore location determines wind patterns, and careful planning is required to optimize the placement of turbines within a wind farm to minimize interference between them.

A large offshore turbine with a capacity of 8 to 12 MW can generate millions of kilowatt-hours (kWh) annually, depending on these factors. For context, the average U.S. household consumes approximately 11,000 kWh per year. While these figures provide a general estimate, actual electricity generation varies throughout the year due to changes in wind conditions.

1. Wind captures blade energy: Wind moves across the turbine's blades, which are designed like airplane wings. This creates a lift, similar to how wings generate lift for airplanes.
2. Lift creates rotation: The lift force generated by the wind is stronger than the drag (resistance), causing the blades to spin around a central rotor.
3. Rotation transfers through drivetrain: The rotating blades are connected to a shaft that runs through the nacelle (the housing at the top of the tower). This shaft may be connected directly to the generator (direct drive) or go through a gearbox (most common) that increases the rotation speed.
4. Generator converts kinetic energy: The rotation of the shaft spins the generator inside the nacelle. This generator is essentially an electromagnet that uses the principle of electromagnetic induction to convert the mechanical energy (movement) of the shaft into electrical energy.

FIGURE 5.4 Wind farm on the cotton farm with cows in the cotton farm in West Texas, USA.

5. Electricity transformation and distribution: The electricity generated by the turbine is often at a lower voltage. A transformer increases the voltage for efficient transmission over long distances before it reaches the power grid and, ultimately, your home or business.

Various turbine types capture this energy. The utility and trade-offs inherent in each design are discussed here.

Onshore wind turbines

Onshore wind turbines, the most ubiquitous and readily recognizable type, are horizontal axis wind turbines (HAWT), and they dominate the wind energy landscape. Fig. 5.4 displays a view of a large wind farm in West Texas, which sits on top of a cotton farm that grows on top of an oil and gas reservoir.

These towering structures, with their spinning blades, are often clustered in wind farms to maximize energy capture. They offer several advantages:

Onshore wind turbines represent a mature technology with a long history of development and refinement. This maturity ensures lower costs, reliable performance, and readily available expertise in manufacturing, installation, and maintenance. Unlike offshore turbines, they benefit from existing infrastructure networks, which simplifies grid integration.

They are highly versatile in terms of site placement, as they can be installed on various land types, including hills, plains, and agricultural areas. This flexibility allows strategic placement in regions with favorable wind resources. Moreover, onshore wind farms can coexist with agricultural activities, such as grazing livestock or low-growing crops, optimizing land-use.

Onshore wind farms are also easily scalable to match the available land area and energy demand. Smaller-scale turbines can even be used for individual homes or businesses, supporting distributed energy generation and empowering communities to contribute to clean energy production.

Accessibility is another key benefit. Onshore turbines are easier to reach for maintenance and repairs compared to offshore installations. Technicians can readily perform routine maintenance or address unexpected issues, resulting in lower operational costs and quicker response times.

Offshore wind turbines

Offshore wind turbines are just emerging as a powerful force in the wind energy sector. These behemoths, often situated miles out at sea, are specifically designed to harness the stronger and more consistent winds that prevail over open water. Many offshore oil platforms can be repurposed as a base for the installation of these structures.

The USA's first commercial-scale offshore wind farm opened in March 2024, a 12-turbine wind farm called South Fork Wind 35 miles (56 km) east of Montauk Point, New York. South Fork can generate 132 megawatts (MW) of offshore wind energy to power more than 70,000 homes. The next large offshore wind farm, Vineyard Wind, is expected to open in 2025 off the coast of Massachusetts, too. When all 62 turbines are spinning, they will generate enough electricity for 400,000 homes and businesses [2]. After China, the UK is the world's next-largest offshore wind market. Scotland's government has described wind power expansion as a "once in a generation economic opportunity," with a pipeline of around 40 GW of offshore wind capacity in addition to 3 GW already in Scottish waters [5].

There are 73 GW of offshore wind capacity under development in the U.S., enough to power 30 million homes, according to the American Clean Power Association. Offshore wind energy offers several distinct advantages. One key benefit is the higher capacity factors achieved by offshore wind farms. These installations benefit from stronger and more consistent wind speeds, allowing turbines to generate electricity closer to their maximum potential. This results in significantly greater energy output per turbine and optimizes electricity production.

Additionally, offshore wind farms reduce visual impact on land-based communities. By being located far from shore, these turbines remain out of sight, addressing aesthetic concerns often associated with onshore wind development. This reduced visibility fosters greater public acceptance of wind energy.

The rapid pace of technological advancements further strengthens the appeal of offshore wind energy. Larger turbines and innovative floating platforms are enabling the deployment of wind farms in deeper waters farther from shore. These advancements open up vast new areas for wind energy generation, expanding the possibilities for sustainable power production.

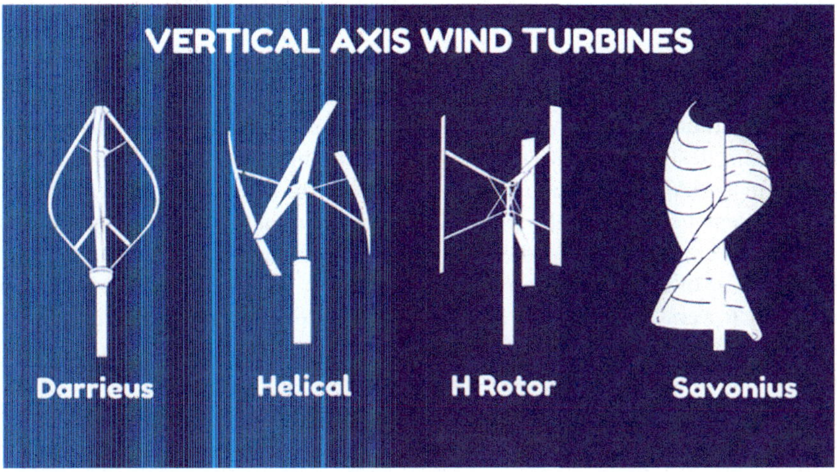

FIGURE 5.5 Types of vertical axis wind turbines [6].

Vertical axis wind turbines

Vertical axis wind turbines (VAWTs) represent a disruptive design in the wind energy sector. Unlike traditional horizontal axis turbines, where the blades rotate horizontally, VAWTs feature rotors with blades that spin on a vertical axis (Fig. 5.5).

One of the key benefits of VAWTs is their ability to capture wind from any direction. This omnidirectional wind capture eliminates the need for constant adjustments to align the turbines with the prevailing wind, making them particularly effective in areas with variable or turbulent wind patterns. Unlike horizontal axis turbines, which require active steering, VAWTs maintain efficient operation regardless of wind direction, simplifying their functionality.

Another advantage is their reduced noise emissions. The vertical orientation and specific design features of VAWTs often result in quieter operation compared to horizontal axis turbines. This characteristic makes them an appealing option for deployment near residential areas, where noise pollution might otherwise pose a challenge.

VAWTs also hold significant potential for urban applications. Their compact footprint and ability to harness wind from any direction make them well-suited for urban environments, where traditional horizontal axis turbines may face space constraints. These turbines can be installed on rooftops or in designated metropolitan areas, efficiently capturing unpredictable wind patterns influenced by surrounding buildings and contributing to sustainable energy solutions in cities.

Vertical axis wind turbines issues

VAWTs generally capture less wind energy compared to HAWTs due to the physics of their blade movement. HAWTs can take advantage of stronger winds at higher altitudes, which VAWTs with blades closer to the ground cannot. Also, the current design of VAWTs makes it challenging to build them as large and powerful as HAWTs, limiting their overall energy production capacity. This may not be ideal for large-scale wind farms. Another factor is turbulence issues. VAWTs operating close to the ground can experience more turbulence from buildings or other structures, reducing efficiency. Mounting them on rooftops or small towers can partially address this. While it is potentially cheaper to manufacture than some HAWTs due to simpler designs, VAWT technology is still under development, and the overall cost per MW-hour of electricity generated can be higher.

Super-sizing offshore wind

A wind turbine's hub height is the distance from the ground to the middle of the turbine's rotor. The hub height for wind turbines has increased significantly and is projected to increase even more, as shown in Fig.5.6. A turbine's rotor diameter, or the width of the circle swept by the rotating blades (the dotted circles in the figure), has also grown over the years. Larger rotor diameters allow wind turbines to sweep more area, capture more wind, and produce more electricity. In addition to getting taller and bigger, wind turbines have also increased in maximum power rating, or capacity. The average capacity of newly installed U.S. wind turbines in 2023 was 3.4 MWh. Higher capacity turbines mean that fewer turbines are needed to generate the same amount of energy across a wind plant—ultimately leading to lower costs.

As of 2025, the largest installed wind turbine was the Haliade-X 12 MW, developed by GE Renewable Energy. It is 260 m (853 feet) high, around 220 m (722 feet) in diameter, and has a 12 MWh capacity. This turbine is installed off the coast of Rhode Island, USA, for testing purposes. The generated electricity can power around 16,000 households per year.

Digitalization and artificial intelligence

Application of machine learning and artificial intelligence could help with anticipating weather, predicting maintenance needs, and optimizing operations, thus boosting output and reliability. This area of windmills is currently the subject of a lot of studies.

Selecting the optimal turbine: A multi-factor decision

While each wind turbine type boasts unique advantages, the optimal choice for a specific location depends on careful consideration of various factors. Site characteristics, wind resource availability, environmental considerations, and economic feasibility all play a crucial role in selecting the most suitable turbine technology.

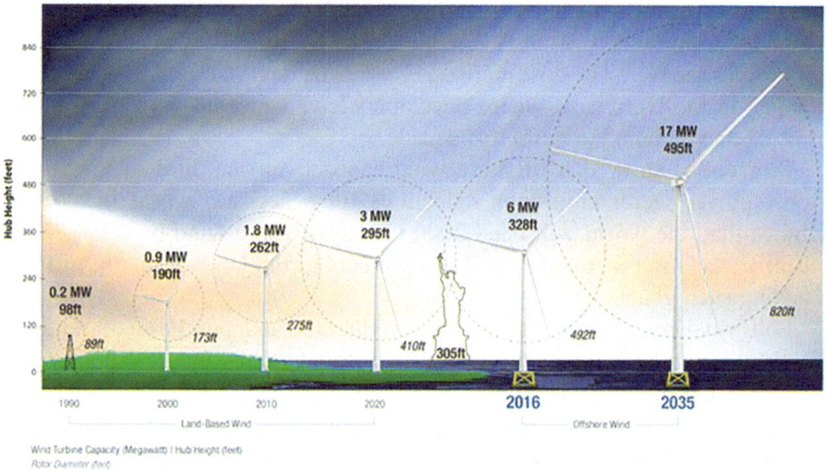

FIGURE 5.6 Trend of turbine sizes between 1990 and 2035 [7].

A thorough assessment of wind speed, direction, and consistency at the potential site is critical before any installation. Onshore wind farms may be viable in areas with moderate wind speeds. At the same time, offshore locations typically require stronger and more consistent winds to justify the higher costs associated with offshore installations. VAWTs can be advantageous in areas with variable wind patterns due to their omnidirectional wind capture capability.

Wind and solar power are often lumped together, but they have important differences that partly explain why one is slowing, and the other is thriving right now. For one, wind power is much more sensitive to location. Wind turbines in a gusty area can generate eight times as much electricity as turbines in an area with just half the breeze. For solar power, the difference between sunny spots and less sunny spots is considerably smaller. That means developers cannot just build wind farms anywhere. In the United States, the best places for wind tend to be in the blustery Midwest and Great Plains. However, many areas are now crowded with turbines, and existing electric grids are clogged with generated electricity from other sources, making it difficult to add more projects.

Onshore wind farms typically have lower upfront costs compared to offshore installations, at about \$31 per MWh on average for new projects, compared to offshore wind, which costs about \$100 for new projects (to be connected to the grid in 2028) [8]. This includes tax credits under the Inflation Reduction Act, which reduced the cost of renewable technologies.

VAWTs are a developing technology, and their cost-competitiveness is still evolving. However, their potential for lower maintenance needs and simpler operation could offer long-term economic benefits. The wind industry has also been hampered by soaring equipment costs after the pandemic wrecked supply chains and inflation spiked. While those factors initially hurt solar, too, the solar industry has adjusted much faster, with China nearly doubling its manufacturing capacity for panels over the last 2 years. Wind supply chains, which are dominated by a few manufacturers in China, Europe, and the United States, have yet to fully recover. The cost increases have been devastating for offshore wind projects in the Northeast, where developers have canceled more than half the projects they planned to build this decade.

Projections show that offshore wind can deliver one-third of the required global power sector emissions reductions for a net-zero world by 2050. To achieve that, the world will need 2000 GW of offshore wind by 2050 [9]. The recent cancelation of many windfarm projects in the USA is an alarm for not reaching such goals.

Policymakers can incentivize the production and purchase of renewable electricity, coupled with support for the development of wind energy infrastructure. This can involve tax breaks, subsidies, and investments in research and development of battery technology and grid modernization. Governments can play a role in encouraging the financial sector to develop strategies and targets that align their financial portfolios with national climate commitments over time. As an example, national development banks can establish ambitious climate-related lending targets, particularly in the offshore wind sector.

Current grids were designed for an era that drew power from large, centralized power plants. These systems are no longer suitable in a world with more decentralized power generation, which will require additional investments for buildout, digitalization, and modernization. Wind power, with its ability to generate electricity across a wide geographic area, can contribute to grid stability and enhance its capacity to support a growing fleet of Electric Vehicles (EVs). Investments in grid infrastructure will be crucial for seamlessly integrating wind power and other renewable energy sources into the grid, ensuring a reliable and clean energy supply for EVs. The International Energy Agency report [10] noted that annual grid investment globally has been stagnant for several years and needs to double within this decade to more than $600 billion.

Life cycle considerations: Carbon footprint of wind turbines

While wind power provides the significant benefit of clean electricity generation during operation, it is important to consider the environmental impact across the entire life cycle of a wind turbine. This includes the resource consumption and emissions associated with manufacturing and construction phases.

A 5 MW wind turbine, for example, requires substantial quantities of raw materials, including approximately 150 tons of steel, 250 tons of concrete for its foundation, and 3 tons of copper. Producing these materials involves significant energy use and associated greenhouse gas emissions.

Additionally, certain components of wind turbines, such as the fiber and polymer coatings used in the blades, often rely on fossil fuels during manufacturing. This creates an initial carbon footprint that must be accounted for when evaluating wind energy's overall environmental impact. Lubricants used

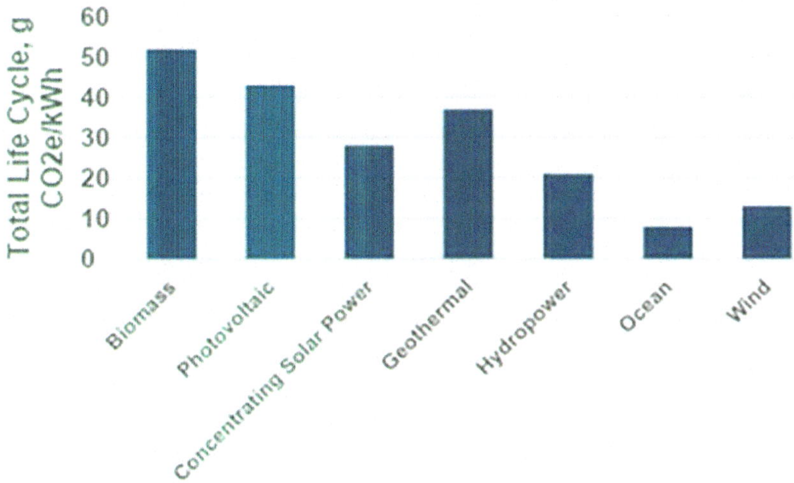

FIGURE 5.7 Life cycle emission factors for electricity generation technologies. (*Data from Ref. [4]*).

in wind turbines may also be fossil fuel-based, further contributing to upfront emissions. Fig. 5.7 presents the median published full life cycle emission factors for renewable energy sources [11]. Taking all of these factors into consideration, wind energy, with only 13 g CO_2 per kWh, ranks as the second-lowest emitter of CO_2, highlighting its relatively low environmental impact over its operational lifetime.

While wind power presents a compelling solution to meet energy needs, its development involves complex challenges. The impact of energy production on wildlife has come into the spotlight again amid US President Trump's plan to pivot the country's supply from renewables such as wind to oil and gas. The current administration has halted work on offshore wind farms until the impacts on the fishing industry, tourism, and whales are reviewed further. Other matters of concern are the intermittency of solar and wind energy and the significant amount of tax incentives that have gone toward renewable projects.

The debate is highlighting one of the most challenging conundrums facing renewable energy projects around the world – to what degree must they balance the impact they can have on local environments with the global effects of climate change? And how do those wind power related impacts compare with the local effects of oil and gas drilling [12]?

The main issue is the impact of wind turbines on bird populations. Collisions with turbine blades can result in bird mortality, especially for migratory birds and raptors with high flight patterns. Certain species, such as night-migrating birds and soaring raptors like hawks and eagles, are particularly vulnerable. Preconstruction studies, including radar monitoring, bird surveys, and collaboration with ornithologists, are essential to understand flight patterns, migration corridors, and nesting areas in proposed wind farm locations. VAWTs, with their slower rotor speeds, may present a lower risk to birds in some areas.

Land-use optimization poses another challenge. Wind farms often compete for land with agriculture, forestry, recreation, and military training activities. Balancing these interests requires identifying areas with lower land-use conflicts and exploring co-location strategies, such as combining wind farms with grazing livestock or integrating them with solar panels to maximize land efficiency.

Thorough environmental impact assessments (EIAs) are vital. These assessments should evaluate risks to wildlife habitats, sensitive ecosystems, and protected areas. Beyond birds, they must consider potential impacts on bats, insects, and plant life. Mitigation strategies, such as establishing buffer zones, adjusting turbine placement, or implementing seasonal shutdowns during critical breeding or migration periods, are necessary to reduce ecological disruption. There is also concern that the widespread deployment of wind turbines could alter wind patterns, potentially affecting climate models and ecosystems. Additional studies are needed to optimize these systems.

Visual impact is another significant issue. Wind turbines, often towering as high as skyscrapers, are more visually intrusive than solar farms and can draw intense opposition from local communities. For example, the Idaho State Legislature opposed a wind farm visible from a historic World War II site, and on Oahu, Hawaii, hundreds of residents protested and blocked the construction of a relatively small wind project. These examples highlight the importance of addressing public concerns and fostering community engagement in wind farm development.

Visual impact can be minimized through thoughtful site selection in areas with existing infrastructure, such as transmission lines, or natural features like hills that help blend wind farms into the landscape. Employing smaller turbines or placing them strategically to reduce visibility from populated areas or scenic viewpoints can also be effective. Additionally, night lighting strategies that limit light pollution and preserve the nighttime sky can help mitigate concerns about aesthetic disruption.

Wildlife impact assessments are also crucial for offshore wind farms. Offshore installations should consider marine ecosystems.

Ongoing research and collaboration among specialists/experts are essential for improving mitigation strategies. Studies of bird flight patterns near turbines, evaluations of different protective measures, and partnerships with ornithologists, wildlife biologists, and developers contribute to best practices. Open communication among stakeholders ensures that solutions are both effective and balanced, minimizing environmental impact while advancing wind energy development.

Community engagement is another cornerstone of successful wind farm projects. Developers must actively involve local communities in the siting process, addressing concerns about noise, visual impact, and property values. Transparent communication about potential effects and a willingness to incorporate community feedback foster trust and a sense of shared ownership over the clean energy benefits of wind power.

Raising public awareness about the environmental and economic benefits of EVs and wind-powered transportation is crucial for driving consumer adoption. Some misinformation has caused discomfort in social media against wind energy, and it must be challenged and corrected. For example, an infamous climate change disinformation artist, Marc Morano, had been talking about whales—specifically, the idea that the higher-than-usual number of dead ones washing ashore along the East Coast is the result of the push to develop 30 GW of offshore wind power by the end of the decade. Regulators and academic researchers say there is no evidence of a link to wind development. Since 2019, hundreds of gray whales have also washed up dead on the West Coast, where offshore wind development is only now getting underway. The clearest common factor is rising ocean temperatures, which disrupt whales' feeding and migration patterns. Fortunately, Morano and his Climate Depot company's mission failed as construction continued on the 12 turbines that now compose one of the largest offshore wind farms built to date in the United States. In December 2023, New York flipped the switch on South Fork Wind, which will deliver renewable power to some 70,000 homes in the state. However, figures like Morano may be gaining ground in a larger mission: Twisting public opinion against renewable energy in other would-be host communities.

A future of flexibility: A blend of technologies

According to Jason Ryan, a spokesperson for the American Clean Power Association, "Energy dominance requires a true 'all of the above' energy approach that unleashes our nation's diverse resources to meet surging demand" [12]. The future of wind energy is likely to witness a strategic blend of these diverse wind turbine technologies. Onshore wind turbines will likely remain a dominant force, particularly in regions with suitable land resources and favorable wind patterns. Offshore wind farms will continue to play an increasingly significant role as technology advances and facilitates the development of even larger and more efficient offshore turbines. VAWTs possess the potential to revolutionize wind energy capture in urban areas and locations with unique needs. By leveraging the strengths of each technology and carefully selecting the most suitable option for each specific application, we can maximize the potential of wind energy to power a sustainable future. Wind turbines provide very cheap electricity, and they often ramp up at night, when solar power is unavailable. This is readily apparent in Texas, where 22% of the electricity is generated by wind, and 8% is generated by solar. Texas generated about 430–465 TWh of electricity in 2025.

As wind energy technology continues to evolve, ongoing research and development efforts will undoubtedly lead to further advancements in design, efficiency, and cost-effectiveness. By embracing a diverse portfolio of wind turbine technologies, we can harness the boundless power of the wind to create a cleaner, more secure, and sustainable energy future for generations to come.

Summary

The future of wind energy points to strong growth. Wind power is expected to contribute nearly a third of the growth in renewable capacity by 2030, second only to solar, with the potential to become the world's leading power source. New technologies such as AI, larger turbines, and floating offshore wind farms are helping wind power supply a growing share of global electricity—potentially more than one-third by 2030–2050. While challenges like higher costs, supply-chain constraints, and permitting delays remain, solutions such as upgrading existing turbines and combining wind with energy storage are strengthening wind energy's role in the clean energy transition.

Beyond the technical aspects, there are challenges associated with wind farm siting, most notably visual impact and bird mortality. Responsible development and mitigation strategies. Collaborative efforts that encompass community engagement, EIAs, and ongoing research are crucial for minimizing potential downsides while maximizing the clean energy benefits of wind power.

Key takeaways

- Wind energy's role in the future energy landscape is expected to grow significantly.
- Challenges associated with wind farm siting, most notably visual impact, bird mortality, and political hurdles, can be overcome with responsible development.
- Offshore wind turbines can benefit from repurposing the offshore oil platforms that have reached the end of their oil production life.

References

[1] Atlas Obscura. (n.d.). Nashtifan windmills. https://www.atlasobscura.com/places/nashtifan-windmills. (Accessed January 19, 2026)

[2] J. McDermott Trump temporarily halts leasing and permitting for wind energy projects. AP News 2025, https://apnews.com/article/wind-energy-offshore-turbines-trump-executive-order-995a744c3c1a2eddb30cacf50b681f13 (Accessed 19 Jan, 2026).

[3] Data from International Energy Agency (IEA)World energy investment 2023, IEA, Paris, 2023 https://iea.blob.core.windows.net/assets/8834d3af-af60-4df0-9643-72e2684f7221/WorldEnergyInvestment2023.pdf (Accessed 19 Jan, 2026).

[4] International Energy Agency (IEA)Renewable capacity growth by technology, main and accelerated cases, 2005–2028, IEA Chart Library, 2024 https://www.iea.org/data-and-statistics/charts/renewable-capacity-growth-by-technology-main-and-accelerated-cases-2005-2028 (Accessed 19 Jan, 2026).

[5] Oil vs wind: which is worse for birds and whales? BBC Future, 2025, https://www.bbc.com/future/article/20250129-oil-vs-wind-which-is-worse-for-birds-and-whales (accessed Jan 19, 2026).
[6] Picture was adapted from Wikipedia vertical-axis wind turbine page; Wikimedia Commons, 2024, https://commons.wikimedia.org/wiki/File:Vertical_Axis_Wind_Turbines.png (Accessed 19 Jan, 2026).
[7] U.S. Department of Energy, Office of Critical Minerals and Energy Innovation. Wind turbines: the bigger, the better. 2024, https://www.energy.gov/cmei/articles/wind-turbines-bigger-better (Accessed 19 Jan, 2026).
[8] Lazard. 2023 levelized cost of energy+; Lazard: New York, 2023 https://www.lazard.com/research-insights/2023-levelized-cost-of-energyplus/ (Accessed 19 Jan, 2026).
[9] International Renewable Energy Agency (IRENA)World energy transitions outlook 2024: 1.5°C pathway, IRENA, Abu Dhabi, 2024 https://www.irena.org/-/media/Files/IRENA/Agency/Publication/2024/Nov/IRENA_World_energy_transitions_outlook_2024.pdf (Accessed 19 Jan, 2026).
[10] International Energy Agency (IEA)Electricity Grids and Secure Energy Transitions: Enhancing the Foundations of Resilient, Sustainable and Affordable Power Systems, IEA, Paris, 2023 https://www.iea.org/reports/electricity-grids-and-secure-energy-transitions.
[11] National Renewable Energy Laboratory (NREL). Life cycle assessment harmonization: life cycle greenhouse gas emissions from electricity generation: update. https://www.nrel.gov/analysis/life-cycle-assessment.html (Accessed 19 Jan, 2026).
[12] J. McDermott Trump tasks congressman with writing executive order he could issue to halt offshore wind. AP News 2025. https://apnews.com/article/trump-offshore-wind-climate-change-new-jersey-wind-turbines-d309a4b859361a5215b89c8242ffe0c4 (Accessed 19 Jan, 2026).

Further readings

European Wind Energy Association.*The economics of wind energy*. EWEA, 2009.
F. Blaabjerg, K.E. Ma, Wind energy systems, Proc. IEEE 105 (11) (2017) 2116–2131.
G.M.J. Herbert, S. Iniyan, E. Sreevalsan, S. Rajapandian, A review of wind energy technologies, Renew. Sustain. Energy Rev. 11 (6) (2007) 1117–1145.
J.F. Manwell, J.G. McGowan, A.L. Rogers, Wind Energy Explained: Theory, Design and Application, John Wiley & Sons, 2010.
R. Burns, Climate-science deniers, right-wing think tanks, and fossil fuel shills are plotting against the clean energy transition, Sierra club. https://www.sierraclub.org/sierra/2024-1-spring/feature/climate-science-deniers-fossil-fuel-shills-plot-against-green-energy.
T. Burton, N. Jenkins, D. Sharpe, E. Bossanyi, Wind Energy Handbook, John Wiley & Sons, 2011.
Y. Kumar, J. Ringenberg, S.S. Depuru, V.K. Devabhaktuni, J.W. Lee, E. Nikolaidis, et al., Wind energy: trends and enabling technologies, Renew. Sustain. Energy Rev. 53 (2016) 209–224.

Chapter 6

Aqua energy: Hydropower, geothermal, and ocean

Synopsis

Snow monkeys and humans have been using Earth-heated waters as baths for ages. We have been captivated by the geysers that intermittently discharge water and steam into the air (see Fig. 6.1), and approximately 1000 such features exist on Earth.

In the United States, the first geothermal plant was built in 1921, in Northern California, in a geyser-filled area that was famously known as "The Gates of Hell." That plant powers a nearby resort hotel and is also still in use. In Sonoma County, California, the Geysers Field stands as the world's largest geothermal power complex, with an energy output of 7000 GWh. Since the 1960s, it has been producing energy by harnessing steam that erupts from the ground at 188°C. The National Oceanic and Atmospheric Administration lists 1660 hot springs in the United States alone that are used for geothermal energy production. Aside from geothermal, numerous dams have been installed that capture hydropower. Hydropower provides about 17% of global electricity, making it the largest single source of low-carbon power. According to the International Energy Agency (IEA) forecast, geothermal and hydropower energy will continue to play significant roles in electricity generation (see Fig. 6.2) [1].

This chapter explores various types of renewable energy generated using water. It begins by describing hydropower energy, followed by a discussion of geothermal systems. Finally, it covers methods of capturing ocean energy.

F.A.C.T.S. Flexibility, Affordability, Conservation, Transition, Sustainability: Our Path to the Future of Energy and the Environment. DOI: https://doi.org/10.1016/B978-0-443-52574-2.00016-1

FIGURE 6.1 Calistoga geyser, California (personal photo).

Hydropower: Technologies for harnessing the power of moving water

Hydropower is the world's largest source of renewable energy. It provided over 14.3% (4578 TWh) of global electricity generation in 2024. It is a mature technology with a proven track record of delivering reliable and baseload electricity. Hydropower uses the energy of moving water to generate electricity. As depicted in Fig. 6.3, dams, canals, turbines, and generators are all part of a hydropower system. At hydropower plants, water flows through a pipe, or penstock, then pushes against and turns blades in a turbine that spins to power a generator to produce electricity. The International Renewable Energy Agency estimates that the global technically feasible potential for hydropower is around 18,000 TWh per year. This section delves into the diverse types of hydropower facilities, each offering unique advantages and applications within a sustainable energy

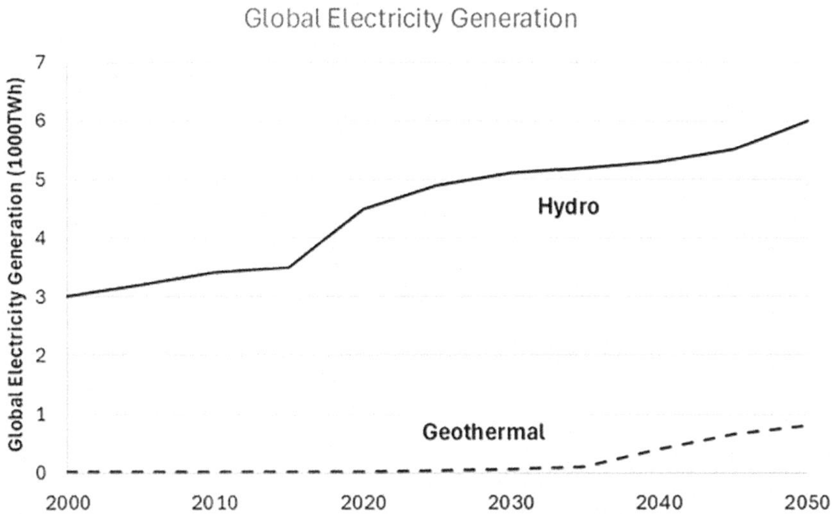

FIGURE 6.2 IEA forecast of electricity generation by geothermal and hydropower [1]. *IEA*, International Energy Agency.

portfolio. Understanding these variations is crucial for optimizing hydropower's role in the evolving energy landscape.

Run-of-river hydropower: A sustainable and scalable approach

Run-of-river hydropower plants are the most common form of hydropower technology, utilizing a straightforward yet effective method to generate electricity from flowing water. These facilities divert a portion of a river's natural flow through a canal, or a large pipe known as a penstock, directing the water to a powerhouse that contains turbines. The force of the moving water spins the turbines, producing electricity.

One of the defining features of run-of-river plants is their minimal infrastructure impact. They typically require little to no dam construction, often taking advantage of existing waterfalls or natural elevation drops. This design significantly reduces environmental disruption and lowers construction costs compared to other types of hydropower systems. However, electricity generation in these plants is closely tied to the river's natural flow, causing power output to fluctuate during droughts or periods of low water levels.

Run-of-river plants are also highly scalable, ranging from small, community-based projects designed to meet local energy needs to larger installations capable of supplying electricity to entire regions. This versatility makes them suitable for various applications.

These plants are well-suited for providing baseload power, offering a steady and reliable source of electricity that supports the core of a power grid's energy

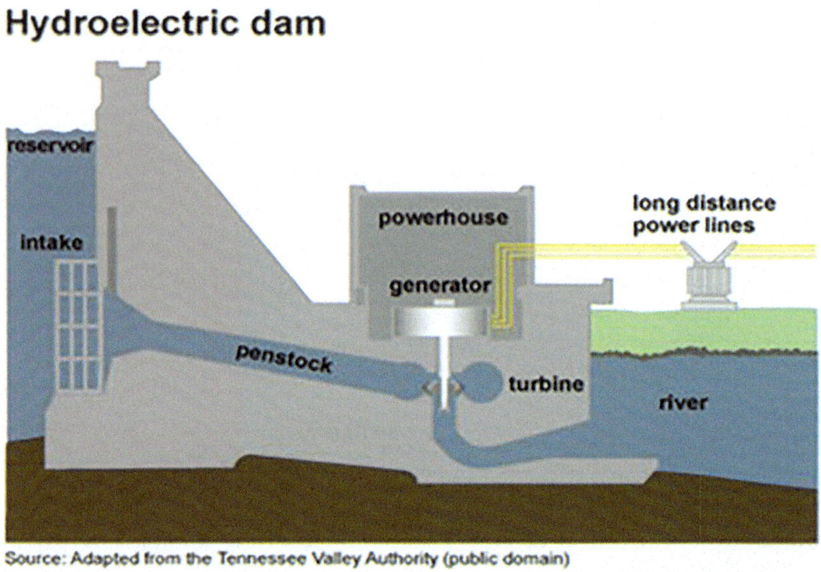

FIGURE 6.3 The hydroelectric dam [2].

supply. As long as there is sufficient water flow, they can operate continuously. Additionally, smaller run-of-river plants contribute to distributed generation by producing electricity closer to where it is consumed, reducing transmission losses and improving grid resilience. For remote communities situated near flowing rivers, run-of-river plants offer a sustainable and dependable energy source, delivering clean electricity in areas where traditional grid connections are impractical.

Pumped hydro energy storage

Pumped hydro energy storage (PHES) plants provide an innovative solution for energy storage and peak power generation, earning the nickname "giant batteries for the grid." These facilities operate using two reservoirs positioned at different elevations. During times of low electricity demand, excess energy from the grid powers pumps that move water from the lower reservoir to the upper one. When energy demand rises, the stored water is released back down through turbines, generating electricity to meet the increased need.

One of the main advantages of PHES is its ability to offer large-scale energy storage. This system allows excess renewable energy produced during low-demand periods to be stored and used later during peak demand, helping to balance fluctuations in supply and demand on the grid. Additionally, PHES plants can quickly ramp up electricity generation to respond to sudden spikes in demand, playing a crucial role in maintaining grid stability and ensuring a consistent electricity supply. However, the effectiveness of pumped storage depends on suitable geographic conditions, specifically locations with a significant elevation difference between the two reservoirs, which can limit its widespread implementation in some regions.

PHES plants are essential for grid stabilization, acting as a buffer to smooth out the supply and demand fluctuations that come with integrating variable renewable energy sources like solar and wind power. They also serve as a reliable source of peak power, reducing the need for costly and polluting Peaker plants that are typically used during high-demand periods. This shift away from fossil fuels for peak power generation contributes to a cleaner energy mix. Moreover, pumped storage supports the integration of renewable energy by storing excess electricity generated during peak production and releasing it when renewable output is low, enhancing the overall efficiency and reliability of renewable energy within the grid.

PHES is a well-established, cost-effective, and low-impact solution for long-duration energy storage. Globally, its potential far exceeds the future storage needs of a fully electrified and decarbonized society. This ensures that energy planners can confidently develop large-scale solar and wind projects, knowing there is a reliable storage option available [3].

Micro hydropower for remote communities

Micro hydropower plants, the smallest type of hydropower facility, harness the power of small streams or flowing water channels to generate electricity. These systems typically have a capacity of less than 100 kilowatts (kW) and are suitable for powering individual homes, remote villages, or small businesses (Fig. 6.4).

One of the primary advantages of micro hydro systems is their minimal environmental impact. Due to their small scale and limited infrastructure needs, they can often be integrated into existing water channels or irrigation systems with little disruption to the surrounding ecosystem. Additionally, micro hydro systems are known for their low maintenance requirements. Once installed, they can operate reliably for long periods with minimal upkeep, making them especially suitable for remote locations where regular maintenance may be challenging. However, the efficiency and success of these systems depend on site-specific factors, such as the water flow rate, the available head (the elevation difference between the water source and the turbine), and the local electricity demands. Proper planning and thorough site assessments are essential to ensure optimal performance.

Micro hydropower systems are highly effective in providing clean and reliable electricity for off-grid applications, powering homes, schools, and clinics in remote areas where connecting to the main power grid is either impractical or too costly. They also contribute to rural development by offering a sustainable and independent energy source for small communities, stimulating local businesses, improving living standards, and fostering energy self-sufficiency. Additionally,

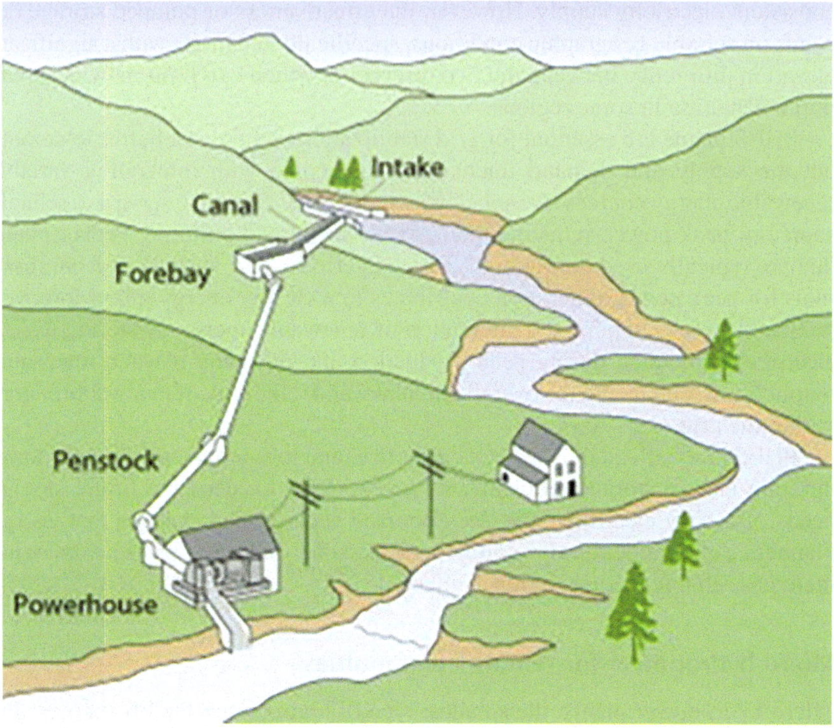

FIGURE 6.4 Micro hydropower plant [4].

micro hydro systems serve as a sustainable energy solution for eco-tourism lodges located near flowing water sources. This not only supports the lodges' commitment to environmental responsibility but also demonstrates the practical application of clean energy practices in the tourism industry.

The future of hydropower: Balancing sustainability and needs

Hydropower remains a crucial source of clean and renewable energy, but its continued development must prioritize responsible practices to minimize environmental impact, especially for large-scale projects. Achieving a balance between maximizing hydropower's energy generation potential and protecting ecosystems and biodiversity is essential for sustainable growth.

One significant challenge is managing sediment buildup in reservoirs, which can disrupt aquatic ecosystems downstream. Developing effective sediment management strategies is vital to maintaining river health and ecosystem stability. Additionally, protecting fish populations requires the implementation of fish passage solutions, such as fish ladders or bypass channels, which allow aquatic species to navigate around dams and continue their natural migration patterns.

Another critical aspect of responsible hydropower development is conducting thorough environmental impact assessments. These evaluations help identify potential ecological risks and guide the implementation of appropriate mitigation measures throughout all stages of a project—from planning and construction to long-term operation.

Environmental and social costs of hydropower

Hydropower is often praised for its clean and renewable energy production, offering a viable alternative to fossil fuels by significantly reducing greenhouse gas emissions. However, large-scale hydropower projects, particularly those involving dam construction, come with considerable environmental and social consequences. Acknowledging these externalities and adopting responsible development practices are critical to ensuring that hydropower contributes to a truly sustainable energy future.

The environmental impacts of large-scale hydropower projects extend beyond the immediate area surrounding the dam, affecting entire river ecosystems. One of the most significant concerns is habitat fragmentation and ecosystem disruption. Dams obstruct the natural flow of rivers, severing the connection between upstream and downstream ecosystems. This disruption hinders fish migration, which is vital for spawning and maintaining healthy fish populations. Additionally, altered water flow disturbs sediment transport, causing erosion in some areas and excessive sediment buildup in others. These changes can smother riverbeds, harm benthic organisms, and disrupt the food chain.

Biodiversity loss is another critical issue. The creation of large reservoirs often requires flooding vast areas of land, permanently destroying terrestrial and aquatic habitats. This submersion can lead to the displacement or extinction of countless species, including plants, insects, and animals that are essential to ecosystem health. For example, rainforests rich in biodiversity may be replaced by stagnant reservoirs, significantly reducing ecological resilience and altering the environmental balance.

Water quality also deteriorates in reservoirs created by dams. Stagnant water changes temperature, oxygen levels, and nutrient composition, often triggering harmful algal blooms that degrade water quality. Decreased oxygen levels stress aquatic life, creating a ripple effect throughout the ecosystem. Furthermore, the decomposition of submerged vegetation releases methane, a potent greenhouse gas, undermining the environmental benefits of hydropower.

Sedimentation poses another challenge. Dams trap sediment that would naturally flow downstream, extending the lifespan of reservoirs but depriving riverbanks and deltas of fertile materials. This loss of nutrient-rich sediment can severely impact agricultural productivity in downstream regions, threatening food security for local communities.

Beyond environmental consequences, large-scale hydropower projects also cause significant social disruptions. One of the most profound impacts is the displacement of communities. Dam construction often forces entire populations to relocate from areas designated for reservoirs. This involuntary resettlement can be deeply traumatic, leading to the loss of ancestral lands, homes, and cultural heritage. Displaced communities frequently struggle to adapt to new environments, facing difficulties in rebuilding their livelihoods, maintaining social networks, and preserving cultural traditions.

Livelihood disruption is another major concern. With more than 550,000 dams in the United States, free-flowing rivers are an endangered species. Changes in water flow and the displacement of communities threaten traditional fishing and farming practices that have sustained populations for generations. Downstream agricultural activities can suffer due to altered irrigation systems and reduced water availability, undermining both food security and economic stability.

Public health risks also emerge as unintended consequences of dam construction. Stagnant reservoir waters create ideal breeding grounds for waterborne diseases such as malaria and schistosomiasis. Additionally, altered water flow can disrupt sanitation systems, increasing the spread of water-related illnesses. Until 2019, there were more than 15,000 "high-hazard" dams in the USA, where a failure would result in the loss of life. Such health threats can overwhelm local healthcare systems, particularly in developing countries where many large-scale hydropower projects are concentrated [5].

Case study: Dam removal on the Klamath River

The previously mentioned environmental reasons have fueled a dam removal movement. Some 2200 dams in the USA, most of them in the past 25 years, have been blasted. The Klamath River dam removal is a case in point.

The Klamath River stretches from Southern Oregon through the Klamath Mountains to the Pacific Ocean in California. This was once the third-largest salmon-producing river on the west coast and teemed with salmon and trout in what seemed to be an inexhaustible supply.

In order to harness hydropower to support a growing region, the Klamath Hydroelectric Project was constructed over a 60-year period beginning in 1902. Six dams were built along a 55-mile stretch of the Klamath River, spanning Southern Oregon and Northern California. These dams provided electricity to an isolated region with minimal power infrastructure.

However, the construction of these dams displaced many Indigenous tribes from their ancestral lands. Before the dams were erected, the tribes had thriving commercial fisheries along the Klamath River. The completion of the downstream dam in 1964 effectively divided the Klamath watershed in half, leading to a sharp decline in the salmon population [6].

The dams disrupted the river's natural flow, creating stagnant, warm pools that became breeding grounds for toxic algae, now a persistent issue in the

Klamath. Over the decades, the dams accumulated more than 13 million tons of sediment, further altering the ecosystem.

The decline of salmon stocks significantly impacted life on the reservation. The commercial sale of salmon, once an economic cornerstone for local tribes, collapsed. However, in other rivers where dams have been removed, salmon populations have rebounded. Additionally, initial expectations that the dams would aid in flood prevention and irrigation were not met.

Tribal advocates and fishermen began calling for dam removal in the 1980s. It took 20 years to build consensus, and even the power company ultimately supported the decision. The dams generated a negligible amount of power, and the cost of upgrading them to modern standards outweighed the expense of removal. Several bills have been introduced in Congress to support the process of removing the dams and restoring the Klamath River's ecosystem [7].

The removal process took approximately four months, during which crews extracted around 1 million cubic yards of Earth from the massive structures. This became the largest dam removal project in U.S. history, reopening 400 miles of river habitat to salmon, trout, and eels for the first time in decades. The draining of reservoirs released accumulated sediment and dead algae into the river, but the drawdown was carefully managed. It is expected that after 1–2 years, the river will get back to normal. An extensive restoration effort is underway on the land beneath the reservoirs and other lands in the project area. The goal is to promote natural, sustainable river and floodplain ecosystems (Fig. 6.5) [8].

By removing the dams, the river and ocean ecosystems were reconnected, allowing fish to access the river's headwaters, its cold-water habitats, and the ocean's rich food supply. The Yurok Tribe and other locals hope this restoration will lead to a healthier river system [10].

This case underscores the need to balance environmental preservation with human needs. Early efforts to provide electricity came at the cost of displacing Indigenous communities, a model that proved unsustainable. In 2019, the Yurok Tribe passed a resolution granting the Klamath River legal personhood, recognizing its right to exist unharmed. This perspective reframes the river not as a resource to be exploited but as a living entity to be cherished, protected, and respected.

Moving forward with hydropower

IEA projects that global hydropower capacity will increase by 17% by 2030, growing from 1360 to 1590 GWh. As hydropower continues to expand, it is vital to prioritize responsible development practices to minimize its environmental and social impacts. A sustainable approach begins with conducting thorough environmental impact assessments during the early planning stages. These assessments must carefully evaluate potential ecological and social disruptions and

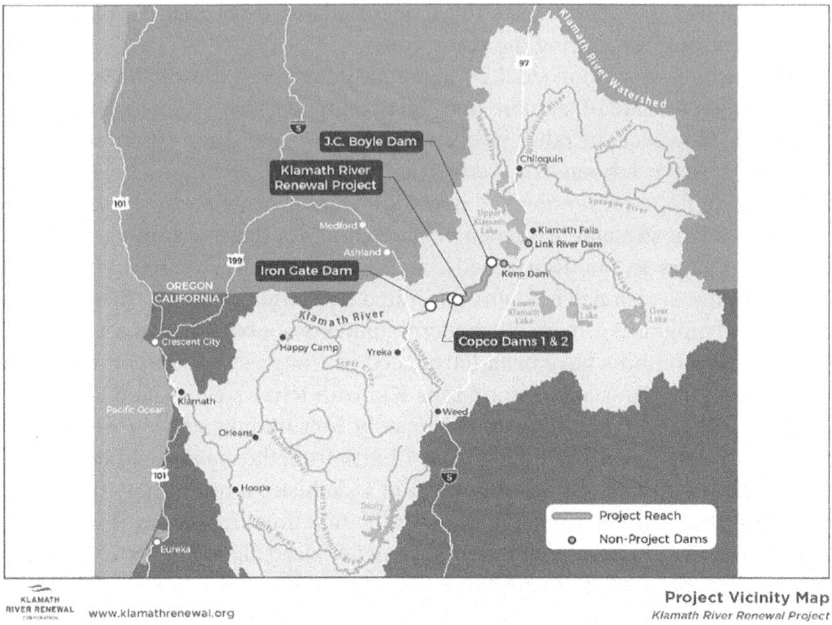

FIGURE 6.5 Klamath River basin and the removed dams. *KRRC*, Klamath River renewal corporation. (*Map lines delineate study areas and do not necessarily depict accepted national boundaries. Courtesy KRRC Ref. [9]*).

guide project design, construction methods, and mitigation strategies. Addressing potential issues proactively allows developers to implement solutions that reduce harm to both the environment and affected communities.

For projects that require the relocation of communities, comprehensive and sustainable resettlement programs are essential. These programs should guarantee fair compensation for lost land and livelihoods, offer sufficient relocation support, and assist in building sustainable livelihoods in new settlements. Such efforts can help restore the well-being of displaced populations and support their long-term stability.

Engaging communities and sharing project benefits are also critical for sustainable hydropower development. Open dialogue and collaboration with affected communities throughout the entire project lifecycle promote transparency and inclusion. Involving local populations in decision-making processes, offering employment opportunities during construction and operation, and investing in community development projects foster trust and empower communities. Prioritizing locally sourced materials and labor further strengthens community ties and enhances economic benefits.

Technological advancements play a key role in reducing the environmental impacts of hydropower projects. Innovations such as improved fish ladders and bypass channels can help protect fish migration patterns, while low-sediment dam designs and effective sediment management strategies can mitigate damage to downstream ecosystems. Continued research and development in these areas can significantly improve the sustainability of hydropower operations.

Diversifying investments across various renewable energy sources is another important strategy. While hydropower is a valuable contributor to clean energy, it is not a one-size-fits-all solution. Expanding investments in solar, wind, geothermal, and tidal energy can reduce dependence on large-scale hydropower projects and help minimize the overall environmental and social footprint of energy production.

In conclusion, hydropower remains a major player in the renewable energy sector despite its environmental and social challenges. By recognizing these challenges and prioritizing responsible development, technological innovation, and community engagement, a more sustainable future for hydropower is possible. The future of hydropower depends on thoughtful development, cutting-edge technology, and a collective effort to build a clean energy future that benefits both people and the planet.

Geothermal energy

The word "Geothermal" comes from the Latin roots, "Geo," defined as "Earth," and "thermal," defined as "heat." As far as today's science can determine, the center of the Earth has been very, very hot for some 3.9 billion years and will continue to be hot for at least that far into the future. This sustainable energy source originates from the natural decay of radioactive materials and residual heat from the planet's formation. The Earth's core and surface have extremely different temperatures, causing thermal energy to flow continuously from the planet's interior to its surface. A portion of the Earth's interior rock melts at temperatures exceeding 4000°C, forming hot, molten rocks known as magma. The mantle, being lighter due to these high temperatures, similar to water or soup in a hot pot, flows upward via convection. Fig. 6.6 displays the energy flow in a geothermal system (Box 6.1).

BOX 6.1 Tech note: High-temperature or high enthalpy?

Industry professionals often use the terms "high enthalpy" and "high-temperature" as synonyms when describing geothermal resources. Enthalpy is a measure of the total energy of a thermodynamic system, including the latent heat of vaporization/condensation. As such, it more accurately describes the energy production potential of a geothermal system that includes both hot water and steam.

Due to tectonics, hot magma intrudes into the Earth's crust. The actual geothermal reservoir is where hot steam or water is trapped under high-pressure

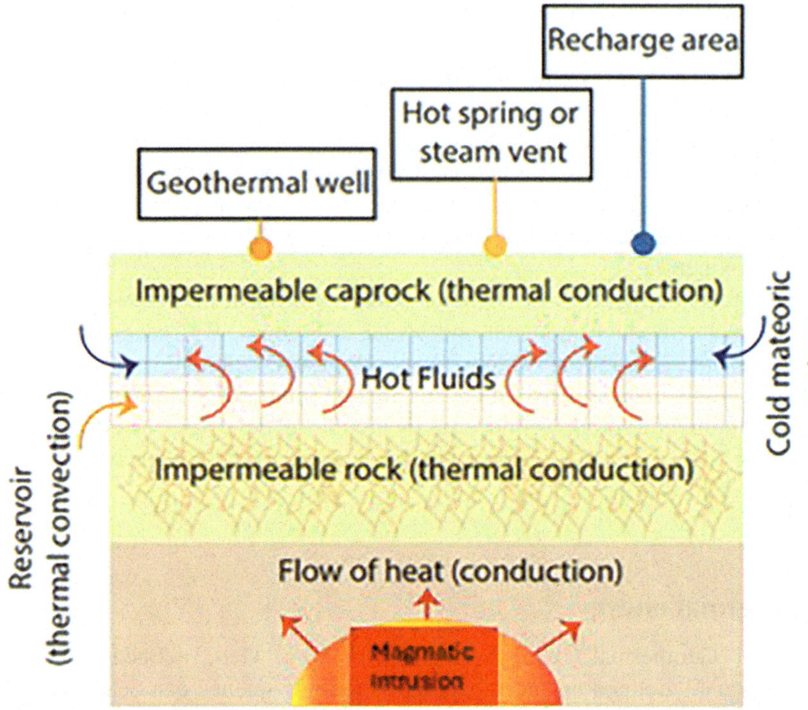

FIGURE 6.6 Heat flow in a geothermal system [11].

beneath a tight rock, and the magmatic intrusion heats it (see Box 6.2). Harnessing geothermal energy is relatively straightforward: For direct heating, relatively shallow wells, typically several hundred meters deep, are used to access natural reservoirs of hot water or steam, which can then be piped into buildings. For electricity generation, wells must reach greater depths where temperatures exceed 150°C. At these temperatures, pressurized steam drives a turbine connected to a generator, converting thermal energy into mechanical energy and then into electricity. Geothermal energy is essentially carbon-free, available at all times regardless of weather conditions, and has a small surface footprint, even though it requires very deep drilling. Recharge areas like lakes, rivers, or the seas provide cold meteoric waters that seep through cracks and faults in the rocks Box 6.2.

BOX 6.2 Tech note: Steam phase diagram

Fig. 6.7 is a phase diagram (pressure-temperature) of water. It represents various phases of water, like ice, liquid, and steam, under various conditions. As shown, the boiling point of water at atmospheric conditions is 100°C. At the end of the liquid-gas curve, there is a point we know as a critical point. The liquid and gas phase equilibrium curve ends at the critical point, i.e., at 218.1 atmosphere (3205 psi) and 374°C. Above these conditions, we have supercritical steam. Supercritical steam is used at modern steam turbine power plants, which are pursuing high efficiency.

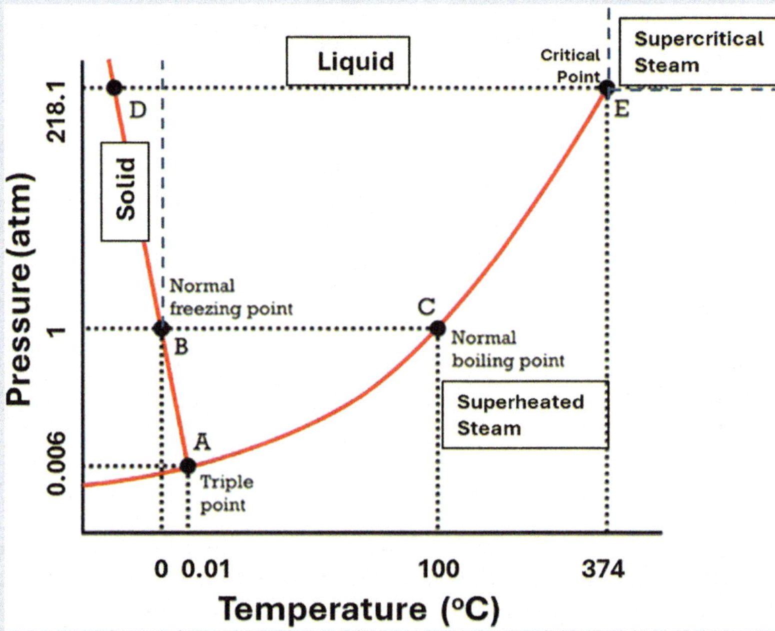

FIGURE 6.7 Water phase diagram [12].

Geothermal energy advantages

Geothermal energy has been increasingly sourcing heat and baseload electricity during the past 100 years. Compared to other renewable energies, such as solar and wind energy, geothermal energy offers several advantages. It boasts high stability (unaffected by season or sunshine) and a high utilization rate (at 75% capacity). This is 5.2 times that of solar photovoltaic power generation and 3.5 times that of wind power generation. Geothermal energy also provides safety, low operating costs, and opportunities for comprehensive integration with the grid [13].

Geothermal power plants impose minimal visual impacts on their surroundings when compared to typical fossil fuel plants. Fig. 6.8 illustrates the land footprint of various electricity-generating technologies in square meters per GWh equivalent. Geothermal power plants have the smallest land footprint and emit low levels of greenhouse gases compared to fossil fuel plants. Additionally, geothermal systems can provide heating and cooling for buildings

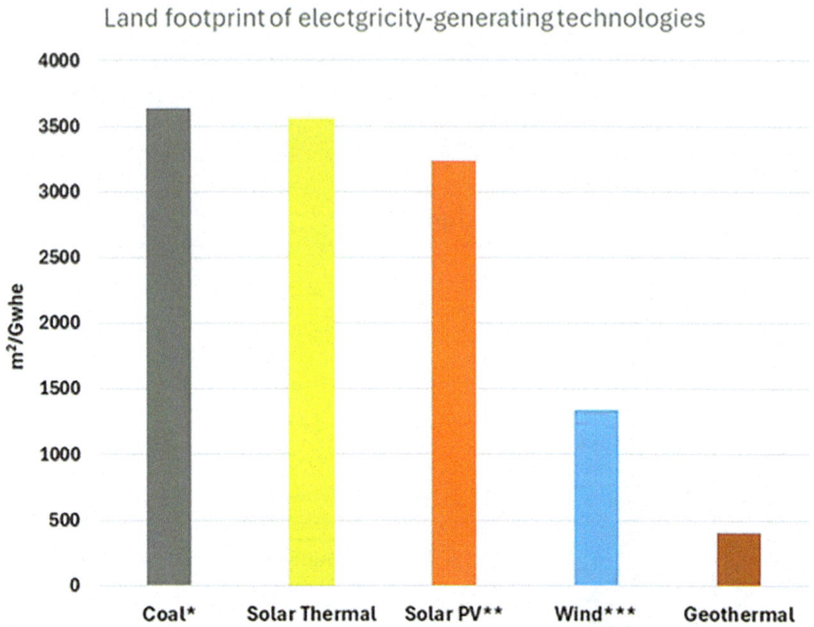

FIGURE 6.8 Land footprint of various electricity-generating technologies. (*Data from Ref. [14]*).

through geothermal heat pumps (GHPs), offering an efficient and environmentally friendly alternative to conventional heating ventillation and air conditioning (HVAC) systems (the latter will be discussed further in the next section).

Geothermal energy is considered environmentally benign. The lifecycle emissions of geothermal binary electricity plants with 100% reinjection are estimated to be as low as 11.3 gm of CO_2 per kWh, and the water consumption of a similar plant is estimated to be 0.66 L/kWh.

Geothermal types

Geothermal resources can be classified into hydrothermal, magmatic, and hot dry rock (HDR) based on their occurrence state. HDR refers to high-temperature rock (about 150°C–650°C) that contains supercritical steam. It is estimated that HDR at depths of 3–10 km holds approximately 30 times the energy of global fossil fuels. Electricity production is more favorable from geothermal resources with medium (90°C–150°C) to high temperatures, while lower temperatures are suitable for direct use in heating and cooling.

Most high-temperature geothermal resources are found around volcanoes in tectonically and volcanically active areas, such as the Pacific Ring of Fire, the mid-Atlantic Ridge, parts of Europe, and the East African Rift. Depending on the local subsurface temperature gradient, these resources can be located at depths ranging from a few hundred meters to several kilometers.

Low- and medium-temperature resources are more widely distributed geographically, with significant resources along faults and fractures in tectonically active areas, as well as deep within sedimentary basins. Because sedimentary basins usually have lower temperature gradients than volcanic areas, deeper drilling (to several kilometers) is often necessary to reach sufficient reservoir temperatures.

Geothermal energy extraction

The primary methods of geothermal energy extraction include dry steam, flash steam, and binary cycle power plants. Dry steam plants utilize steam directly from geothermal reservoirs to run turbines. Flash steam plants, the most common type, draw high-pressure hot water from the ground, which vaporizes into steam as it reaches the surface. Binary cycle power plants transfer the heat from geothermal water to another liquid that vaporizes at a lower temperature, driving the turbines.

Conventional hydrothermal geothermal resources are limited, with the current capacity in the U.S. around 40 GWh. However, the next generation of geothermal systems has a potential capacity of about 5500 GWh. These advanced systems are discussed in the following section.

Enhanced geothermal systems

The low permeability and porosity of natural HDR restrict heat extraction, leading to the development of enhanced geothermal systems (EGS). EGS artificially stimulates HDRs using hydraulic, chemical, and thermal methods. This technology is possible due to advances made in the oil industry, such as drilling horizontal wells at exceptional depths and hydraulic fracturing (fracking), which the geothermal industry refers to as hydro shearing. These techniques could expand the use of reliable and low-emissions renewable geothermal energy, but there are serious environmental concerns to consider.

The EGS systems could be either conductive or convective, depending on the nature of the heat flow underground [15]. An example of a conductive EGS is shown in Fig. 6.9. An injection well injects high-pressure cold fluids into the artificially fractured reservoir (indicated by the blue line). This working fluid absorbs thermal energy as it passes through the reservoir's fractures and returns to the surface through a production well (indicated by the red line). The exchanged energy at the conversion plant on the surface is used for power generation, and the cooled working fluid is reinjected into the thermal reservoir and forms a closed-loop.

Theoretical calculations indicate that a 20°C reduction in the temperature of 1 km^3 of HDR can provide enough thermal energy to operate an electric generator with a 10 MWh capacity for 20 years, equivalent to 1.3 million tons of oil. While

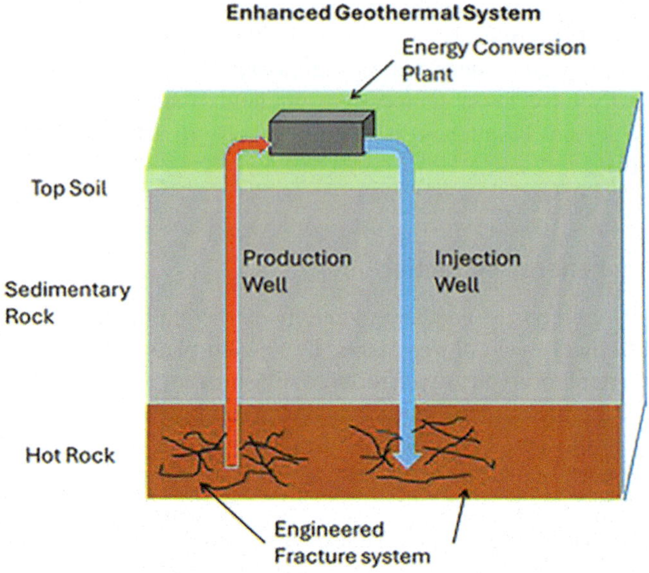

FIGURE 6.9 Enhanced geothermal system.

conventional EGS uses water as the working fluid, there is a 10%–20% water loss during EGS stimulation and operation.

Adaptation of advanced drilling techniques is enabling an increase in scale and a decrease in the cost of EGS projects. In 2024, the levelized cost of electricity generation in the western regions of the USA was around 30–50 $/MWh [16]. In some regions, it is preferred to use direct heating instead of electricity generation. The direct utilization of low-temperature geothermal resources is generally associated with a lower carbon footprint and fewer environmental issues.

By making geothermal generation flexible and dispatchable, EGS can provide additional grid stability to the electrical grid, which depends on large contributions of intermittent generation, such as from wind and solar energy.

With all its benefits, EGS has some environmental impacts. As was discussed in Chapter 2, hydraulic fracking has been associated with small earthquakes, and even when conducted thousands of feet below drinking water sources, it carries a potential risk of contaminating water with toxic chemicals from the propping agent (if they are used). The occurrence of seismic events can be considered during the project hazard analyses, monitored, and accommodated. Maintaining the thermal reservoir's pressure to ensure the successful and continuous operation of the EGS equipment requires a significant amount of water (about 870–15,898 L/MWh). Fortunately, EGS can use lower-quality water sources than traditional power plants, benefiting dry areas where groundwater and fresh surface water are inevitably consumed.

Advanced geothermal systems

These systems are deep, large, artificial closed-loop circuits in which a working fluid circulates and is heated by subsurface rocks through conductive heat transfer, similar to domestic-based GHPs. As advanced geothermal systems do not require a water-bearing reservoir with good permeability, they have the potential to be applicable in almost any location worldwide. However, substantially longer well bores are needed to increase the surface area for heat transfer, which may result in higher drilling costs. Another option is to drill multilateral wells with several branches downhole to increase the contact area between the hot rock and the working fluid. As we move into deeper hot rocks, new drilling tools are required to enable drilling at high temperatures.

Supercritical geothermal systems

These systems are characterized by very high temperatures and a natural reservoir containing fluid in the supercritical state. For pure water, this means a temperature of at least 374°C and a pressure of at least 221 bar. Such supercritical fluids can be found deep within volcanic hydrothermal systems in locations like Iceland, Japan, Kenya, Mexico, and New Zealand. The productivity of these supercritical systems could be significantly greater than that of conventional high-temperature geothermal systems due to the higher energy content of the fluid. However, their utilization presents several technological challenges, such as managing corrosive fluids, which have been the subject of recent investigations.

Low-temperature geothermal heat pumps

A GHP can extract the heat content of low-temperature resources (less than 90°C or 194°F). These heat pumps utilize the stable temperature conditions found within the top layers of soil to heat and cool buildings, using underground pipes buried at depths where temperatures range between 5°C and 30°C to transport warm and cool air directly into buildings during winter and summer, respectively. Fig. 6.10 shows a typical GHP. Widely used across North America, Europe, and East Asia, GHP systems are highly energy-efficient, consuming 25%–50% less electricity than air-source heat pumps or conventional air conditioning systems while emitting less greenhouse gas pollution. Installing an air-cooling system that uses ground transference and heating systems is roughly three times more costly than installing a ventilation system.

Geothermal energy challenges

Like other energy sources, geothermal energy has its limitations. In certain resource areas, geothermal fluids or steam contain significant amounts of hydrogen sulfide (H_2S) and other non-condensable gases (NCGs), such as carbon dioxide (CO_2), which can negatively impact the environment if released into the

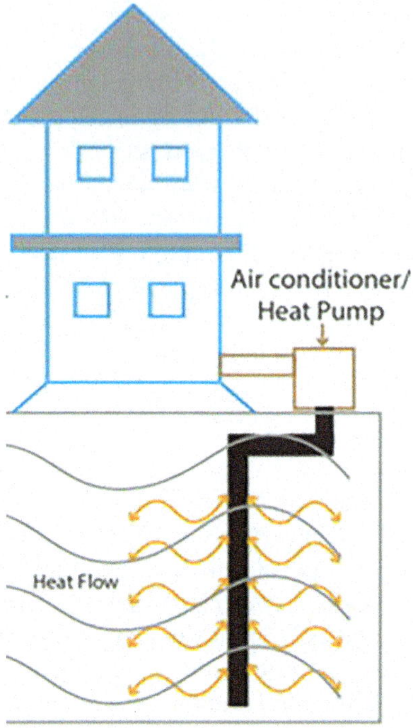

FIGURE 6.10 Ground geothermal heat pump [9].

atmosphere. Since NCGs must be removed from the steam before it enters the turbine, geothermal fields with high concentrations of these gases are unsuitable for power generation. Additionally, geothermal resources are often located in remote regions, necessitating the construction of transmission connections and other infrastructure to access these sites, which increases the indirect land requirements or rights of way. Developing geothermal energy in areas with high scenic value can further complicate the process by increasing the licensing burden for companies.

Moreover, geothermal steam fields demand sophisticated maintenance to ensure efficient operation. Periodic drilling of make-up wells is often necessary to replace older wells that have diminished steam production, leading to additional costs. In some regions, scale buildup in production lines also presents operational challenges, potentially affecting system efficiency and longevity.

For example, Indonesia has the world's largest geothermal resource, with the potential for 29 GWh of capacity. However, less than 5% of the potential has been utilized. Regulatory barriers, inadequate feed-in tariffs, lack of financing, and early-stage exploration risks have frustrated the exploitation of the country's large geothermal potential. Geothermal resources are geographically constrained, and the initial costs of drilling and plant construction can be high. The required capital investment for geothermal energy ranges from $3000 to $6000 per kWh, whereas solar and terrestrial wind energy range from $1700 to $2100 per kWh. Additionally, there are environmental concerns related to the release of gases and the potential for induced seismicity. Despite these challenges, advancements in technology and growing interest in sustainable energy solutions continue to drive the growth and development of geothermal energy.

Leasing prime geothermal land is also more challenging than leasing similar land for oil and gas development. In the U.S., most areas with subsurface characteristics best suited for EGS deployment are situated on federal land, and the approval process is very slow.

Extracting minerals: A new way of utilizing geothermal resources

Granite rock in geothermal systems is typically rich in lithium, and hot water has a greater capacity to absorb lithium. As a result, geothermal water often contains high concentrations of lithium, making it a promising source for lithium extraction. The process of extracting lithium from geothermal water is comparable to removing salt from seawater through evaporation.

One advanced method for lithium extraction is direct lithium extraction (DLE), a promising technique with several variations. DLE employs specialized materials such as nano-filters or ion exchange resins that function like chemical sieves, selectively capturing lithium chloride from the brine while leaving other dissolved minerals behind. The captured lithium chloride is then processed and purified, often into lithium hydroxide, a critical component in battery production.

Another method, electrodialysis, operates similarly to ion exchange by using electrically charged membranes to separate lithium ions from the geothermal brine. This process can be conducted in stages to concentrate lithium content and is sometimes combined with DLE to improve efficiency.

Compared to traditional lithium extraction methods like evaporation ponds used in salt lakes, both DLE and electrodialysis offer significant advantages. They are generally more environmentally friendly, requiring less water and land, and producing less ecological impact.

However, these technologies are still in development, with ongoing research focused on improving efficiency, reducing costs, and ensuring environmentally responsible practices. The combined potential of energy generation and lithium production is positioning geothermal energy as an increasingly attractive option for investors.

Expanding geothermal energy for a sustainable future

Unlocking the vast potential of geothermal energy for a clean and secure energy future requires a comprehensive five-point strategy. The first step is diversifying geothermal resources by focusing on the development of abundant yet underutilized low—and medium-temperature resources. This includes exploring volcanic zones and sedimentary basins and repurposing existing infrastructure, such as abandoned mines and oil or gas wells. Additionally, integrating shallow geothermal systems, GHPs, and underground thermal energy storage can expand the range of geothermal applications. Geothermal systems for heating and cooling individual homes remain somewhat expensive to install, but they last for decades, reduce energy bills up to 50%, and avoid reliance on the electrical grid.

Positioning geothermal energy as a climate champion is equally important. This involves highlighting its competitive advantages, such as clean electricity generation, sustainable heating and cooling, and minimal greenhouse gas emissions. Industry stakeholders should actively promote these benefits, along with geothermal energy's potential for extracting critical minerals like lithium used in battery production. Public awareness campaigns and educational initiatives targeting policymakers are crucial for fostering broader adoption.

Enabling investment is another key component. Addressing the high upfront costs and exploration risks of geothermal projects requires the development of tailored risk-mitigation strategies, streamlined licensing procedures, and supportive energy policies that recognize geothermal energy's distinct characteristics. These measures will encourage investment and support its integration into existing energy systems.

Driving innovation through research and development funding is vital for unlocking geothermal energy's full potential. This includes exploring advanced technologies such as hot, dry rock systems, closed-loop systems, green hydrogen production, supercritical geothermal resources, and mineral extraction from geothermal brines. Pilot projects demonstrating the technical feasibility of these innovations can build confidence among industry stakeholders and accelerate deployment.

Lastly, leveraging the expertise and infrastructure of the oil and gas industry can significantly speed up geothermal development. By repurposing abandoned wells for geothermal plants, the industry can contribute to clean electricity generation and sustainable heating applications. Implementing this strategic approach will unlock geothermal energy's vast potential and pave the way for a sustainable and secure energy future. Geothermal means jobs in drilling and production for engineers, geologists, and project managers in the petroleum industry.

Case study: Iceland geothermal energy

Until the early 1970s, Iceland utilized imported fossil fuels for nearly three-quarters of its energy. Despite the presence of many hot springs, lava domes, and bubbling mud pots, the country's geothermal resources remained largely untapped. The oil crisis of 1973, when prices more than tripled, became a turning point. Alongside other efforts to develop domestic energy sources, public investment funds provided loans to support geothermal projects, whose upfront costs were substantial. By the early 1980s, almost all Icelandic homes were heated using geothermal energy. Today, more than a quarter of the country's electricity comes from geothermal sources, one of the highest shares in the world, with most of the remainder supplied by hydropower. Part of this success is the proximity of hot rocks to the surface, where temperatures exceeding 150°C can be reached at depths of roughly 1–2 km [17].

In 2008, Iceland's three largest energy companies launched a joint research initiative known as the Iceland Deep Drilling Project (IDDP). The project aimed to drill much deeper, "superhot rock" (SHC) near Krafla, northeast Iceland, to access steam at temperatures approaching 400°C [18]. In general, the hotter the well, the greater and more efficient the energy output. At a depth of about two km, however, IDDP engineers unexpectedly struck magma, with temperatures near 900°C. Under these conditions, water becomes supercritical, meaning it is neither liquid nor gas and can produce 5–10 times more power per well than standard geothermal wells. They had created the hottest and most powerful geothermal well in the world. Despite the presence of corrosive gases, this well flowed (around 160 kg/s) of superheated steam at 452°C and 142 bars, i.e., an electric potential of 36 MWe. A conventional geothermal well often delivers 50–100 kg/s for ~5–7 MWe. However, it was plugged in 2012 due to several surface failures. IDDP drilled a second well in 2016, which reached a depth of 4.6 km and a temperature of 600°C. The casing failure has stopped production from this well.

Drilling these wells identified several technical and geologic challenges that must be overcome to enable the success of superhot rock wells. A major achievement of the IDDP wells was to demonstrate that it is possible to drill into a supercritical geothermal reservoir. Also, with fewer wells, it is possible to generate more energy. A third well has been planned to be drilled in 2026. This project has global relevance, especially for volcanic regions such as Japan, Indonesia, Kenya, Italy, and the western United States.

Beyond energy production, geothermal resources are used extensively in Iceland for direct applications, including greenhouse heating, fish farming, snow melting, and tourism, most famously at sites like the Blue Lagoon. Iceland has also become a center for geothermal research and innovation, exporting expertise, technology, and project development know-how worldwide.

Overall, geothermal energy has enabled Iceland to achieve one of the world's cleanest energy systems, combining energy security, affordability, and very low greenhouse gas emissions while demonstrating the long-term potential of geothermal power when favorable geological conditions exist.

Case study: Quidnet energy – Innovating long-duration energy storage

As the global transition to renewable energy accelerates, the need for cost-effective, long-duration energy storage has become a critical challenge. Quidnet energy, a Texas-based company, has developed an innovative energy storage solution inspired by traditional hydroelectric power but adapted for widespread deployment. This case study explores Quidnet's technology, its potential impact on the energy sector, and the challenges it faces in scaling up.

Quidnet energy's proprietary technology, known as geomechanical pumped storage, stores energy by using excess electricity to pump water into underground rock formations. The system pressurizes the water within naturally occurring geologic formations, effectively converting them into energy storage reservoirs. When electricity demand rises, the stored water is released to drive turbines and generate power, similar to conventional hydroelectric systems but without the need for large dams or significant topographical constraints.

This approach offers several advantages over traditional battery storage and pumped hydro. First, it provides long-duration storage at a lower cost compared to lithium-ion batteries, addressing the intermittency of renewables such as wind and solar. Second, it utilizes subsurface rock formations that are widely available, making it scalable across various geographies. Lastly, Quidnet's storage system boasts a long operational lifespan with minimal environmental impact, as it does not require hazardous materials or extensive land use modifications.

Despite its promise, Quidnet faces challenges in regulatory approval, market adoption, and competition from established battery technologies. Gaining trust from utilities and grid operators will be essential for widespread implementation. Additionally, proving the long-term durability and efficiency of its geomechanical reservoirs will be key to commercial success. However, with increasing demand for grid resilience and sustainable storage solutions, Quidnet is well-positioned to play a significant role in the evolving energy landscape.

In summary, Quidnet energy represents a pioneering effort in the long-duration energy storage sector. Leveraging geomechanical storage provides an innovative alternative to batteries and conventional hydro. If successfully scaled, this technology could become a critical component of a resilient, renewable-powered grid, supporting global decarbonization efforts.

Summary

Geothermal energy, a reliable and clean source of heat and electricity, stands alongside wind and solar as a crucial player in the clean energy transition. Its widespread availability, from volcanic zones to sedimentary basins, complements the variable nature of wind and solar power, stabilizing electricity grids. Furthermore, geothermal technology has transcended its focus on electricity, now offering sustainable heating and cooling solutions. These features make geothermal a cost-effective and weather-independent source of renewable energy.

Looking ahead, advancements in geothermal technology, advances in superhot rock, collaboration across energy sectors, and the growing use of geothermal heating and cooling will fuel accelerated deployment.

Ocean power: An analysis of emerging technologies

The gravitational pull of the moon and sun, along with the rotation of the Earth, create tides in the oceans. In some places, tides cause water levels near the shore to rise and fall up to 40 feet. People in Europe harnessed this movement of water to operate grain mills for more than a millennium. The amount of power available in the movement of ocean waves has inspired many engineers to develop machines and devices to capture and use this free oceanic power. This section delves into some of the most promising ocean energy technologies, meticulously evaluating their potential to contribute to a sustainable energy future while acknowledging the hurdles that need to be overcome for their widespread adoption.

Ocean energy technologies

Ocean energy can be captured through three technologies as described:

Wave energy: Capturing the kinetic energy of motion

Wave energy involves capturing the kinetic energy generated by the motion of ocean waves. As waves pass a fixed point, the sea surface rises and falls, causing water near the surface to move and gradually lose its kinetic and potential energy. This movement also impacts the pressure beneath the surface. Due to the periodic and oscillatory behavior of ocean waves, various wave energy devices have been developed to harness this energy and convert it into electricity.

One common method is the use of overtopping devices, which resemble ramps or sea walls and are strategically placed to capture the energy of waves as they surge over the structure. The water is collected in a reservoir positioned at a higher elevation, and as it is released, it drives turbines to generate electricity. Although overtopping devices are relatively simple in design, their efficiency can be reduced by fluctuations in wave activity and the risk of damage during storms [19].

Another approach is the oscillating water column (OWC), which utilizes a partially submerged chamber to capture wave energy. The motion of the waves compresses air trapped above the water column, forcing it through turbines connected to generators. OWCs can be more efficient than overtopping devices and operate effectively under a broader range of wave conditions. However, their complex design and maintenance requirements, along with the impact of chamber size and shape on efficiency, present challenges that need further refinement [20].

Point absorbers offer a different solution by floating on the water's surface and moving up and down with the waves. This motion powers internal pistons or hydraulic systems that generate electricity. These devices can be deployed flexibly in deep offshore waters, but they tend to involve higher development and

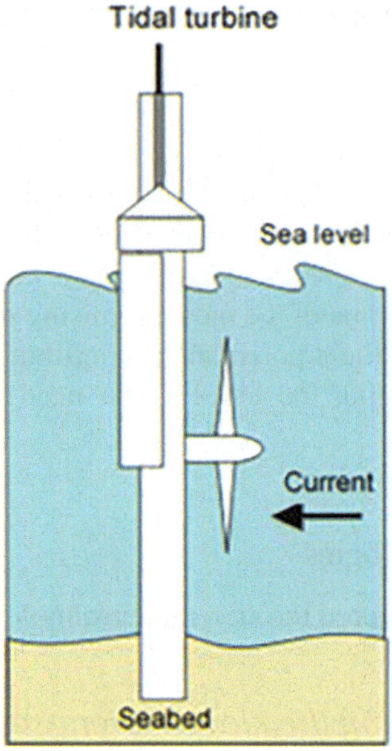

FIGURE 6.11 Tidal turbine source *(Modified from National Energy Education Development Project, EIA [22])*.

maintenance costs. Additionally, their potential effects on marine ecosystems require careful evaluation [21].

Wave energy has significant potential due to its high energy density. Some estimates suggest that the theoretical global wave energy capacity could surpass the world's current electricity consumption, making it an appealing renewable energy source. However, the technology remains in its early development stages, facing obstacles such as high costs, durability in harsh ocean environments, and environmental concerns related to marine life. Ongoing research and development are essential to overcoming these challenges and fully realizing the potential of wave energy.

Tidal energy: Channeling the predictable power of currents

Tidal energy harnesses the predictable movement of ocean tides to generate electricity. This form of renewable energy takes advantage of the consistent ebb and flow of tides, offering a reliable and steady power source. One prominent technology in this sector is the tidal stream generator. These submerged turbines function similarly to wind turbines but are placed underwater to capture the kinetic energy of tidal currents (Fig. 6.11). Their predictable and dependable energy output stems from the well-understood behavior of tides. However, their deployment is limited to locations with strong tidal currents, and both installation and maintenance are complex and costly, requiring a thorough cost-benefit analysis.

Another method of generating tidal energy involves tidal barrages, which function like dams built across estuaries or bays where tidal currents are strong. Water passing through turbines within the barrage generates electricity. While tidal barrages can produce substantial amounts of clean energy, they also pose significant environmental risks. By disrupting natural water flow and altering marine habitats, these structures can negatively impact ecosystems. Additionally, the high initial construction costs and extended building periods present considerable challenges, making it crucial to assess their feasibility carefully [23].

Tidal energy holds immense potential, with some estimates indicating it could supply up to 20% of the United Kingdom's electricity demand. The predictability of tidal patterns makes this energy source exceptionally reliable. Nonetheless, the availability of suitable locations with strong tidal currents is limited, and the environmental and economic impacts of large-scale infrastructure, such as tidal barrages, must be carefully evaluated. Addressing these challenges is essential to fully realizing the benefits of tidal energy.

Ocean thermal energy conversion: Exploiting thermal differences for power generation

Ocean thermal energy conversion (OTEC) capitalizes on the temperature difference between warm surface seawater and cold deep seawater to generate electricity through a thermodynamic cycle. This process operates through two main systems: open-cycle and closed-cycle OTEC.

Open-cycle OTEC uses warm surface water to evaporate a working fluid, typically one with a low boiling point such as ammonia. The resulting vapor drives a turbine to produce electricity, and the vapor is then condensed using cold deep seawater, completing the cycle. This method provides a stable baseload power supply in tropical and subtropical regions where temperature differences are consistent. However, open-cycle OTEC systems are complex and energy-intensive due to the vast amounts of seawater that must be processed. Additionally, their use is geographically limited to areas with significant thermal gradients between surface and deep waters.

Closed-cycle OTEC operates similarly but uses a secondary fluid like ammonia in a closed-loop. Warm surface seawater heats this fluid, causing it to vaporize and drive a turbine. The vapor is then cooled and condensed by deep seawater, allowing the cycle to repeat. Closed-cycle OTEC systems are more environmentally friendly than open-cycle systems because they do not require the extraction and discharge of large volumes of seawater, reducing harm to marine life. However, they tend to be less efficient due to the additional heat transfer step and remain in the early stages of development.

OTEC presents a promising, clean, and continuous energy source, especially for tropical and subtropical regions. Despite this potential, both open- and closed-cycle systems face challenges related to efficiency, high operational costs, and the need for further technological advancements. Environmental concerns, such as the potential release of harmful substances during biofouling prevention and the disruption of marine ecosystems due to seawater extraction in open-cycle systems, also require thorough evaluation.

Despite the promise of ocean energy in creating a sustainable energy future, significant obstacles hinder its commercial viability. High costs are a major challenge, as ocean energy technologies are still in their early development stages. Research, development, and deployment expenses are substantial compared to established renewable sources like solar and wind. Additional factors, such as corrosion and biofouling—the accumulation of marine organisms on equipment—also contribute to higher operational costs.

Limited infrastructure and supply chains further complicate the development of ocean energy. Unlike solar and wind energy, the infrastructure for ocean energy technologies is underdeveloped, inflating project costs and slowing deployment. A reliable supply chain for specialized materials, components, and installation vessels is essential for scaling up. Moreover, a workforce skilled in managing offshore devices is necessary to support ongoing operations and maintenance in challenging marine environments.

Ocean energy: Transition to the future

Ocean energy stands as a promising frontier in the global transition to sustainable energy. According to IEA [24], the estimated annual electricity production potential from various ocean energy sources is substantial: wave power could generate between 8000 and 80,000 TWh, OTEC could produce around 10,000 TWh, salinity gradient power offers up to 2000 TWh, and tides and currents could contribute approximately 1100 TWh. These figures are particularly striking when compared to the global electricity production of 28,000 TWh in 2023.

Despite this vast potential, ocean energy remains in its early stages of development, with few commercially operating plants. As of 2024, global electricity generation from ocean energy is only a small fraction of its potential. However, overcoming commercialization challenges, reducing environmental impacts, and advancing technology could unlock widespread adoption of this resource. Collaborative efforts among researchers, developers, policymakers, and the private sector, combined with responsible environmental practices, are critical to realizing this potential. By prioritizing innovation and sustainability, ocean energy can become a cornerstone of a clean energy future. Harnessing the immense power of waves, tides, and thermal gradients must be done responsibly to protect ocean ecosystems for future generations.

The journey toward integrating ocean energy into the global energy mix is not without challenges. However, advancements in technology, supportive policies, and the growing need to transition from fossil fuels are paving the way for ocean energy to play a more prominent role. The global market outlook for ocean energy is poised for growth, driven by several key factors. Heightened concerns about climate change are accelerating the shift toward clean energy, making ocean energy an appealing alternative due to its minimal greenhouse gas emissions. Rising global energy demand necessitates the exploration of diverse renewable energy sources, and the depletion of fossil fuel reserves underscores the importance of sustainable and long-term solutions like ocean energy.

Technological innovation will be central to the expansion of ocean energy. Efforts to reduce costs will focus on innovative manufacturing techniques, cost-effective materials, and achieving economies of scale through broader deployment. Standardized designs and modular components will help streamline production and maintenance. Enhancing efficiency is equally critical, with ongoing research aimed at improving device designs, exploring advanced materials, and developing sophisticated control systems to maximize energy capture from waves, tides, and thermal gradients.

The feasibility and effectiveness of ocean energy technologies vary by region due to differences in geography, wave patterns, tidal currents, and thermal gradients. Europe leads global ocean energy development, leveraging its extensive coastlines and strong currents. The European Union has set ambitious goals for ocean energy deployment, with the United Kingdom, France, and Portugal actively advancing projects in this sector.

In the Asia-Pacific region, countries like China, Japan, and South Korea present enormous potential for ocean energy development. These nations have vast coastlines and growing energy demands. China is investing heavily in ocean energy research and development, Japan is applying its advanced marine engineering expertise, and South Korea is exploring tidal energy to meet its increasing energy needs.

North America also offers significant opportunities due to its diverse oceanographic conditions. The United States and Canada are pursuing wave, tidal, and OTEC technologies. The U.S. focuses on the Pacific Northwest's wave energy and the Atlantic Coast's tidal resources, while Canada is exploring tidal energy in the Bay of Fundy. Government support, particularly from the U.S. Department of Energy, and state-level policies have been instrumental in driving ocean energy initiatives.

For developing countries and remote island communities, ocean energy presents a pathway to clean and reliable energy access. Small scale waves and tidal technologies can provide sustainable energy solutions for isolated areas. Tropical island nations could benefit significantly from OTEC due to the natural temperature differences in their surrounding waters. However, these regions face financial and infrastructural challenges, requiring international collaboration, technology transfer, and innovative financing to support large-scale deployment.

The future of ocean energy is filled with promises. As technologies mature, costs decrease, and policy support strengthens, ocean energy can become a major contributor to the global energy supply. Responsible development, including thorough environmental impact assessments and ongoing monitoring, will be crucial to ensure that ocean energy deployment aligns with environmental protection goals.

Adopting a region-specific approach that considers the unique characteristics and challenges of each area will help unlock ocean energy's full potential. This strategy can foster a future where diverse regions worldwide benefit from clean, renewable ocean energy.

Summary

The "*Aqua Energy*" chapter provided a review of the key technologies for generating energy from geothermal systems, hydropower, and ocean. These technologies utilize the Earth's static and dynamic energies and the natural movement of water to generate electricity. Although their effectiveness is influenced by geographic conditions, they do have their own environmental considerations to be dealt with.

Key takeaways

- Hydropower is a key contributor to the future production of renewable energy.
- Large-scale hydropower projects, particularly those involving dam construction, come with considerable environmental and social consequences.
- Geothermal, hydropower, and ocean energy have immense potential to contribute to clean energy generation, especially in regions where natural conditions are favorable.
- Ocean energy stands as a promising frontier in the global transition to sustainable energy.

References

[1] Data from International Energy Agency (IEA). Renewable Electricity; in Renewables 2025, IEA, Paris, 2025 (Accessed Jan. 19, 2026).

[2] U.S. Energy Information Administration (EIA). Hydropower Explained; Energy Explained, 2023 (Accessed Jan. 19, 2026)

[3] A. Blakers, Z. Weber, I. Silalahi, Pumped hydro energy storage: a global opportunity, Prog. Energy 7 (2) (2025) 022004. https://doi.org/10.1088/2516-1083/adaabd.

[4] U.S. Department of Energy. Microhydropower systems. https://www.energy.gov/energysaver/microhydropower-systems (Accessed Jan. 19, 2026).

[5] T. Lohan, Undammed: Freeing Rivers and Bringing Communities to Life, Island Press, Washington, DC, 2025.

[6] A river reconnected. Sierra Magazine (Accessed Jan. 19, 2026).

[7] P.A. Sheikh, C.V. Stern *Klamath River Dam Removal and Restoration*; CRS In Focus IF11616; Congressional Research Service: Washington, DC, 2024. (Accessed Jan. 19, 2026).

[8] Klamath River Renewal Corporation (KRRC). Restoration *brochure*; 2024. (Accessed Jan. 19, 2026).
[9] Klamath River Renewal Corporation (KRRC). Project *M*aterials (webpage). (Accessed Jan. 19, 2026).
[10] *Sierra* (Sierra Club). Klamath Dams: Closer to Coming Down *(digital edition article)*. https://digital.sierramagazine.org/publication/?i=837107&article_id=4901012&view=articleBrowser (Accessed Jan. 19, 2026).
[11] L. Kumar, M.S. Hossain, M.E.H. Assad, M.U. Manoo, Technological advancements and challenges of geothermal energy systems: a comprehensive review, Energies 15 (23) (2022) 9058. https://doi.org/10.3390/en15239058.
[12] W. Wagner, A. Pruß, The IAPWS formulation 1995 for the thermodynamic properties of ordinary water substance for general and scientific use, Journal of Physical and Chemical Reference Data 31 (2) (2002) 387–535. https://doi.org/10.1063/1.1461829.
[13] A. Kagel, D. Bates, K. Gawell, A Guide to Geothermal Energy and the Environment, Geothermal Energy Association, Washington, DC, 2007 http://www.charleswmoore.org/pdf/Environmental%20Guide.pdf (Accessed Jun. 20, 2018).
[14] P. Brophy, Environmental advantages to the utilization of geothermal energy, Renew. Energy 10 (2–3) (1997) 374.
[15] R. Horne, A. Genter, M. McClure, W. Ellsworth, J. Norbeck, E. Schill, et al., Enhanced geothermal systems for clean firm energy generation, Nat. Rev. Clean Technol. 1 (2025) 148–160. https://doi.org/10.1038/s44359-024-00019-9.
[16] M.J. Aljubran, R.N. Horne, Supply characterization of baseload and flexible geothermal power, Sci. Rep. 14 (2024) 17619. https://doi.org/10.1038/s41598-024-68580-8.
[17] R. Galchen The heat of the moments. The New Yorker 2025, https://www.newyorker.com/magazine/2025/11/24. (Accessed Jan. 19, 2026).
[18] G.O. Friðleifsson, B. Pálsson, A. Albertsson, B. Stefánsson, E. Gunnlaugsson, J. Ketilsson, et al. The IDDP Success Story – Highlights. Proc. World Geothermal Congress 2020+1; Reykjavik, Iceland, 2021. https://www.worldgeothermal.org/pdf/IGAstandard/WGC/2020/37000.pdf. (Accessed Jan. 19, 2026).
[19] T. Brekken, B. Batten, E.A. Amon, From blue to green [Ask the Experts], IEEE Control Syst. Mag. 31 (5) (2011) 18–24. https://doi.org/10.1109/MCS.2011.941960.
[20] R. Vertechy, M. Fontana, G.P. Rosati Papini, M. Bergamasco, Oscillating-water-column wave-energy-converter based on dielectric elastomer generator, in: Proceedings of the SPIE, 8687, 2013, p. 86870I. https://doi.org/10.1117/12.2012016.
[21] J. Xie, L. Zuo, Dynamics and control of ocean wave energy converters, Int. J. Dyn. Control 1 (3) (2013) 262–276. https://doi.org/10.1007/s40435-013-0025-x.
[22] U.S. Energy Information Administration (EIA). Tidal *P*ower. https://www.eia.gov/energyexplained/hydropower/tidal-power.php. (Accessed Jan. 19, 2026).
[23] WTS Energy. Tidal *E*nergy. https://www.wtsenergy.com/glossary/tidal-energy/. (Accessed Jan. 19, 2026).
[24] Columbia Climate School. Tapping into Ocean Water (Ocean Power). https://news.climate.columbia.edu/2017/02/14/tapping-into-ocean-power/. (Accessed Jan. 19, 2026).

Further readings

Alternative Energy Tutorials, https://www.alternative-energy-tutorials.com/wave-energy/wave-energy-devices.html#google_vignette.

R. Baria, L. Mortimer, G. Beardsmore, Engineered geothermal systems, development and sustainability of, in: R.A. Meyers (Ed.), Encyclopedia of Sustainability Science and Technology, Springer, New York, 2012, pp. 3501–3514.

A. Bartle, Hydropower potential and development activities, Energy Policy 30 (14) (2002) 1231–1239.

D.W. Brown, A hot, dry rock geothermal energy concept utilizing supercritical CO_2 instead of water, in: Proceedings of the 25th Workshop on Geothermal Reservoir Engineering, 2000.

Climate Policy InitiativeUsing Private Finance to Accelerate Geothermal Deployment: Sarulla Geothermal Power Plant, Climate Policy Initiative, Indonesia, 2015 https://www.climatepolicyinitiative.org/publication/using-private-finance-to-accelerate-geothermal-deployment-sarulla-geothermal-power-plant-indonesia/.

F.R. Førsund, Hydropower Economics, International Series in Operations Research & Management Science, 217, 2nd, Springer, 2015 https://doi.org/10.1007/978-1-4899-7519-5.

G.Ó. Friðleifsson, W.A. Elders, A. Albertsson, The concept of the Iceland deep drilling project, Geothermics 49 (2014) 2–8. https://doi.org/10.1016/j.geothermics.2013.03.004.

I. Stober, K. Bucher, Geothermal Energy: From theoretical models to exploration and development, 2nd, Springer Science & Business Media, 2021 https://doi.org/10.1007/978-3-030-71685-1.

L.L Guo, Y.J Zhang, Z.W Yu, Z.J Hu, C.Y Lan, T.F. Xu, Hot dry rock geothermal potential of the Xujiaweizi area in Songliao Basin, northeastern China, Environ. Earth Sci. 75 (2016) 470.

C.B Harto, J.N Schroeder, R.M Horner, T.L Patton, L.A Durham, D.J Murphy, et al., Water use in enhanced geothermal systems (EGS): geology of U.S. stimulation projects, water costs, and alternative water source policies. 2014. http://www.ipd.anl.gov/anlpubs/2014/10/108702.pdf. Accessed Apr. 3, 2019

S.R. Miller, Has geothermal's moment finally arrived? A new generation of innovators is gambling on its future. Sierra. https://www.sierraclub.org/sierra/2024-1-spring/feature/geothermal-renewable-energy-moment-finally-arrived.

K. Kaygusuz, Hydropower and the world's energy future, Energy Sources 26 (3) (2004) 215–224.

Killingtveit, Å. (2019). Hydropower. In T. M. Letcher (Ed.), Managing global warming: An interface of technology and human issues (pp. 265–315). Academic Press/Elsevier. https://doi.org/10.1016/B978-0-12-814104-5.00008-9.

I. Kougias, G. Aggidis, F. Avellan, S. Deniz, U. Lundin, A. Moro, et al., Analysis of emerging technologies in the hydropower sector, Renew. Sustain. Energy Rev. 113 (2019) 109257.

X.-Z. Li, Z.-J. Chen, X.-C. Fan, Z.-J. Cheng, Hydropower development situation and prospects in China, Renew. Sustain. Energy Rev. 82 (2018) 232–239.

E.F. Moran, M.C. Lopez, N. Moore, N. Muller, D.W. Hyndman, Sustainable hydropower in the 21st century, Proc. Natl. Acad. Sci. 115 (47) (2018) 11891–11898.

O. Paish, Micro-hydropower: status and prospects, J. Power Energ 216 (1) (2002) 31–40.

Pathway to commercial liftoff: next-generation geothermal power, US DOE, 2024, LIFTOFF_DOE_NextGen_Geothermal_v14.pdf (energy.gov)

T. Reinsch, P. Dobson, H. Asanuma, E. Huenges, F. Poletto, B. Sanjuan, Utilizing supercritical geothermal systems: a review of past ventures and ongoing research activities, Geotherm. Energy 5 (2017) 1–25. https://doi.org/10.1186/s40517-017-0075-y.

The World Bank ESMAP Technical report 002/2012, Geothermal Handbook: Planning and Financing Power Generation.

US DOE, Energy Efficiency and Renewable Energy, A history of Geothermal Energy Research and Development in the United States, https://www1.eere.energy.gov/geothermal/pdfs/geothermal_history_1_exploration.pdf.

B. Wagner, C. Hauer, H. Habersack, Current hydropower developments in Europe, Curr. Opin. Environ. Sustain. 37 (2019) 41–49.

X.X Wang, N.Y Wu, Z. Su, Y.C. Zeng, Progress of the enhanced geothermal systems (EGS) development technology, Progr. Geophys. 27 (2012) 355–362.

X.X Wang, N.Y Wu, Z. Su, Y.C. Zeng, Development of enhanced geothermal systems (EGS)-with soultz geothermal farm in france serving as an example, J. Eng. Therm. Energy Power 27 (2012) 631–636.

C. Xu, P. Dowd, Q. Li, Carbon sequestration potential of the Habanero reservoir when carbon dioxide is used as the heat exchange fluid, J. Rock Mech. Geotechn. Eng. 8 (2016) 50–59.

T.F Xu, Y.L Yuan, Z.J Jiang, Z.Y Hou, B. Feng, Hot dry rock and enhanced geothermal engineering: international experience and china prospect, J. Jilin Univ. 46 (2016) 1139–1152.

W. Yu, L. Pan, The potential of coupled carbon storage and geothermal extraction in a CO_2-enhanced geothermal system: a review, Geotherm. Energy 8 (19) (2020). https://doi.org/10.1186/s40517-020-00173-w.

I. Yüksel, Development of hydropower: a case study in developing countries, Energy Sources, Part B 2 (2) (2007) 113–121.

I. Yüksel, Hydropower for sustainable water and energy development, Renew. Sustain. Energy Rev. 14 (1) (2010) 462–469.

Chapter 7

Emerging electrical energy sources

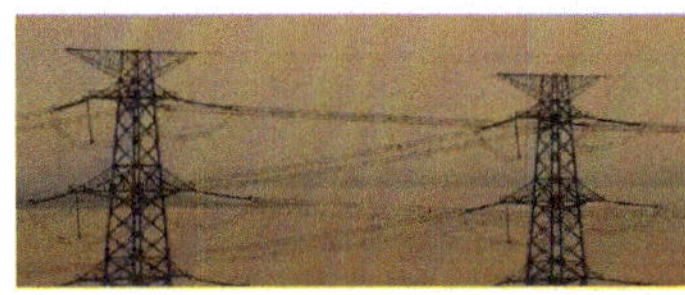

Synopsis

Energy demand globally increased by 2.2% in 2024, well above the previous decade's average. Electricity use rose by nearly double that amount, 4.3%. This was due to heatwaves, growing electrification, industrial expansion, and the growth of data centers to accommodate energy-hungry generative artificial intelligence (AI) processing. Our insatiable demand for electricity, the lifeblood of our modern world, necessitates a continuous exploration of diverse and sustainable energy sources. This chapter first delves into the technologies and resources that will help us toward electrification. We will describe the critical role of electrical batteries and how they revolutionized our relationship with portable electronics and hold immense promises for large-scale energy storage applications. Then, we will discuss electric cars and how their number is increasing exponentially on the roads. However, the narrative of batteries is intricately interwoven with a complex web of mineral extraction methods and their environmental impacts, including ecosystems, water sources, and air pollution. Moving forward, research into alternative battery chemistries that utilize less resource-intensive materials or boost greater recycling potential is crucial for a truly sustainable future.

Next, we cover electricity generation via biomass. Like any other energy source, biomass has its own challenges, including land-use competition for food production and assurance of its sustainability. Finally, hydrogen (H_2) emerges as a clean and versatile energy carrier with the potential to revolutionize various

F.A.C.T.S. Flexibility, Affordability, Conservation, Transition, Sustainability: Our Path to the Future of Energy and the Environment. DOI: https://doi.org/10.1016/B978-0-443-52574-2.00007-0

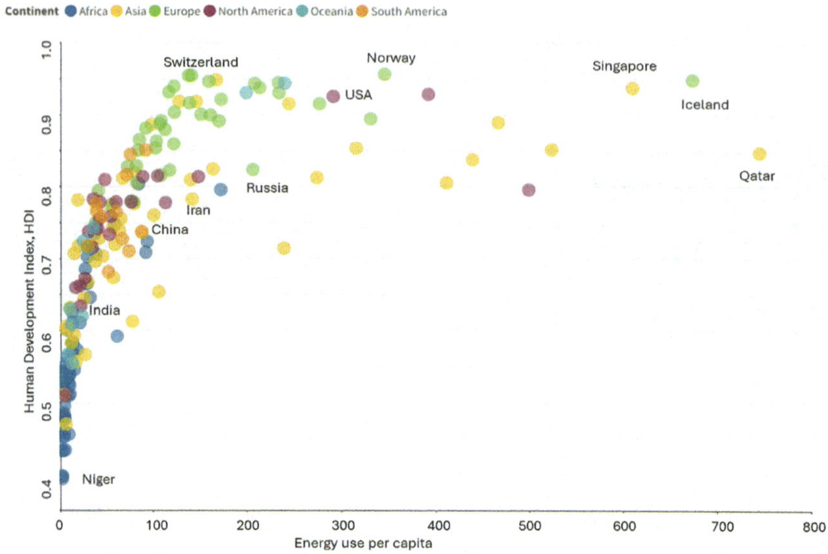

FIGURE 7.1 Human development index versus per capita electricity consumption for different countries [1].

sectors, from transportation to industrial applications. We will discuss different methods of hydrogen generation, including renewable sources like electrolysis powered by solar or wind energy. Then, the significant challenges facing the hydrogen industry will be covered, namely, transportation and storage. Additionally, ensuring the cost-effectiveness of hydrogen generation compared to traditional fossil fuels and developing efficient fuel cell technologies that convert hydrogen back into electricity are crucial for economic viability.

The importance of electricity

Access to electricity is a fundamental driver of progress, shaping living standards, education, and health outcomes. Approximately 1 billion people globally still live their daily lives without electricity, the majority of whom are in Sub-Saharan Africa. Fig. 7.1 shows countries based on their "human development index (HDI) score" alongside their per capita electricity use. HDI dimensions are health (life expectancy at birth), education (mean years of schooling), and living standards (gross national income per capita). A clear correlation emerges as countries develop; their electricity consumption rises. Forecasts predict this trend to continue, with developing nations outside the Organization for Economic Co-operation and Development (OECD) expected to see a 60% increase in electricity use by 2040, compared to a global rise of 45%.

Some have challenged the validity of such correlations as a true measure of HDI since it ignores the improvement in energy efficiency [2].

The high levels of energy use in many developed countries, such as those in Western Europe and North America, could be described as the extravagant use of energy. This generally corresponds to the high levels of land-use, water, and materials, the release of waste, and the emission of greenhouse gases and other pollutants. Through more efficient use of energy, high-consumption lifestyles could create more space for people trying to climb the human development ladder.

The growing reliance on electricity presents a significant challenge: Balancing our growing energy needs with the urgent need to reduce greenhouse gas emissions that contribute to climate change. The electric power sector is a prime target for decarbonization efforts, and in the near future, our electricity will likely continue to come from a mix of sources—fossil fuels, hydropower, variable renewables (solar and wind), and nuclear energy [3].

Nuclear energy currently provides around 11% of the world's electricity and is a major source of low-carbon generation in the United States, Europe, and globally. However, its future role is uncertain due to several factors—primarily rising costs and, to a lesser extent, lingering challenges like spent fuel disposal, safety concerns, and nuclear proliferation risks [4]. For details, see Chapter 3.

The specific mix of sources used for electricity generation varies depending on location, resource availability, and government policies. The trend globally is moving toward a larger share of renewable energy sources to reduce reliance on fossil fuels and combat climate change. While not a source of energy in itself, electricity stored in batteries can be a secondary source used to power electric vehicles or provide backup power during grid outages.

Electric cars

In the early 1890s, Henry Ford was serving as the chief engineer of Detroit's Edison Illuminating Company, a local electricity plant, while devoting his spare time to a "quadricycle" with an internal combustion engine that ran on gasoline at a time when most motorized transport cars were powered by lead-acid batteries or steam.

According to an account by Michael Brian Schiffer, [5] in 1896, Henry Ford attended a banquet for Edison's power-plant officers at the Oriental Hotel in Coney Island, Brooklyn, New York. After dinner, as Edison, then nearly 50, and his managers were discussing electric vehicles, Alex Dow, Ford's boss at the Detroit plant, introduced Henry to Edison. The 33-year-old Ford sat with Edison and discussed his invention. Soon, Edison commented, "Young man, that is the thing; you have it. Keep at it! Electric cars must be near power stations. The storage battery is too heavy. Your car is self-contained—carries its own power plant."

Edison had identified the problem with EVs then, and it is still the problem now. Gasoline has vastly more energy density than the best battery. About 20 gallons of gasoline, which weighs 100 and 20 pounds or so, will convey a Ford F-150 around 400 miles: close to twice the target range of the Ford Lightning's standard 1800-pound battery. Part of the early appeal of the automobile, along with the everyday freedoms it offered, lay in "touring"—the ability to go long distances on a whim. Today, even though most people use their cars for shorter trips and fly longer distances, the mystique of touring remains.

By 1903, Edison had apparently changed his mind again about the power source for automobiles. "Electricity is the thing," he told the Automobile Association that year. "There are no whirring and grinding gears with their numerous levers to confuse. There is not that almost terrifying, uncertain throb and buzz of the powerful combustion engine. There is no water circulating system to get out of order, no dangerous and evil-smelling gasoline, and no noise."

In January 1914, in an interview with the New York Times, Henry Ford discussed a vehicle that he and Edison were building together. "Within a year, I hope, we shall begin the manufacture of an electric automobile," he said. Edison's new nickel-iron batteries, supplied by the Edison Storage Battery Company, promised a range of up to a hundred miles. Finally, the electric vehicle was expected to have a low price and high production quantity.

Only a few Ford-Edison cars were ever made. During the First World War, access to the world's oil reserves pushed the combustion engine forward, and Edison's batteries did not work as advertised.

Electric cars, or battery electric vehicles (BEVs), operate using a fundamentally different principle than traditional gasoline cars. Instead of an internal combustion engine, they utilize an electric motor powered by a large battery pack. When you plug an EV into a charging station or home outlet, electricity flows into the battery, storing energy. This stored energy is then released to power the electric motor, which in turn drives the wheels, propelling the car forward. Unlike gasoline cars, EVs produce zero tailpipe emissions, contributing to cleaner air and reducing reliance on fossil fuels.

The battery pack in an EV is a complex system, typically composed of hundreds or thousands of individual lithium-ion cells. These cells are similar to those found in laptops and smartphones, but much larger and designed for the high-power demands of a vehicle. The battery's energy is managed by a sophisticated battery management system (BMS), which monitors temperature, voltage, and other factors to ensure safe and efficient operation. The electric motor itself is a marvel of engineering, converting electrical energy into mechanical energy with high efficiency. Many EVs also incorporate regenerative braking, which captures energy during deceleration and feeds it back into the battery, further improving efficiency. Beyond the core components of the battery and motor, electric cars also feature advanced electronics and software systems [6].

The simplicity of the electric powertrain compared to a gasoline engine also translates to fewer moving parts, potentially leading to lower maintenance costs and increased reliability. As battery technology continues to advance, electric cars are becoming increasingly competitive with gasoline vehicles in terms of range, performance, and affordability, paving the way for a cleaner and more sustainable transportation future.

One needs to be aware of the fact that many car manufacturers, when reporting the sales of their electric vehicles, combine the following types:

BEVs: These are fully electric vehicles that run solely on battery power and must be plugged in to charge.
Plug-in hybrid electric vehicles: These vehicles have both an electric motor and a gasoline engine. They can be driven in electric-only mode for a certain range, and then the gasoline engine takes over when the battery depletes. They can also be plugged in to charge, while the gasoline engine requires gasoline.
Hybrid electric vehicles: These also have both an electric motor and a gasoline engine, but they cannot be plugged in. The battery is charged through regenerative braking and by the gasoline engine. They have a smaller electric motor and battery compared to PHEVs and primarily rely on gasoline engines. They are not considered "electric vehicles" in the same way as BEVs and PHEVs. However, they do use less gasoline and are therefore better for the environment than traditional vehicles.

Issues with electric vehicles

A key concern for many drivers is the vehicle range, i.e., how far the car can drive on one charge. One major issue with EVs is their loss of range during cold seasons. On average, they lose 20% of their range in the winter. Of course, when it is even colder than freezing, the range may drop more. In winter, cold temperatures slow down the chemical reactions that power the battery, so the car cannot get as much energy out of it, or it cannot get as much energy as quickly.

Another reason for this drop is due to the fact that the energy that keeps the inside of the car warm for the driver and passengers is coming from batteries. In a gas-powered vehicle, heat is basically free. It comes from the waste heat due to fuel combustion in the engine [7].

Other concerns are the long charging time, the ease of access to charging stations, battery life, and the environmental footprint of battery materials and their later disposal.

Electrical batteries

Batteries have emerged as a critical component in the global shift toward a low-carbon energy landscape. Their capacity to store energy from diverse sources and deliver it on demand has positioned them as a cornerstone for grid stability, transportation electrification, and renewable energy integration.

The past decade has witnessed an unprecedented surge in global battery storage capacity increase and price drop (Fig. 7.2). Driven by technological advancements, falling costs of component materials, and the increasing integration of renewable energy sources, the market for battery storage has expanded

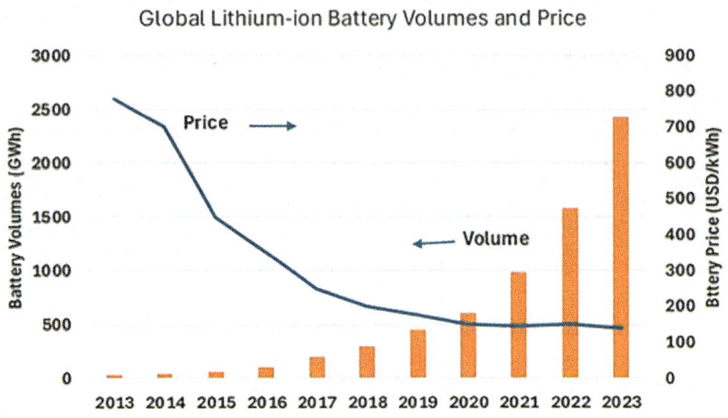

FIGURE 7.2 Global battery volume and price, 2010–2023 [8].

exponentially. Between 2014 and 2024, there was about 93% drop in the cost of battery storage projects. From modest beginnings, the energy storage landscape has transformed into a dynamic sector, with utility-scale projects, residential batteries, and electric vehicle batteries contributing to the overall growth. While lithium-ion batteries have dominated the market, research and development efforts are exploring alternative technologies to address challenges related to cost, sustainability, and performance. The future of battery storage holds immense promises, with the potential to revolutionize the energy landscape and pave the way for a decarbonized future.

Batteries serve a multifaceted role in the energy transition. They can be used for grid-scale energy storage during periods of low demand, thus ensuring a consistent power supply. This is particularly crucial for integrating variable renewable energy sources like solar and wind power. They can be used for peak shaving, i.e., by discharging stored energy during periods of high electricity consumption, thereby mitigating the need for additional power plants. Finally, they can rapidly respond to fluctuations in grid frequency, helping to maintain system stability and prevent blackouts.

In 2024, Hawaii replaced its last coal plant with a grid-scale battery energy storage system (BESS). These are rechargeable batteries that can store energy from various sources and discharge it when necessary, with a total capacity of 185 megawatts of instantaneous discharge. The BESS market is expected to grow to over USD 15 billion in 2027 [9].

The mineral dependence of electrical batteries

Electrical batteries are fundamentally transforming the way we interact with electricity. However, this transformative technology is intricately interwoven with a complex web of mineral dependence. Lithium (Li), nickel (Ni), cobalt (Co), and copper (Cu) are considered the "critical minerals" that endow batteries with their remarkable energy densities and functionalities. This section delves into the critical role these minerals play in battery technology and the challenges associated with their extraction.

Lithium: This lightweight alkali metal serves as the cornerstone of most rechargeable batteries. Li-ion batteries, the dominant technology today, leverage lithium ions that shuttle between electrodes during charge and discharge cycles. This enables efficient energy storage and release. The specific form of lithium used and its processing method significantly impact battery performance and cost. The race to find lithium is driven by expectations that fast-rising electric car sales will make the lithium required for batteries in those vehicles a valuable commodity. Brine deposits found in concentrated salt lakes generally offer a more environmentally friendly extraction method compared to traditional hard-rock mining (e.g., from lithium-rich granitic rocks, such as feldspar). Recently, there has been a rush to extract lithium from underground brines and produced water in the oil fields. However, brine extraction can still have a significant impact on local water resources. Large volumes of water need to be processed to extract a reasonable volume of lithium. In one project in Arkansas, with a lithium concentration of 437 mg/l, 3 billion gallons of brine needed to be processed annually—more than 1 million gal/D—to produce 30,000 tons of lithium carbonate a year [10]. This can support between 600,000 and 900,000 EVs.

Nickel (Ni) & Cobalt (Co): These versatile elements play a crucial role in enhancing the performance of Li-ion batteries. Cathode (positive electrode) chemistries utilize varying amounts of nickel, leading to distinct performance characteristics. Lithium cobalt oxide (LCO) batteries, with high cobalt content, prioritize stability and lifespan but offer lower energy densities. Lithium nickel manganese cobalt (NMC) batteries, with a higher nickel content and lower cobalt content, provide a well-balanced combination of energy density, power output, and cost. High-nickel NMC batteries boast the highest energy densities but may come at the expense of lifespan and safety.

Copper (Cu): This highly conductive metal forms the foundation of the electrical pathways within a battery. Copper current collectors efficiently channel electrons within the battery, ensuring optimal performance. Additionally, copper wiring plays a vital role in connecting batteries to external circuits, enabling the flow of electricity. The purity and form of the copper used can impact battery performance and cost.

The specific combination and ratio of these minerals determine the characteristics of a battery, such as its energy density, power output, lifespan, and safety profile. For instance, batteries designed for electric vehicles prioritize high energy density for extended range, while those used in power tools prioritize high-power output for rapid bursts of energy. Furthermore, the overall design of the battery, including the separator that keeps the electrodes apart and the electrolyte that facilitates the movement of lithium ions, plays a crucial role in overall performance and safety (Box 7.1 and 7.2).

BOX 7.1 Tech note—Lithium measurement

Lithium concentrations in brine are measured in two ways: Parts per million (ppm) and milligrams per liter (mg/l). In general, 100 ppm and 100 mg/l are equivalent measures, but not in engineering work. This is because lithium is found in fluids with high levels of dissolved solids, which can vary widely, leading to significant differences between ppm and mg/l. The ppm should be understood as mg/kg (mg of lithium per kilogram of fluid). A volumetric measure does not reflect variability in fluid density.

BOX 7.2 Tech note—Mineral usage in electric cars

Every electric vehicle contains about 75 kg of copper, or three times more than a conventional vehicle. A single wind turbine generally contains 500 kg of nickel. Note that the amount of nickel requires 100 tons of steelmaking coal to be refined. Every crystalline silicon solar panel contains 20 g of silver paste. It takes 80 metric tons of silver to generate approximately a gigawatt of solar power. (In power terms, that is equivalent to 9000 Nissan Leafs) [11].

Environmental and social considerations

The surging demand for Li-ion batteries, fueled by the proliferation of electric vehicles and electronics, has intensified the focus on the extraction of these critical minerals. However, mining activities associated with Li, Ni, Co, and Cu raise significant environmental and social concerns.

An all-electric future heavily depends on copper, and looming supply shortfalls could hamper reaching net-zero emissions by 2050. The S&P Global report [12] forecasts copper demand will nearly double to 50 million metric tons by 2035. By 2050, demand will reach more than 53 million metric tons. This figure is "more than all the copper consumed in the world between 1900 and 2021." Supplying even 5% of the world's electricity demand with cadmium telluride (CdTe) or copper indium gallium selenide (CIGS) solar cells would require directing today's entire worldwide production of key elements (tellurium, indium, and gallium) to PV fabrication [13]. The rates of growth in metal production that would be necessary to support higher levels of CdTe or CIGS penetration are unprecedented.

Mineral demand for clean energy technologies will almost triple by 2030, and quadruple by 2040 in the net-zero emission scenario, reaching nearly 40 Mt.

Given this trajectory, the development of diverse, resilient, and sustainable clean energy supply chains for critical minerals is an essential task [14].

The difficulties inherent in turning the intermittent output of PV installations into a reliable source of electric power are likely to remain a constraint on large-scale PV deployment in the future.

Declining ore quality complicates this picture. The world's industrial machine has already exploited the easiest metal reserves to extract. As ore quality declines, the industry must use more energy to mine it. Recent studies show that the average ore grade of copper mines has decreased by about 25% in just 10 years. That means more fossil fuels must be burned to haul and crush more rock. Notice that ramping up electricity production would require a massive investment in new green power plants or alternative energy sources.

Mining operations can disrupt ecosystems through habitat destruction, deforestation, soil erosion, and water pollution. The use of chemicals for processing the ore can further contaminate water sources and contribute to air pollution. For instance, lithium extraction from brine deposits can lead to increased salinity and disruption of local water tables. Hard-rock mining for lithium and other battery metals can generate large volumes of waste rock and tailings, which can leach harmful chemicals into the surrounding environment if not properly managed. Dust from mining operations can cause respiratory problems, while water contamination can lead to a variety of health issues.

Mineral prices can swing dramatically, making profitability hard to lock in, especially after a boom. Lithium is a clear example: after rising roughly eightfold in 2021–2022 (relative to early-2021 levels), spot prices fell ~75% in 2023 and stayed ~80% + below the 2022 peak into 2024–2025, highlighting why projects need designs that remain viable across price cycles. The cost of lithium extraction from produced water is yet to be confirmed by high-volume commercial operations. The lithium price forecast is expected to drop due to emerging alternative battery technologies potentially impacting the lithium-ion EV battery market.

In some mining operations, unethical labor practices, including child labor, can occur, particularly in countries with lax regulations. Additionally, local communities may be displaced or experience negative impacts on their health and livelihoods due to mining activities. Indigenous communities with ancestral ties to the land may be particularly vulnerable to the social and environmental impacts of mining. Most of the rare Earth metals in the past 25 years came from Bayan Obo, the largest rare Earth mineral deposit in the world. Once a sacred mountain in Mongolia, the Chinese government has been extracting its minerals and, thus, dominating the rare Earth metal markets [11].

Promoting responsible sourcing practices that prioritize ethical labor standards, minimize environmental impact, and ensure transparency throughout the supply chain. This could involve collaborating with reputable suppliers who adhere to ethical sourcing guidelines and environmental regulations. Certification schemes can play a role in verifying responsible sourcing practices.

Developing efficient and cost-effective methods for recycling batteries to recover critical minerals and reduce reliance on virgin ore extraction. Currently, battery recycling rates are relatively low due to the complex nature of battery chemistry and the lack of established infrastructure. Research and development efforts are underway to improve battery recycling technologies and make the process more economically viable [15].

Up to 95% of the materials used in phones can be recovered and reused. In 2024, the Australian recycling scheme, MobileMuster, an industry initiative, collected 109 tons of phones and recovered materials worth 2.5 million USD>. Experts suggest that the world's disused phones contain up to 16 billion USD of metals and enough cobalt to make 20 million EV batteries. The global eWaste is expected to grow, making recycling batteries and phones a priority [16].

Advancements in battery technology

Continued advancements in battery technology can help reduce the overall amount of critical minerals required for the same level of performance. This could involve optimizing electrode designs, improving energy densities, and extending battery lifespans. Additionally, advancements in battery management systems can help to optimize battery performance and lifespan, further reducing the need for frequent replacements.

Research and development efforts focused on exploring alternative battery chemistries that utilize less resource-intensive materials or offer greater recycling potential. This could involve exploring sodium-ion batteries, which utilize readily available sodium instead of lithium as the primary element. Lithium carbonate costs around 20,000–80,000 USD per ton, while sodium salts cost less than 500 USD per ton. Although they are bulkier, sodium batteries do not need cobalt or nickel. Per kWh, the cost difference is 40–60 USD for sodium versus 100–140 USD for Lithium. Because of their lower efficiency, they are better suited for stationary applications. More work is required to scale-up sodium battery production to an industrial level. Another option is using solid-state batteries, which offer potential advantages in terms of safety and energy density while potentially relying on less resource-intensive materials.

Alternative methods of energy sources for generating electricity include using compressed air or pumped water.

Compressed air can be stored underground as a renewable energy source. In this technology, known as compressed air energy storage (CAES), electricity is used to compress air and inject it into underground reservoirs or salt domes during periods of low energy demand. During peak energy demand hours, the stored air is released into a piping system and mixed with natural gas for combustion in expanders, which rotate a generator to produce electricity. Conventional CAES successfully stores energy for later use, but limitations include wasting most of the heat generated during compression. In addition, storage of the compressed air in salt caverns limits storage to certain geographic locations and a duration of only a few hours. Notice that CO_2 can be used instead of air, making geologic CO_2 sequestration more economically viable and adding a pathway for reducing CO_2 emissions.

Emerging challenges related to artificial intelligence and cryptocurrency operations

This section explores both the exciting possibilities of renewable energy technologies and the obstacles we need to overcome, with a specific focus on the burgeoning energy demands of cryptocurrency mining and AI operations.

The energy-intensive proof-of-work consensus mechanism used in some popular cryptocurrencies raises concerns about their long-term environmental impact [17]. Proof-of-work relies on vast networks of computers solving complex mathematical problems to verify transactions and secure the network. This process requires significant computing power, which translates to high energy consumption. The cryptocurrency mining in the USA was estimated to be consuming between 0.6% and 2.3% of the total electricity usage in 2024. Some of the mining operations have moved to less developed nations due to their cheaper electricity. This has caused the occurrence of frequent blackouts in these locations. Of course, such usage brings a significant carbon footprint, especially when fossil fuel is used for electricity generation. Alternatively, less energy-intensive consensus mechanisms such as proof-of-stake are under development, but their widespread adoption remains uncertain [18].

Annually, more than 1 trillion semiconductors are sold and used in smartphones, cars, and most modern equipment. Over 11000 data centers are operational, with new ones opening every month. They are handling everything from streaming to 2 billion+ prompts sent daily to AI models. This sector will continue growing strongly, driven by AI, cloud computing, high-performance electronics, and innovations like quantum computing [19].

The immense computational power required for complex AI tasks, such as machine learning and ChatGPT, translates to high energy consumption [20]. Training AI models often involves running algorithms on powerful computers for extended periods [21]. When a data center comes online, retail customers typically help foot the electric bill. American utilities sought nearly $30 billion in retail-rate increases in the first half of 2025. A Bloomberg analysis estimated that in areas near data centers, wholesale electricity costs have risen by more than 200% over the past five years — and rates are expected to keep climbing. Power plants cannot produce nearly enough electricity to meet surging demand. Eric Schmidt, former CEO of Google, has stated that the United States must add 92 GW of power to the national grid to satisfy data center demand alone. Renewable energy sources such as wind and solar can currently meet only a fraction of that need. In the near term, new data centers will largely be powered by fossil fuels, with developers purchasing land near natural gas deposits to secure easier access to this energy. The rush to build data centers echoes the great railroad expansion of the 19th century. The current administration has made the construction of data centers a national priority. Research into energy-efficient AI algorithms and hardware is crucial for ensuring the sustainability of this rapidly evolving field. This includes developing hardware specifically designed for AI tasks, as well as optimizing algorithms to reduce the amount of computing power required to achieve desired results (Box 7.3).

BOX 7.3 Tech note—AI processors

High-performance graphics processing units (GPUs) and tensor processing units (TPUs) are vital to AI processing. Unlike traditional CPUs, GPUs and TPUs are designed to handle parallel processing tasks, making them ideal for the matrix operations central to AI algorithms. However, this advanced processing capability comes at a cost: these units require a lot of electricity to function effectively. Data centers, where AI models are often trained and deployed, house thousands of servers – each containing multiple GPUs and TPUs, all working in tandem. The energy consumption of these data centers is substantial. According to some estimates, data centers account for about 1% of global electricity demand, a figure that is expected to rise as AI adoption grows.

Balancing growth with sustainability

As the world becomes more electrified, the increasing energy demands necessitate a multipronged approach to ensure long-term sustainability, as shown as follows:

Encouraging the use of energy-efficient mining equipment and AI hardware is crucial to reducing overall energy consumption [22]. This can involve developing regulations or standards for energy efficiency in these sectors, as well as promoting research and development in this area.

Crypto mining and AI operations can become more sustainable by migrating to renewable energy sources. This could involve utilizing green energy certificates, which represent the environmental benefits of renewable energy production, or locating facilities near sustainable energy generation plants, such as solar farms or wind parks [23].

Ongoing research and development in both renewable energy technologies and energy-efficient approaches for crypto mining and AI operations are essential for a sustainable future. This includes continued advancements in renewable energy storage solutions, as well as ongoing research into alternative consensus mechanisms for cryptocurrencies and energy-efficient AI algorithms and hardware.

Unveiling the power of biomass

Biomass, encompassing organic matter derived from biological processes, presents a compelling avenue for producing more environmentally friendly fuels,

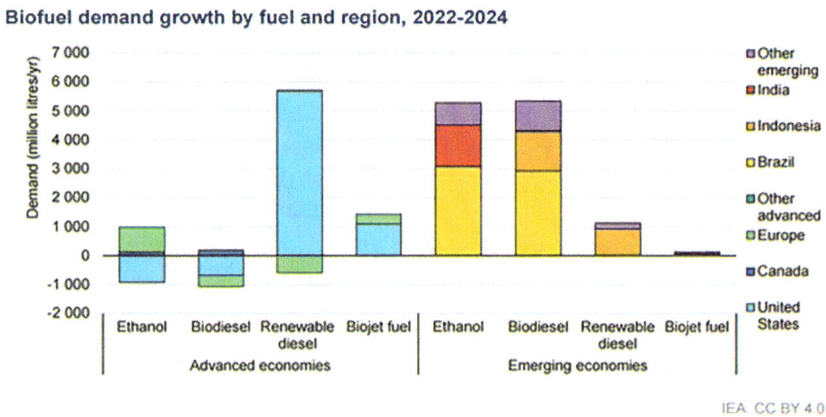

FIGURE 7.3 Biofuel demand growth by fuel and region, 2022–2024 [25].

as shown in Fig. 7.3. Nearly two-thirds of biofuel demand growth will occur in emerging economies, primarily India, Brazil, and Indonesia. Modern bioenergy is the largest source of renewable energy globally today, accounting for 55% of renewable energy and over 6% of the global energy supply [24]. This section delves into the various sources of biomass, exploring their potential as renewable energy resources and the considerations for their sustainable utilization.

Forests offer a readily available source of biomass in the form of trees, branches, and wood chips. These residues can be directly combusted in efficient power plants for heat and electricity generation or processed into wood pellets for convenient transportation and storage. However, the sustainability of forest biomass hinges on implementing responsible forestry practices. Selective harvesting, prioritizing reforestation, and upholding biodiversity are crucial to safeguard ecosystems and prevent deforestation. Certifications like the Forest Stewardship Council (FSC) can help ensure responsible wood sourcing practices.

Agricultural activities generate a significant amount of crop residues, including corn stover, wheat straw, and cobs. These residues represent a valuable source of biomass that can be converted into biofuels or electricity through processes like gasification. Utilizing crop residues for energy production offers farmers an additional revenue stream while promoting waste reduction. Nevertheless, it is critical to ensure that residue removal does not deplete soil nutrients or negatively impact soil health. A portion of the residues should be left on the field to maintain soil fertility and organic matter content, promoting long-term agricultural sustainability. Techniques like no-till farming, which minimizes soil disturbance, can further contribute to soil health.

Organic waste, including food scraps, yard trimmings, and paper products, constitutes a significant portion of MSW generated from households and commercial establishments. These organic wastes under anaerobic conditions in landfills generate biogas, which consists mostly of methane. Diverting this organic waste from landfills for conversion into biogas offers a sustainable waste management solution while generating renewable energy [26]. Additionally, wastewater treatment plants produce sewage sludge, which can also be a source of biomass for biogas production. Effective waste segregation and collection systems are essential for maximizing the utilization of organic waste as a biomass source. Public awareness campaigns and educational initiatives can play a role in encouraging waste segregation at the source.

Cultivating dedicated energy crops specifically for biofuel production can be a viable option, but considerations are necessary. Fast-growing grasses like switchgrass or miscanthus offer high yields and require fewer inputs compared to traditional food crops, potentially utilizing marginal lands unsuitable for food production. However, careful planning and lifecycle assessments are essential to ensure that energy crop cultivation does not lead to indirect land-use change or negatively impact local ecosystems.

Biofuels

Biofuels are mostly derived from organic materials like plants, algae, or animal waste. The choice of biofuel depends on various factors, including feedstock availability, production costs, environmental impact, and compatibility with existing infrastructure. Biofuels can be categorized into generations based on the feedstock and production process. A breakdown of different biofuels is provided in the following section.

First-generation biofuels: These are the most common biofuels currently in use, but they raise concerns about competition with food production and potential land-use changes. Ethanol is made by fermenting the sugars or starches of crops like corn or sugarcane. It is blended with gasoline to reduce reliance on fossil fuels. Biodiesel is produced from vegetable oils, animal fats, or used cooking grease. It is a cleaner-burning alternative to petroleum-based diesel fuel.

Second-generation biofuels: These are made in response to the limitations of first-generation biofuels by using nonfood sources like Cellulosic Ethanol, which is made from the cellulose content of plant materials like wood chips, switchgrass, or agricultural residues. Cellulose requires complex pretreatment processes to break it down into fermentable sugars. Biodiesel from Waste Fats utilizes used cooking oil, animal fats, or other waste greases for biodiesel production, diverting waste from landfills and creating a sustainable fuel source.

Third-generation biofuels: These are advanced biofuels under development, focusing on more efficient and sustainable feedstocks. Algae biofuels use algae that are fast-growing organisms that can be cultivated in various environments and produce oils that can be converted into biodiesel or jet fuel. Biogas is produced from the anaerobic digestion of organic waste materials like food scraps, manure, or sewage sludge. Biogas is a methane-rich fuel that can be used for electricity generation or transportation.

Sustainable biomass utilization

While the availability of various biomass sources is promising, ensuring the sustainability of this renewable energy option requires careful consideration of several factors:

Large-scale cultivation of crops specifically for biofuel production can compete with food production for land-use, potentially leading to deforestation, particularly in developing countries, and contributing to rising food prices. Sustainable biomass utilization should prioritize the use of waste residues, nonfood crops grown on marginal lands, and energy crops cultivated on land unsuitable for food production.

Biomass is often considered carbon neutral as it releases the carbon dioxide absorbed by the plants during growth through combustion. However, a holistic lifecycle assessment of biomass energy production is needed. The emissions associated with land-use changes, transportation, processing of biomass, and potential emissions from inefficient combustion technologies need to be factored in. Sustainable practices that minimize these emissions are essential for maximizing the environmental benefits of biomass energy. Techniques like carbon capture and storage (CCS) can be explored to further reduce the lifecycle emissions associated with biomass energy production.

Biomass harvesting activities can disrupt ecosystems and reduce biodiversity. Practices that minimize habitat destruction and ensure the conservation of natural habitats are crucial. Additionally, focusing on harvesting techniques that promote the regeneration of biomass sources, such as coppicing (cutting back the trees and shrubs to a stump at or near ground level to cause growth of multiple new sprouts) and maintaining biodiversity within energy crop plantations, can support long-term sustainability.

Technological advancements for biomass utilization and conversion processes

Biomass can be converted into usable energy through various conversion technologies:

Direct burning of biomass in boilers or power plants generates heat, which can be used for electricity production or industrial processes. Advanced combustion technologies with high efficiency and stringent emission controls are essential for minimizing air pollution. Additionally, cofiring biomass with fossil fuels can be a transitional strategy to reduce reliance on traditional fuels while ensuring efficient use of existing infrastructure.

This process thermochemically converts biomass into a combustible gas mixture (syngas) through partial combustion with limited oxygen. Syngas can then be used for various applications, including electricity generation, the production of biofuels like biogasoline or biomethanol, or industrial heat. Gasification offers the advantage of being a more versatile technology compared to combustion, as it can handle a wider range of biomass feedstocks and produce cleaner syngas with lower emissions. Additionally, gasification can be integrated with advanced technologies like combined cycle power plants for improved efficiency.

Hydrogen can be produced from solid biomass through biological, thermochemical, or electrochemical conversion. The hydrogen yield is the highest for thermochemical processes like gasification, pyrolysis, and reforming, which can produce 40–190 kg H_2 per ton of feedstock, while biological processes like fermentation and anaerobic digestion only achieve around one-quarter of those yields (5–50 kg H_2 per ton of feedstock). The energy efficiency of most of these processes is 40%–67% (lower heating value). Thermal and thermochemical conversion, which have the highest TRL, can achieve costs as low as USD 1.2–2.4/kg H_2 but, on average, are closer to USD 4.5–7/kg H_2[27].

In summary, biomass presents a promising renewable energy option with the potential to contribute significantly to a sustainable energy future. However, responsible sourcing, careful planning, and a commitment to minimizing environmental impacts are crucial for maximizing the benefits of this resource. By prioritizing the use of waste residues, cultivating energy crops on marginal lands, and employing efficient conversion technologies with robust emission control measures, we can unlock the vast potential of biomass while safeguarding the environment for generations to come. Technological advancements in areas like the pretreatment of biomass feedstocks to improve conversion efficiency and developing cleaner gasification technologies will further enhance the viability of biomass as a sustainable energy source.

Furthermore, fostering international collaboration and knowledge sharing can accelerate the development and deployment of sustainable biomass utilization practices. Additionally, creating supportive policies and economic incentives can encourage investment in biomass energy technologies and infrastructure, paving the way for a future powered by clean and renewable biomass energy.

Hydrogen fuel

There is a consensus that the decarbonized energy system of the future will be mostly electric. The primary sources of energy will be renewable energy, such as wind and solar power, which are already mature technologies and are cost-competitive with fossil fuels in a growing number of markets worldwide. However, there are many activities today that cannot be undertaken simply by substituting hydrocarbon molecules for electrons. They include industrial processes such as steelmaking, long-distance travel via land, sea, and air, and energy storage when renewables fail to meet electricity demand for days or weeks. For these hard-to-abate sectors, what is needed is fuel that does not generate a carbon footprint. One of the most promising options is clean hydrogen.

Hydrogen is the universe's most abundant element, yet it is remarkably scarce in Earth's atmosphere, existing at only 500 parts per billion. Primarily, we find hydrogen bound with oxygen in water (H_2O) and with carbon in hydrocarbons. Unlike its hydrocarbon counterparts, hydrogen combustion produces no greenhouse gases, making it a promising, low-carbon energy source.

The potential of "clean" hydrogen—hydrogen produced with minimal carbon emissions—to contribute to global climate goals has generated immense excitement. Headlines about progress toward a hydrogen economy are commonplace. However, the reality is that clean hydrogen remains a rare and costly commodity.

The basics of hydrogen energy

Hydrogen is not a primary energy source but rather an energy carrier, akin to a battery storing energy generated from other sources. When utilized in a fuel cell, hydrogen, upon reacting with oxygen, produces only water vapor, making it a remarkably clean-burning fuel with close to zero GHG emissions (ignoring the amplifying effect of produced water vapor as discussed in Chapter 8) (Box 7.4).

BOX 7.4 Tech note—Hydrogen properties

Hydrogen has a high energy density. This is true on a per unit mass basis but not on a volumetric basis (see Table 7.1). The volumetric energy density of liquid hydrogen is 3 times lower than that of gasoline. This is because of the low density of liquid hydrogen compared to gasoline. The volumetric energy density is what matters for the use of liquid fuels for transportation. Also, to liquify hydrogen, it must be cooled to a very low temperature and compressed to high-pressure, and to do this, it uses nominally 25% of the energy contained in the hydrogen.

TABLE 7.1 Heating values for various fuels.
(HHV and LLV are higher and lower heating values, respectively).

Fuel	State at ambient temperature and pressure	LHV (MJ/kg)	HHV (MJ/kg)	Volumetric (MJ/liter)
Hydrogen	Gas	120.1	141.8	10.8–12.7
Methane	Gas	50	55.5	23.4–25.9
Ethane	Gas	47.5	51.9	26.6–29.2
Gasoline	Liquid	41.9	44.4	30.7–32.7
Diesel	Liquid	40.3	42.7	33.6–35.8
Methanol	Liquid	17.	22.7	13.4–17.9

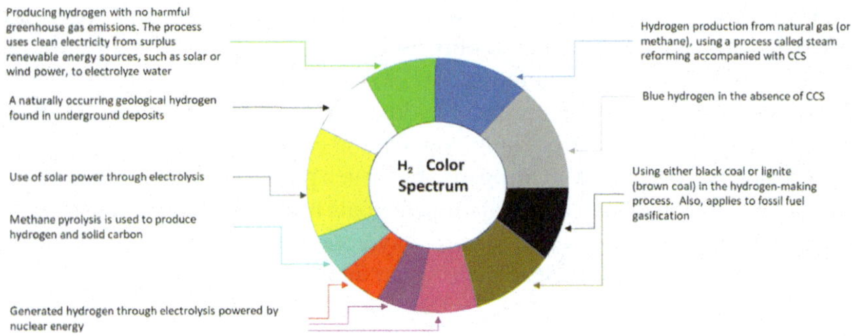

FIGURE 7.4 Hydrogen color spectrum.

Hydrogen generation

Depending on its production method, hydrogen comes in many colors [28]. Fig. 7.4 displays different color schemes. We will only address a few of them in this section.

Hydrogen production methods vary significantly in their environmental impact. White (or gold) hydrogen is a nascent concept involving naturally occurring underground hydrogen reservoirs [29]. Humans have encountered free hydrogen since ancient times. The continuously burning gas seep at Chimaera, near Antalya, Turkey, has 7.5%–11.3% H_2. Other occurrences have been observed in many locations. Fig. 7.5 shows the habitat of natural hydrogen and how it is generated. The total annual flow of hydrogen from geologic sources has been estimated at around 23 million tons/year. The origin of natural hydrogen is either the Earth's mantle or mostly due to chemical reactions between water and iron-rich minerals at high temperatures within the Earth's crust. The generated hydrogen is either trapped in a deep geological structure or seeps to the surface through faults and cracks. Currently, exploration for natural hydrogen is in its infancy, where the search for hydrogen shows on the surface may point the way to identifying larger reserves in the subsurface. This is analogous to the situation at the end of the 19th century when hydrocarbon exploration was just beginning. The development of appropriate exploration techniques will reveal the potential of yet unknown accumulations and seepages. The natural processes producing hydrogen have been active since primordial times and will continue for millions of years in the future. Because of this, natural hydrogen can be classified as a renewable source of energy. Large-scale commercialization of natural hydrogen is in its early stages.

Grey hydrogen is the predominant method of hydrogen generation, derived from fossil fuels via steam methane reforming (SMR) or partial oxidation.

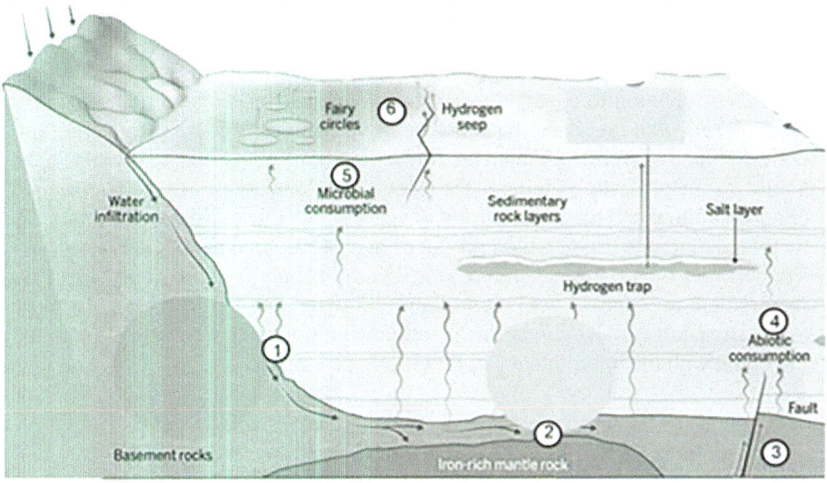

FIGURE 7.5 **Underground natural hydrogen generation.** (Notes on Fig. 7.5: (A) Refers to radiation from radioactive rocks that can split water and generate hydrogen. (B) Refers to the reaction between iron-rich rock and water to generate hydrogen. (C) Refers to the seepage of hydrogen from Earth's mantle. (D) Refers to the consumption of hydrogen by rocks and gases to form water and methane. (E) Consumption of hydrogen at shallower depths by microbes that generate methane. (F) refers to hydrogen seeping to the Earth's surface via fractures. Shallow depressions, also known as fairy circles, are the outcome of this seepage.) (*Modified from Ref. [30].*).

While efficient, with around 76%–77% of input energy converted to hydrogen, it generates substantial greenhouse gas emissions. Experimental in situ methods also exist but require further development.

Blue hydrogen aims to reduce grey hydrogen's carbon footprint by coupling SMR with carbon capture and storage (CCUS). This requires both low-emission natural gas sources and effective CCUS implementation.

Green hydrogen is produced through electrolysis, splitting water into hydrogen and oxygen. This is an established and well-known method, constituting the most effective technique for water splitting. The reaction, however, is very endothermic; it requires the use of electricity. Its environmental impact hinges on the electricity source. Renewable power yields truly green hydrogen, while grid-based electrolysis often relies on a mix of renewable and fossil fuel energy. Commercialized electrolyzers have an efficiency of 40%–60%.

There are other methods of hydrogen generation, such as biological fermentation, that are addressed in many technical references [28]. Underground conversion of fossil fuel into hydrogen is a nascent technology that helps with achieving net-zero emission targets (Box 7.5 and 7.6).

BOX 7.5 Tech note—Water electrolysis

The principle of electrolysis is quite simple. A direct current is injected into an electrolytic solution to separate the water molecules into hydrogen and oxygen. Electrolyzers, like fuel cells, have an anode, a cathode, and an electrolyte, and come in different types. Polymer Electrolyte Membrane electrolyzers (PEMEL) have a solid-specialty plastic material as the electrolyte. Water reacts at the anode (which contains iridium and titanium) to form oxygen, which moves across to the cathode (where platinum is used), which produces hydrogen. Alkaline electrolyzers use a similar process, but the electrolyte is a liquid alkaline solution, and there is no requirement for catalysts such as iridium and platinum. The technology is more established, but PEMELs can ramp up and down more quickly and are more suited to coping with the intermittency of renewable power.

BOX 7.6 Tech note—Low-emission hydrogen

Low-emissions hydrogen includes hydrogen, which is produced through water electrolysis with electricity generated from a low-emission source (renewables, e.g., solar, wind turbines, or nuclear). Hydrogen produced from biomass or fossil fuels with CCUS technology is also counted as low-emissions hydrogen.

Fuel cells

In order to generate electricity, Hydrogen can be used either directly as a gas or as a fuel cell. Some companies have already made engines and turbines that work with hydrogen or blends of natural gas and hydrogen. Although it requires some retrofitting, the existing gas turbines in power plants can use hydrogen gas safely [31].

Hydrogen fuel cells hold immense potential as a versatile power source across various sectors. Functioning similarly to a battery, they utilize an anode, cathode, and electrolyte to convert hydrogen fuel into electricity. Fig. 7.6 illustrates a hydrogen fuel cell's operation, highlighting inputs and outputs.

Hydrogen gas enters the anode, while air is fed to the cathode. A catalyst separates hydrogen molecules into protons and electrons, generating electricity. Protons pass through the electrolyte to the cathode, where they combine with oxygen from the air to produce water and heat as byproducts.

Several fuel cell types are emerging, each utilizing different electrolytes and catering to specific applications. Proton exchange membrane fuel cells (PEMFCs) are particularly promising for transportation due to their ability to operate at low temperatures and rapidly adjust their output. Other fuel cell types include alkaline and solid oxide fuel cells.

These fuel cells can be used in:

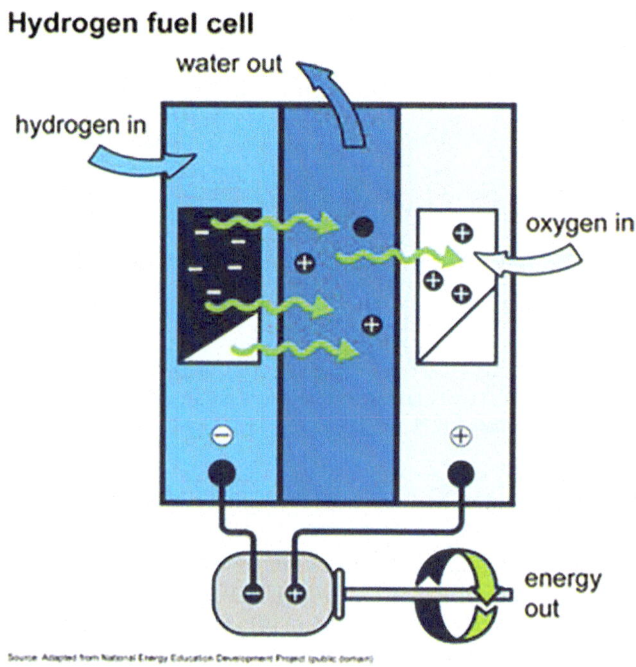

FIGURE 7.6 Hydrogen fuel cell [32].

Transportation: Hydrogen-powered vehicles are being developed as a clean alternative to gasoline-powered cars. They boast advantages like long driving range and fast refueling times.
Electricity generation: Fuel cells can generate electricity for homes and businesses, especially in areas with intermittent renewable energy sources like solar and wind. They offer a reliable backup or primary power source.
Industrial applications: High-temperature industrial processes can potentially be powered by hydrogen fuel cells, significantly reducing reliance on fossil fuels.

Clean hydrogen challenges

While promising, hydrogen energy still faces some hurdles:

Cost challenges

Depending on production technologies, the cost varies widely. Fig. 7.7 shows the historical and estimated production costs in 2030. Costs for both "with" and "without" CCUS application are shown.

There were close to 1200 hydrogen refueling stations in operation globally in 2023, but they have increased only marginally lately. Cost has been a key

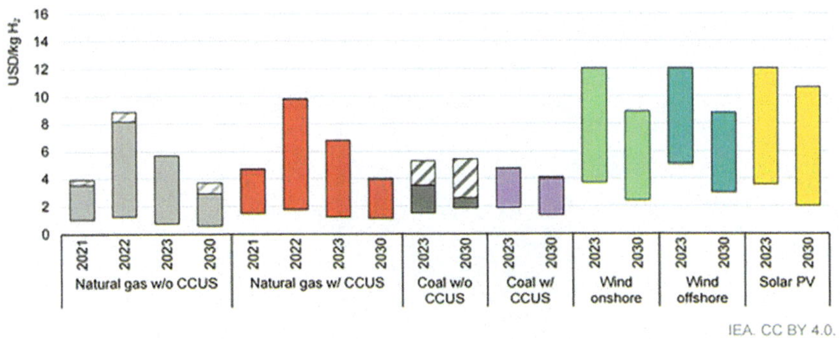

FIGURE 7.7 Hydrogen production cost by pathway, 2023, and in the net-zero emissions by 2050 scenario, 2030 (IEA, 2024) (The dashed area represents the CO_2 price impact, based on USD 15–140/t CO_2 for the NZE Scenario) [27].

hurdle. In 2019, the cost of hydrogen production without and with CCUS ranged between 1.5 and 8 USD/kg of hydrogen. In California, where sales have been more substantial, hydrogen pump prices increased by over 100% between 2021 and the end of 2023 (15 USD/kg H_2 to 34 USD).

Filling up a 5-kg tank would have cost USD 82.55 before the price jump, i.e., USD 0.206 per mile compared to USD 0.06 for gasoline and USD 0.03 for EVs (source: recharge news.com). Even green hydrogen, produced with renewable energy, would likely be more expensive.

While large-scale hydrogen production exists for industrial use, it is often not considered clean. SMR from natural gas (grey hydrogen) or coal/lignite gasification (brown hydrogen) are dominant methods, generating significant carbon emissions. Production is just one piece of the cost puzzle. Transporting hydrogen adds substantial expense. Two potential solutions exist: on-site production or the use of repurposed natural gas pipelines for distribution.

Blue versus green hydrogen

The choice between blue hydrogen (fossil fuel-based with carbon capture) and "green" hydrogen (renewable energy-based) depends on several factors:

- Can captured carbon be effectively stored?
- Is affordable renewable energy readily available?
- Is natural gas abundant and cheap?
- Does a carbon tax incentivize cleaner production methods?

Markets with abundant, cheap natural gas and carbon storage options might favor blue hydrogen. Conversely, regions with plentiful, affordable renewable energy, high carbon pricing, and limited natural gas might prioritize green hydrogen. This suggests a strong regional influence on clean hydrogen production strategies.

Notice that when it comes to electricity generation, direct conversion of solar power to electricity (PV) is much more efficient than using it to generate hydrogen and then using hydrogen fuel cells to generate electricity (18% vs 6% on average). The advantages of green hydrogen are energy density, continuity, and transportability.

Hydrogen transportation

Beyond production, the distribution and storage of hydrogen pose significant logistical challenges. Hydrogen is typically compressed to between 35 and 150 bars (~500 to ~2200 psi) for pipeline transmission, whereas the distribution system that provides gas to many end users typically operates at pressures less than ~7 bar (~100 psi). The optimal transport method depends on distance and existing infrastructure. While hydrogen pipelines totaled over 5000 km in 2023, primarily in Europe and the U.S., this pales in comparison to the 3 million kilometers of natural gas pipelines. Repurposing natural gas pipelines or blending hydrogen with natural gas offer short-term solutions, but dedicated hydrogen pipelines are essential for long-term growth. Note that blending reduces the energy content of the gas stream due to hydrogen's lower energy density compared to natural gas, but it also lowers CO_2 emissions.

Transporting hydrogen over long distances often requires conversion, shipment, and reconversion. Conversion into liquefied hydrogen (LH_2), liquefied organic hydrogen carriers (LOHCs), or ammonia (NH_3) increases volumetric energy density and improves handling for maritime transport.

Liquefied hydrogen (LH_2) requires extreme cooling (−253°C), incurring high energy costs and storage losses. But, with a density of 70.8 kg/m^3 is highly suitable for applications requiring greater storage capacities compared to gaseous forms. While reconversion is straightforward, the lack of specialized transport ships limits its current viability.

LOHCs offer safe, ambient temperature hydrogen handling and can leverage existing oil infrastructure. However, the need to return empty carriers and energy-intensive reconversion increases costs and complexity.

Ammonia (NH_3) is a well-established chemical with extensive production infrastructure. Converting hydrogen to ammonia enables a stable supply for large-scale industrial use. However, the energy-intensive ammonia cracking process for hydrogen recovery remains a technological challenge. Its corrosive nature also hinders its wide application.

Methanol is another hydrogen carrier and is primarily produced from coal or natural gas. Hydrogenation of CO_2 is another way of producing methanol. However, it is a hazardous substance with significant toxicity if ingested, inhaled, or absorbed through the skin, posing risks to both human health and the environment.

In 2021, European gas transmission operators outlined plans for a 39,700-km European Hydrogen Backbone (EHB) by 2040. Approximately 69% of this network would repurpose existing natural gas infrastructure. Estimated transportation costs of 0.11–0.21 cents per kilogram of hydrogen over 1000 km make the EHB a promising cost-effective solution for long-distance hydrogen transport.

Hydrogen storage

For storage, hydrogen is typically compressed to more than 350 bar (~5000 psi). At ambient temperature and atmospheric pressure, 1 kg of H_2 gas occupies a volume of 11 m^3. With a low density of 0.09 kg/m^3, H_2 storage has become one of the key barriers restricting its widespread use. Also, hydrogen possesses only about one-third the energy density of natural gas, necessitating larger storage volumes. Natural gas can be stored indefinitely, making geologic reservoirs attractive for large-scale storage. Similar principles govern hydrogen and carbon dioxide storage, though hydrogen's smaller molecular size increases the risk of leakage through micropores. Salt caverns, with their impermeable caprocks, are particularly well-suited for hydrogen storage. Leveraging existing natural gas storage facilities in salt caverns can expedite and reduce the cost of hydrogen infrastructure development.

While the U.S. announced development of three industrial-scale subsurface hydrogen storage facilities in 2023 in Texas salt caverns, the growing demand for low-carbon energy solutions will undoubtedly require expanded storage capacity.

Environmental issues

Hydrogen is an indirect greenhouse gas that has stronger warming effects, but is shorter-lived than other greenhouse gases. If leaked, it causes atmospheric warming. If burned, hydrogen generates nitrogen oxides pollutants (NO_x), which contribute to climate change. For example, hydrogen's Global Warming Potential (GWP) peaks around 7 years after emission, and is 40 times more damaging than CO_2 [33].

- Hydrogen reacts with hydroxyl radicals, which are the main way methane is decomposed in the atmosphere. This increases the lifetime of methane.
- Hydrogen radicals in the troposphere react with oxygen, thereby producing more ozone (which has a warming effect).
- Hydrogen increases water vapor in the stratosphere (which has a warming effect). The Global Warming Potential of H_2 is around 12–20.
- In power plants, there is also likely to be an increased level of Nitrous Oxide (NO_x) production due to an increased level of flame temperature [34]. NO_x can cause human respiratory illness and contributes to climate change.

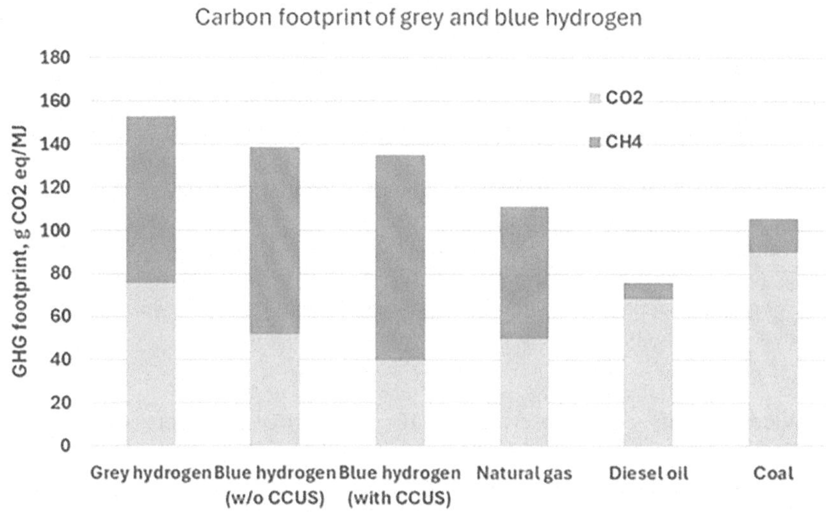

FIGURE 7.8 Comparison of carbon dioxide equivalent emissions from gray hydrogen, blue hydrogen, natural gas, diesel oil, and coal (The methane leakage rate is assumed to be 3.5%) [35].

CO_2 emissions from hydrogen utilization come from the production, processing, and transport of hydrogen. Hydrogen storage and transmission systems may require specialized high-pressure equipment and will require a significant amount of energy for compression. Carbon accounting must, therefore, track the CO_2 emissions at each conversion step in the supply chain. To effectively regulate hydrogen, policymakers must consider the full lifecycle emissions of both electrolysis and natural gas-based production methods. Fig. 7.8 compares carbon dioxide equivalent emissions from gray, blue, and green hydrogen, as well as conventional fuels like natural gas, diesel, and coal. This analysis includes direct emissions and fugitive methane emissions (for diving deeper into how to estimate hydrogen emissions , please refer to the end of this chapter).

Researchers have indicated that even with low methane leakage rates, blue hydrogen often emits more greenhouse gases than simply burning natural gas. Furthermore, neither current blue nor green hydrogen production methods achieve true net-zero emissions without additional CCUS [36].

Water footprint of hydrogen production

Water consumption associated with hydrogen production varies significantly across different production pathways. The specific mix of technologies employed in each region is crucial in determining the overall water intensity. Approximately 9 L of water is required to produce 1 km of hydrogen, yielding 8 km of oxygen as a byproduct, which can be utilized in healthcare or industrial applications (Box 7.7).

BOX 7.7 Tech note: Water footprint of H_2 production

Blue hydrogen production, involving the conversion of natural gas with heated water, consumes between 12 and 19 kg of water per kilogram of hydrogen. Moreover, the quality of water needed varies across production methods. Electrolyzers demand high-purity water, unlike other process components. Inadequate water treatment can damage electrolyzers and hinder operations. Desalinating water for electrolysis using reverse osmosis consumes 3–4 kWh of electricity per cubic meter of water.

The International Energy Agency [27] projects substantial water demand for hydrogen production in 2050: 3500 gigalitres for blue hydrogen, 4500 gigalitres for green hydrogen, and 250 gigalitres for mineral extraction.

Water scarcity could pose a challenge to hydrogen projects, especially in renewable-rich but water-stressed areas. Regions that currently do not face water shortages might be at risk in the future due to increases in droughts caused by climate change. For this reason, it will be important for hydrogen projects to use different water sources and take care of their sustainable management. Based on the announced projects, around 40% of the global low-emissions electrolytic hydrogen production by 2030 is located in water-stressed regions. Making hydrogen from seawater can be tricky because the salt is corrosive, and the process can split the salt and create toxic chlorine gas. The latter can corrode the electrodes. Developers in various regions are therefore considering large-scale desalination of seawater, or the use of treated wastewater, to ensure an adequate water supply for their projects. Desalination plant costs represent only a minor portion (less than 2%) of overall hydrogen project expenses. The process of reverse osmosis seawater desalination requires around 3–6 kWh of electricity per m^3 of water, resulting in a cost for hydrogen production of around USD 0.05/kg H_2. In certain cases, developers could oversize desalination capacity to ensure an ample freshwater supply for the project and surrounding areas, benefiting agriculture, industry, and domestic users. This approach provides community benefits with limited impact on the economics of the hydrogen production project itself. Around 70% of the low-emissions hydrogen production in 2030 from announced projects using water electrolysis is within 100 km from the coast, so seawater desalination is a viable option [27].

Researchers are investigating ways to mitigate water scarcity. For example, new electrodes have been devised that can split ocean water to make hydrogen without the need for desalination. The process uses an electrolyzer that splits seawater into four "streams"—useful oxygen and hydrogen, and acidic and alkaline streams that are recycled back into the ocean. A chlorine-blocking layer on the anode allows water to flow through while stopping the salt [37]. The impact of the acidic and alkaline streams needs further investigation.

Safety considerations

Like any energy carrier, hydrogen presents inherent health and safety risks when deployed on a large-scale. Effective communication and management of these risks are crucial for widespread adoption. While established energy sources like gasoline, diesel, natural gas, electricity, and coal have associated risks, including flammability and toxicity, these risks are familiar and have been mitigated through regulations and safety practices.

Hydrogen's unique properties, such as high flammability, wide ignition range, and invisibility, require specific safety measures. Its ability to embrittle certain metals also necessitates careful material selection. Ammonia, another potential hydrogen carrier, presents additional challenges due to its toxicity, flammability, and corrosive nature. However, extensive industry experience in handling both substances provides a foundation for robust safety protocols.

Ultimately, the successful integration of hydrogen into the energy landscape hinges on a comprehensive understanding and proactive management of its safety profile.

Infrastructure challenges

The widespread adoption of hydrogen-powered transportation requires a robust network of fueling stations and storage facilities. However, the initial infrastructure costs are substantial. Hydrogen pumps cost at least USD 2 million, significantly higher than the USD 40,000 price tag for EV fast-charging points. Ongoing operational expenses are also more significant due to the complexities of hydrogen handling, including transportation, storage, compression, and dispensing. Unless these costs are drastically reduced, the overall operating cost of a hydrogen fuel cell vehicle is likely to exceed that of an electric vehicle at such high levels that it is not yet practical.

Infrastructure planning and development, especially in long-lead projects like power transmission, should begin immediately to support future hydrogen production. Based on announced projects, by 2030, Latin America could produce more than 7 Mtpa of hydrogen with a low-carbon intensity. However, achieving this potential in full would require a significant increase in electricity generation capacity—equivalent to 20% of the region's current power output—and substantial investments in enabling infrastructure, such as transmission lines.

Hydrogen cars

Initially, automakers were excited about hydrogen fuel cells because they are more efficient compared to traditional gasoline engines. They can convert up to 60% of the energy in hydrogen into usable power, while internal combustion engines typically achieve only about 30% efficiency. This means that hydrogen

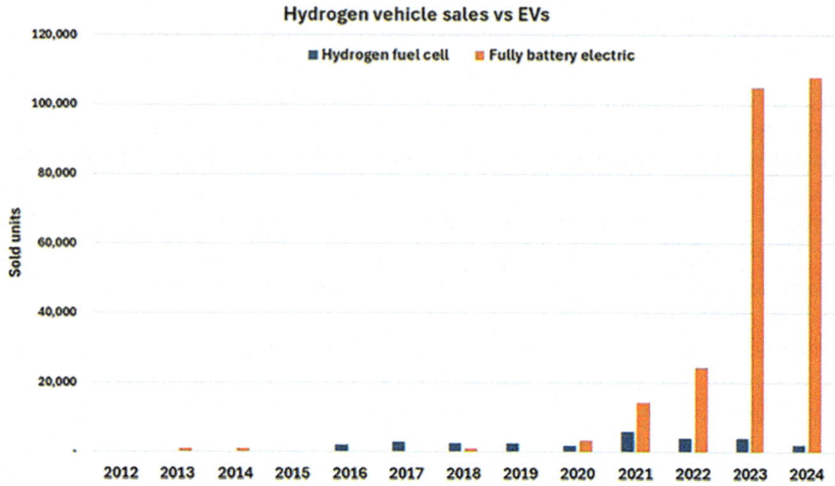

FIGURE 7.9 Toyota sales of hydrogen vehicles versus EVs [38].

vehicles can potentially offer better fuel economy and lower operating costs over the vehicle's lifetime.

While sales of EV vehicles have increased substantially during the last decade, sales of hydrogen fuel vehicles have faced serious hurdles. This is evident from the graph of car sales shown in Fig. 7.9.

Lack of fueling stations, high cost of building new refueling stations, higher fuel cost, and higher car prices have discouraged the public from buying hydrogen cars. Overall, producing hydrogen, compressing it, transporting it, and then converting it to electricity in the fuel cell involves multiple energy conversion steps, each with inherent losses. This makes hydrogen vehicles less energy-efficient compared to BEVs, where electricity is stored directly in the battery. Also, "Green hydrogen" is still in the early stages of development and faces scalability challenges.

However, Hydrogen is poised to play a pivotal role in the global transition to clean energy. According to the IEA, hydrogen demand could surge nearly eightfold by 2050, reaching almost 700 million tons annually and contributing up to 24% of global energy needs. Realizing this potential will require substantial investments—over USD 11 trillion—in production, storage, and transport infrastructure. Hydrogen sales could exceed USD 700 billion by 2050, with billions more invested in end-use applications.

The United States now accounts for almost half of low-emissions hydrogen production from fossil fuels, with CCUS announced for 2030. This could increase low-emissions hydrogen production 18-fold, from around 0.6 Mtpa H_2 in 2023 to 11 Mtpa H_2 by 2030. Other countries are also working toward hydrogen production facilities. In 2024, the World Bank approved a USD 1.5 billion loan to India for promoting hydrogen from renewable energy, scaling up renewable electricity, and enhancing finance for low-carbon energy investments.

As technology matures and costs decline, hydrogen will become increasingly viable for various sectors. Its decarbonization potential is particularly promising for industries like iron and steel, chemicals, and heavy-duty transport. Leveraging existing hydrogen applications in petroleum refining and ammonia production can already prevent millions of tons of CO_2 emissions annually. The heavy-duty transportation sector, including maritime, is actively exploring hydrogen as an alternative fuel. Regarding hydrogen-based vehicles, the two alternatives under consideration are either direct storage of H_2 provided by refueling stations or indirect, on-board production of H_2 from methanol.

Methanol ships are leading the way in terms of alternatively fueled vessels [excluding liquefied natural gas (LNG)], due to their higher technological and commercial readiness, and greater ease of handling when compared to ammonia. Ammonia-fueled vessels are at an earlier stage of development than methanol-fueled vessels, with safety regulations for the use of ammonia as a fuel for ships still to be established. Ammonia-fueled engines for maritime applications have been tested, and vessel trials are already underway or are being planned [39].

Based on announcements, the United States will remain the frontrunner in hydrogen production from fossil fuels with CCUS, with 5.6 Mtpa H_2 by 2030, followed by Europe with more than 3.5 Mtpa H_2 by 2030. However, many projects are now facing diverse challenges that are hindering progress in the sector.

The transition to a hydrogen economy must prioritize minimizing the wider environmental impact. Reducing emissions and water consumption in the mining and processing of materials crucial to hydrogen production is essential for a sustainable future.

Hydrogen development barriers

The current emphasis on affordability means that some alternative sources, such as green hydrogen and some other sustainable fuels, may not be competitive with traditional fuels in the near term. Some of the barriers to the way of deploying new hydrogen projects are [40]:

- *Unfamiliar technology*: Hydrogen fuel cell technology is less well-understood by the general public compared to BEVs. This lack of familiarity can lead to skepticism or concerns about safety and reliability.
- *BEV competition*: Battery electric vehicles have seen significant improvements in range, charging speed, and cost in recent years. This rapid progress has made BEVs a very attractive alternative, making it harder for hydrogen vehicles to compete.
- *Regulatory uncertainty*: Project developers require clear and stable regulatory frameworks to secure investments. Although many governments are

making progress, regulations are often not yet clear, which can lead to companies halting their hydrogen programs or postponing projects while awaiting clarity.

- *Demand uncertainty*: The slow development of the market for low-emissions hydrogen and insufficient demand mean that project developers cannot secure off-takers to underpin investments.
- *Financial hurdles and incentive delays*: Low-emissions hydrogen production and use cases remain more expensive than the fossil fuel incumbents they could replace.
- *Licensing and permitting issues*: Like the developers of renewable energy projects in the past, several low-emissions hydrogen project developers have struggled to get necessary approvals from authorities, including access to water supplies and grid connections.
- *Operational challenges*: Although several technologies to produce low-emissions hydrogen are commercially mature, they are now facing the scale-up stage. The change in scale implies that new project configurations need to be tested and perfected through a process that presents technical and operational hurdles for developers.

Summary

In this chapter, we reviewed different pathways of electrification that are needed to reduce the carbon footprint of energy consumption by 2050. Batteries, hydrogen, and biomass provide means of generating electricity. However, they inherently carry with them issues that could be problematic ethically and environmentally. By embracing a holistic approach that prioritizes responsible sourcing, sustainable practices, technological innovation, and collaborative efforts, these sources of energy can transition from promising concepts to a cornerstone of a clean and sustainable energy future.

Key takeaways

- The limited supply of rare Earth minerals may hinder the propagation of EVs unless new technologies unlock the lithium battery issues.
- Sodium batteries could be the next wave in making cheaper EVs.
- A key concern for many drivers is the limited EV range, i.e., how far the car can drive on one charge. More work is required to remove this hurdle.
- Electricity generation from Biomass competes with the food supply and needs proper attention.
- Some alternative sources, such as green hydrogen, are not competitive with traditional fuels in the near term.
- Hydrogen fuel cell technology is less well-understood by the general public compared to BEVs. This lack of familiarity can lead to skepticism or concerns about safety and reliability.

- The full life cycle of any generated kWh of electricity needs to be fully transparent to the consumers.

Extra science

Methodology for estimating the carbon footprint of hydrogen production technologies (based on Howarth and Jacobson, 2021 [35]):

According to the overall SMR reaction:

$CH_4 + 2H_2O \rightarrow CO_2 + 4H_2$

1 mole of carbon dioxide and 4 moles of hydrogen gas (H_2) are produced per mole of methane consumed. Based on the gross calorific heat content of hydrogen, 0.286 MJ per mole, or, 3.5 moles H_2 per MJ, we arrive at the following::

(3.5 moles H_2/MJ) * (1 mole CO_2/4 moles H2) = 0.875 moles CO_2 per MJ

The amount of carbon dioxide (molecular weight 44.01 g per mole) produced during the SMR process is 38.51 g CO_2 per MJ (Table 7.2). The amount of methane consumed is given by:

(3.5 moles H_2/MJ) * (1 mole CH_4/4 moles H2) = 0.875 moles CH_4 per MJ (or 14.04 g per MJ)

For this reaction to occur, significant input of energy, between 2.0 and 2.5 kWh per m^3 of hydrogen is needed. This energy comes almost entirely from NG when producing gray hydrogen. Using a mean value of 2.25 kWh per m^3 of hydrogen, the energy in NG required to produce a mole of hydrogen is:

(2.25 kWh/m^3 of H_2) * (3.6 MJ/kWh) * (1 m^3/1000 L) * (22.4 L/mol) = 0.1814 MJ per mole H_2

When burning NG for heat, 50 g CO_2 per MJ in emissions are produced, using gross calorific values. Note that higher carbon dioxide emission values are reported when using net calorific values. Therefore,

(0.1814 MJ/mole H_2) * (50 g CO_2/MJ) = 9.07 g CO_2 per mole H_2

38.5 g CO_2 per MJ from the SMR process + 31.8 g CO_2 per MJ from heat = 70.3 g CO_2 per MJ. Additionally, it takes energy to produce, process, and transport the natural gas used to generate hydrogen. The indirect upstream emissions are approximately 7.5% of the direct carbon dioxide emissions for NG, or an additional 5.3 g CO_2 per MJ (7.5% of 70.3 g CO_2 per MJ) [41]. Therefore, the total quantity of carbon dioxide produced is 75.6 g CO_2 is produced per MJ (Table 7.2).

TABLE 7.2 Carbon footprint of hydrogen production technologies [35].

-	Gray H_2	Blue H_2 (w/o flue-gas capture)	Blue H_2 (w/flue-gas capture)
SMR process			
CH_4 consumed (g CH_4/MJ)	14	14	14
CO_2 produced (g CO_2/MJ)	38.5	38.5	38.5
Fugitive CH_4 emissions (g CH_4/MJ)	0.49	0.49	0.49
Fugitive CH_4 emissions (g CO_2eq/MJ)	42.1	42.1	42.1
Direct CO_2 emissions (g CO_2/MJ)	38.5	5.8	5.8
CO_2 capture rate	0%	85%	85%
Energy to drive SMR			
CH_4 consumed (g CH_4/MJ)	11.6	11.6	11.6
CO_2 produced (g CO_2/MJ)	31.8	31.8	31.8
Fugitive CH_4 emissions (g CH_4/MJ)	0.41	0.41	0.41
Fugitive CH_4 emissions (g CO_2eq/MJ)	35.3	35.3	35.3
Direct CO_2 emissions (g CO_2/MJ)	31.8	31.8	11.1
CO_2 capture rate	0%	0%	65%
Energy to power carbon capture			
CH_4 consumed (g CH_4/MJ)	0	3	6
CO_2 produced (g CO_2/MJ)	0	8.2	16.3
Fugitive CH_4 emissions (g CH_4/MJ)	0	0.11	0.21
Fugitive CH_4 emissions (g CO_2eq/MJ)	0	9.5	1
Direct CO_2 emissions (g CO_2/MJ)	0	8.2	16
Indirect upstream CO_2 emissions (g CO_2/MJ)	5.3	5.9	6.5
Total CH_4 consumed (g CH_4/MJ)	25.6	28.6	31.6
Total CO_2 emitted (g CO_2/MJ)	75.6	51.7	39.7
Total fugitive CH_4 emissions (g CO_2eq/MJ)	77.4	86.9	95.4
Total emissions (g CO_2eq/MJ)	153	139	135

Note: *The methane leakage rate is assumed at 3.5%.*

Therefore, total methane consumption to generate gray hydrogen is the sum of 14.04 g CH_4 per MJ (from SMR) and the heat generation

11.6 g CH_4 per MJ which is 25.6 g CH_4 per MJ.

When using NG, about 0.8% is vented due both to leaks and to purposeful emissions. Combined with the 2.6% for field-level emissions, 3.4% of production is emitted to the atmosphere overall (fugitive CH_4). Note that in addition to some methane being lost between production and consumption due to leaks, methane is also burned by the natural gas industry to power natural gas processing and transport. Based on 2015 data, natural gas production in the United States was 817 billion m^3, while consumption was 771 billion m^3 .The methane emission as a percentage of gas consumption is estimated to be:

(3.4% of production) $*$ ($817 \times 109\ m^3/771 \times 109\ m^3$) = 3.5% of consumption.

Thus, the estimated upstream emissions of methane is:

(3.5% of consumption) $*$ (consumption of 25.6 g CH_4 per MJ) =0.90 g CH_4 per MJ

Notice that the half-life of methane in the atmosphere is about 12 years, far less than that of carbon dioxide.

Based on a recommended 20-year time frame, the Global Warming Potential (GWP) is at 86. Thus, the upstream methane emissions associated with the production of gray hydrogen can be estimated as:

(0.90 g CH_4 per MJ) $*$ (86 g CO_2eq/g CH_4) = 77.4 gCO_2eq per MJ

Gray hydrogen emission = carbon dioxide emissions (75.6.0 g CO_2 per MJ) + unburned methane (77.4 g CO_2eq per MJ) =153 g CO_2eq per MJ (Table 7.2).

References

[1] Adapted from Our World in Data. United Nations Development Programme, Boston University Institute for Global Sustainability | visualizingenergy.org | CC BY 4.0. (Accessed Jan 19, 2026).

[2] Boston University. Does more energy use increase the level of human development? Visualizing Energy 2022. https://visualizingenergy.org/does-more-energy-use-increase-the-level-of-human-development/. (Accessed Jan 19, 2026).

[3] U.S. Energy Information Administration (EIA)International Energy Outlook 2017, EIA, Washington, DC, 2017 https://www.eia.gov/outlooks/ieo/pdf/0484(2017).pdf (Accessed Jan 19, 2026).

[4] Massachusetts Institute of Technology Energy Initiative (MITEI)The Future of Nuclear Energy in a Carbon-Constrained World: An Interdisciplinary MIT Study, MIT, Cambridge, MA, 2018 https://energy.mit.edu/research/future-nuclear-energy-carbon-constrained-world/.

[5] M.B. Schiffer, Taking Charge: The Electric Automobile in America, second ed., Smithsonian Books, Washington, DC, 2010.

[6] U.S. Department of Energy. How Do All-Electric Cars Work? Alternative Fuels Data Center. https://afdc.energy.gov/vehicles/how-do-all-electric-cars-work

[7] GreenCars. Range (GreenCars 101). https://www.greencars.com/greencars-101-topics/range.

[8] Data from International Energy Agency (IEA)Batteries and Secure Energy Transitions, IEA, Paris, 2024 https://www.iea.org/reports/batteries-and-secure-energy-transitions.

[9] CarbonCredits.com. Zero13 and XTCC Reveals $100B Climate Finance for Net Zero at Davos (WEF). https://carboncredits.com/zero13-and-xtcc-reveals-100b-climate-finance-for-net-zero-at-davos-wef/.

[10] Society of Petroleum Engineers. It Is Easy to Find Lithium. Turning a Profit Is Hard. https://jpt.spe.org/it-is-easy-to-find-lithium-turning-a-profit-is-hard.

[11] A. Nikiforuk, The rising chorus of renewable energy skeptics. Resilience 2023, https://www.resilience.org/stories/2023-04-10/the-rising-chorus-of-renewable-energy-skeptics/.

[12] S&P Global. Copper on the Rise. https://www.spglobal.com/en/research-insights/special-reports/copper-on-the-rise.

[13] Massachusetts Institute of Technology Energy Initiative (MITEI)The Future of Solar Energy: An Interdisciplinary MIT Study, MIT, Cambridge, MA, 2015 https://energy.mit.edu/research/future-solar-energy/.

[14] International Energy Agency (IEA). Critical Minerals (Topic page). 2022. https://www.iea.org/topics/critical-minerals.

[15] The New York Times. Can a Sustainable Mining Experiment in New Caledonia Power Tesla's Ambitions? 2021. https://www.nytimes.com/2021/12/30/world/asia/tesla-batteries-nickel-new-caledonia.html.

[16] World Economic Forum. What Happens When You Recycle Your Phone? (Video). https://www.weforum.org/videos/what_happens_recycle_phone/.

[17] U.S. Energy Information Administration (EIA). Today in Energy (post). https://www.eia.gov/todayinenergy/detail.php?id=61364.

[18] Frontiers in Blockchain. 2023. https://www.frontiersin.org/journals/blockchain/articles/10.3389/fbloc.2023.1151724/full.

[19] World Economic Forum. Nature Positive: The Role of the Technology Sector (publication page). https://www.weforum.org/publications/nature-positive-role-of-the-technology-sector/.

[20] Foley & Lardner LLP. AI Requires Lots of Power. https://www.foley.com/p/102j836/ai-requires-lots-of-power/.

[21] Hyperstack. What You Need to Know About Large AI Model Training. https://www.hyperstack.cloud/blog/thought-leadership/what-you-need-to-know-about-large-ai-model-training

[22] International Monetary Fund (IMF). Carbon Emissions from AI and Crypto Are Surging and Tax Policy Can Help. 2024, https://www.imf.org/en/Blogs/Articles/2024/08/15/carbon-emissions-from-ai-and-crypto-are-surging-and-tax-policy-can-help.

[23] Simple.app. Crypto and Climate Change. https://simple.app/blog/crypto-and-climate-change/.

[24] International Energy Agency (IEA). Bioenergy. 2024. https://www.iea.org/energy-system/renewables/bioenergy.

[25] International Energy Agency (IEA). Biofuel Demand Growth by Fuel and Region, 2022–2024 (Chart). 2023. https://www.iea.org/data-and-statistics/charts/biofuel-demand-growth-by-fuel-and-region-2022-2024.

[26] Environmental and Energy Study Institute (EESI). Fact Sheet: Biogas: Converting Waste to Energy. https://www.eesi.org/papers/view/fact-sheet-biogasconverting-waste-to-energy.

[27] International Energy Agency (IEA)Global Hydrogen Review 2024, IEA, Paris, 2024 https://www.iea.org/reports/global-hydrogen-review-2024.

[28] P. Nikolaidis, A. Poullikkas, A comparative overview of hydrogen production processes, Renew. Sustain. Energy Rev. 67 (2017) 597–611. https://doi.org/10.1016/j.rser.2016.09.044.

[29] V. Zgonnik, The occurrence and geoscience of natural hydrogen: a comprehensive review, Earth-Sci. Rev. 203 (2020) 103140. https://doi.org/10.1016/j.earscirev.2020.103140.

[30] *Science*. Hidden Hydrogen: Does Earth Hold Vast Stores of a Renewable, Carbon-Free Fuel? (PDF reproduction). https://www.ribelvalvole.it/wp-content/uploads/2023/08/science.adh1477.pdf.

[31] GE Vernova. Hydrogen-Fueled Gas Turbines. https://www.gevernova.com/gas-power/future-of-energy/hydrogen-fueled-gas-turbines.

[32] U.S. Geological Survey (USGS). Hydrogen Fuel Cell (illustration); Energy Resources Program, 2022. https://www.usgs.gov/media/images/hydrogen-fuel-cell

[33] I. B. Ocko, S. P. Hamburg, Climate consequences of hydrogen emissions, Atmospheric Chemistry and Physics 22 (14) (2022) 9349–9368. https://doi.org/10.5194/acp-22-9349-2022.

[34] GE Vernova. Hydrogen for Power Generation; 2022. https://www.gevernova.com/content/dam/gepower-new/global/en_US/downloads/gas-new-site/future-of-energy/hydrogen-for-power-gen-gea34805.pdf.

[35] R.W. Howarth, M.Z. Jacobson, How green is blue hydrogen? Energy Sci. Eng. (2021). https://doi.org/10.1002/ese3.956.

[36] C. Bauer, K. Treyer, C. Antonini, J. Bergerson, M. Gazzani, E. Gencer, et al., On the Climate Impacts of Blue Hydrogen Production, Sustain Energ Fuels (2021). https://doi.org/10.1039/D1SE01508G.

[37] J. Dinneen, Special Electrodes Can Split Seawater to Produce Hydrogen Fuel, New Scientist (2024). https://www.newscientist.com/article/2448643-special-electrodes-can-split-seawater-to-produce-hydrogen-fuel/.

[38] Data adopted from Financial Times. (Toyota data as reported). https://www.ft.com/content/69422656-7a25-4bc7-b626-1e69d4da953c.

[39] International Energy Agency (IEA). Global Hydrogen Review2024 (PDF). https://iea.blob.core.windows.net/assets/89c1e382-dc59-46ca-aa47-9f7d41531ab5/GlobalHydrogenReview2024.pdf.

[40] GreenCars. The Rise and Fall of Hydrogen for Passenger Cars. https://www.greencars.com/greencars-101/the-rise-and-fall-of-hydrogen-for-passenger-cars.

[41] Santoro, R. L., Howarth, R. W., & Ingraffea, A. R. (2011). Indirect emissions of carbon dioxide from Marcellus Shale gas development. Agriculture, Energy, & Environment Program, Cornell University. http://www.eeb.cornell.edu/howarth/Marcellus.htm.

Further readings

COP 26 | 'Widespread Use of Green Hydrogen in Heating and Cars is not Only Stupid — it's Practically Impossible' | Recharge (rechargenews.com).

Department of Energy: https://www.energy.gov/eere/fuelcells/hydrogen-fuel-basics

S. Griffith, Electrify: An Optimist's Playbook for Our Clean Energy Future, MIT Press, 2021, doi:10.7551/mitpress/14054.001.0001.

Gas for Climate press release, 2021, European Hydrogen Backbone grows to 40,000 km, covering 11 new countries. https://gasforclimate2050.eu/news-item/european-hydrogen-backbone-grows-to-40000-km/.

Hydrogen News Intelligence, 2023, Massive Underground Reservoir of Natural Hydrogen in Spain 'could deliver the cheapest H2 in the world' https://www.hydrogeninsight.com/innovation/massive-underground-reservoir-of-natural-hydrogen-in-spain-could-deliver-the-cheapest-h2-in-the-world/2-1-1431515

IEA, 2021. Global Hydrogen Demand by Sector in the Net Zero Scenario, 2020-2030, IEA, Paris https://www.iea.org/data-andstatistics/charts/global-hydrogen-demand-by-sector-in-the-netzero-scenario-2020-2030.

International Energy Agency, "Global Hydrogen Review 2021," 2021. https://www.iea.org/reports/global-hydrogen-review-2021

IEA, 2023, Towards Hydrogen Definitions Based ON Their Emissions Intensity, www.iea.org

IEA, 2019, The Future of Hydrogen - Report prepared by the IEA for the G20, Japan, Seizing today's opportunities

Kearny and Uniper, 2021, Competitiveness of green hydrogen import pathways for Germany in 2025, www.kearny.com

I. Moretti and M. Weber, 2021, Natural Hydrogen: A Geological Curiosity or the Primary Energy Source for a Low-Carbon Future? https://www.renewablematter.eu/idrogeno-naturale-curiosita-geologica-o-futura-fonte-di-energia

National Grid Group: https://www.nationalgrid.com/stories/energy-explained/what-is-hydrogen

National Renewable Energy Laboratory: https://www.nrel.gov/hydrogen/

World Bank Group, 2022, Sufficiency, Sustainability, and Circularity of Critical Materials for Clean Hydrogen Climate-Smart Mining Facility, WB-Hydrogen-Report-2022.pdf

Part IV

On the path toward net carbon zero

Synopsis

In Zoroastrianism, one of the world's oldest faiths, four of the seven sacred elements hold central importance: fire, representing divine energy; water, the source of life; earth, the sacred provider; and air, the protective shell of the world. Because these elements are holy, a central tenet of the faith is protecting and honoring them.

Building upon the analysis of energy sources in Parts I–III, Part IV explores specific methodologies for sustaining and preserving these venerable elements.

In Chapter 8, we discuss methods of reducing and capturing emissions from the atmosphere. Then, we focus on how to achieve sustainable water, soil and minerals in Chapter 9.

Chapter 8

Clean air—Breathing new life into a threatened atmosphere

Synopsis

Life sustainability on Earth depends on the air we breathe. This gas mixture not only fuels respiration but also regulates global temperatures and shields us from harmful solar radiation. However, human activities have demonstrably disrupted the delicate equilibrium of the atmosphere, leading to a multitude of environmental challenges. This chapter delves into the multifaceted factors contributing to air pollution, analyzes the associated environmental and health consequences, and illuminates the pathways toward achieving sustainable air quality for a thriving future.

Air minus pollution

During the COVID-19 pandemic, wearing masks became a necessity to protect us against contracting a potentially deadly virus. Similarly, during wildfires, we need to wear masks to safeguard against smoke inhalation. The wildfires that ravaged Canada during the summer of 2023 were among the most severe in its history, consuming an astonishing fifteen times more land than the usual annual impact, totaling 11 million acres—an area twice the size of New Jersey. Subsequently, as smoke blanketed cities like New York and its surroundings, air quality plummeted to "very unhealthy" levels, obscuring skylines with a

F.A.C.T.S. Flexibility, Affordability, Conservation, Transition, Sustainability: Our Path to the Future of Energy and the Environment. DOI: https://doi.org/10.1016/B978-0-443-52574-2.00002-1

thick haze. With increasing frequency and severity, wildfires in Canada and across the western United States have undermined nearly half of the air-quality improvements achieved through the Clean Air Act in recent years. In this region, smoke has emerged as a significant source of pollution, rivaling, if not surpassing, fossil fuels. Nationwide, the number of Americans exposed to at least one day of "extreme smoke" annually has surged twenty-seven-fold since 2006 [1].

When air quality deteriorates, studies have established a correlation with a rise in crime rates, a decline in student test scores, an increase in erroneous umpire decisions, and more investment blunders. Exposure to air pollution has been correlated with a range of health issues, from asthma and emphysema to Alzheimer's and Parkinson's diseases, from cancer and strokes to depression and suicide, from miscarriages and premature births to infant mortality. Annually, air pollution contributes to as many as 10 million deaths worldwide. Given the magnitude of the damage, maintaining the status quo represents a profound failure of mobilization and an astonishing normalization of the crisis.

Air pollution encompasses a complex mixture of gaseous and particulate contaminants originating from a diverse range of anthropogenic sources. These are listed below in the order of most noxious to least polluting:

- Fossil fuel combustion: The continued reliance on fossil fuels for electricity generation, transportation, and industrial processes remains a significant contributor to air pollution. Incomplete combustion releases a suite of pollutants, including nitrogen oxides (NOx), sulfur oxides (SOx), particulate matter (PM), and volatile organic compounds (VOCs).
- Industrial processes: Manufacturing facilities, particularly those involved in metal smelting, chemical production, and cement manufacturing, contribute significantly to air pollution by releasing pollutants such as SOx, heavy metals, and persistent organic pollutants.
- Agricultural practices: The intensive use of fertilizers in agriculture leads to the emission of ammonia, a precursor to PM formation. Additionally, agricultural burning practices and the decomposition of organic matter contribute to air pollution. Furthermore, some farming activities, such as livestock production and cultivation practices, are a source of significant methane emissions.
- Waste management: Open burning of waste, prevalent in developing nations, releases a myriad of pollutants, including PM, VOCs, and dioxins. Landfills also contribute to air pollution by emitting methane, a potent greenhouse gas.

Indoor air pollution: While often overlooked, indoor air pollution poses a significant health risk. Sources include building materials, combustion appliances, cleaning products, and cigarette smoke. The limited exterior ventilation designed into modern buildings can exacerbate indoor air pollution concerns.

Greenhouse gas emissions

Gases that trap heat in the atmosphere are called greenhouse gases (GHGs). They are composed of carbon dioxide, methane, nitrous oxide, and fluorinated gases. Other GHGs not counted in US or international GHG inventories, are water vapor and ozone (Box 8.1). Water vapor is the most abundant GHG. Still, most scientists believe that water vapor produced directly by human activity contributes very little to the amount of water vapor in the atmosphere. Therefore the US Energy Information Administration (EIA) does not estimate emissions of water vapor. However, the GHG effect of human-generated water vapor needs to be evaluated more closely and considered in environmental policy. Ozone is technically a GHG, but ozone can be helpful or harmful, depending on where it is in the Earth's atmosphere. The ozone layer naturally occurs at higher elevations in the atmosphere (the stratosphere), where it blocks ultraviolet (UV) light that is harmful to plant and animal life from reaching the Earth's surface. The protective benefits of stratospheric ozone outweigh its contribution to the greenhouse effect. The United States and countries all around the world ban and control the production and use of several industrial gases that destroy atmospheric ozone and create holes in the ozone layer [2].

According to the Environmental Protection Agency (EPA) and related data from 2023 to 2024, CO_2 accounted for roughly 79%–80% of US GHG emissions, with methane around ~11%, nitrous oxide ~6%, and fluorinated gases ~3% of the total in CO_2-equivalent terms [3]. In this Chapter, we will focus on the two major gases, namely, CO_2 and CH_4.

In the United States, large emitters are required to report their carbon footprint. To achieve this, they need to measure their GHG emissions or create an emissions inventory. Emission scope types are discussed in Box 8.2.

BOX 8.1 Tech Note: Climate change and CO_2 emission

The phenomenon of climate change is intricately linked to the greenhouse effect, a well-established scientific principle. The key mechanisms are shown below:

- Incoming solar radiation warms the Earth's surface. The planet then naturally emits a portion of this absorbed energy back toward space in the form of infrared radiation (heat).
- Specific atmospheric gases, including carbon dioxide (CO_2), water vapor, methane, and nitrous oxide, function as a radiative transfer agent. These

(continued)

BOX 8.1 Tech Note: Climate change and CO_2 emission—*cont'd*

gases absorb and re-emit infrared radiation within the atmosphere. For every degree Celsius that Earth's atmospheric temperature rises, the amount of water vapor in the atmosphere can increase by about 7%, according to the laws of thermodynamics.

- While some infrared radiation escapes back into space, a portion is trapped by GHG, causing a net warming effect within the atmosphere and at the Earth's surface. This phenomenon is known as the greenhouse effect.

BOX 8.2 Tech Note: Emission scope types

GHG emissions are categorized using one of the three scopes, as shown in Fig. 8.1 Scope 8.1 emissions are direct GHG emissions that occur from sources that are owned or controlled by the company, for example, emissions from combustion in owned or controlled boilers, furnaces, and vehicles; emissions from chemical production in owned or controlled process equipment. Scope 2 emissions account for GHG emissions from the generation of purchased electricity and steam consumed by the company. Scope 2 emissions physically occur at the facility where electricity is generated. Scope 3 is a reporting category that allows for the treatment of all other indirect emissions. Scope 3 emissions are a consequence of the activities of the company but occur from sources not owned or controlled by the company [4].

Fig. 8.2 displays the key features of this phenomenon.

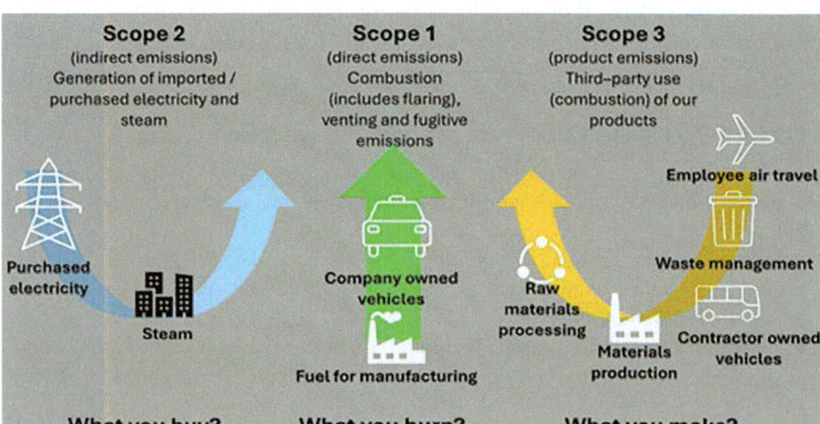

FIGURE 8.1 Different emission categories [5].

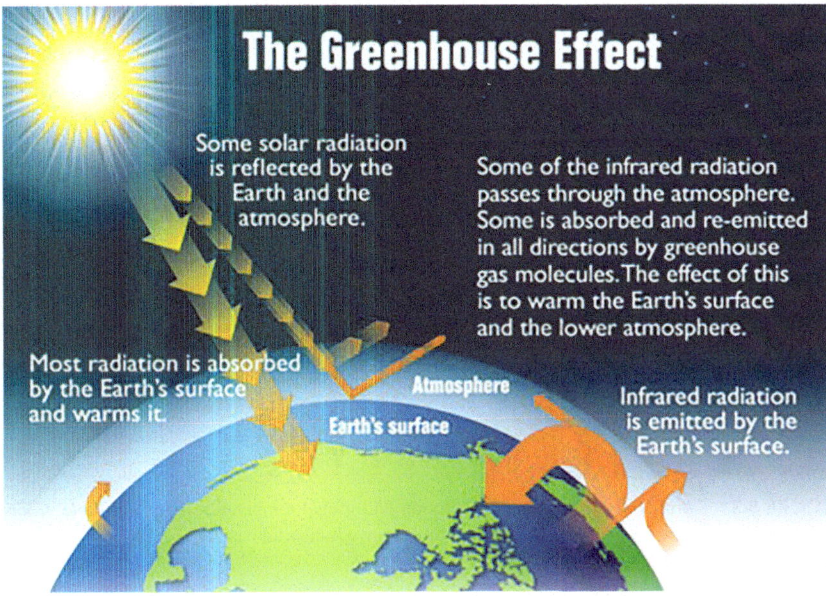

FIGURE 8.2 Greenhouse gas effect [6].

Typical emission values

To put the emissions values in proper perspective, an average passenger vehicle emits about 4.6 metric tons of carbon dioxide (CO_2) per year. Units of industrial CO_2 emission are generally measured in million tons per annum (MTPA). This is equivalent to about 217,400 average cars on the road. One gigaton per year is equivalent to 1000 MTPA. On the other hand, in 2024, human activities emitted around 37.8 $GtCO_2$ globally (EIA), with less than half of these emissions directly contributing to the rise in atmospheric CO_2 concentrations (the remainder were absorbed by natural carbon removal processes, countering the human disruption of the carbon cycle). Notice that 2024 and 2025 were the hottest years on record.

It is crucial to recognize that a natural level of GHG plays a vital role in maintaining Earth's habitable temperature. Without these gases, the average global temperature would be a frigid $-18°C$, unsuitable for life as we know it. The Intergovernmental Panel on Climate Change (IPCC) report mentioned that "it is extremely likely that more than half of the observed increase in global average surface temperature from 1951 to 2010 was caused by the anthropogenic (manmade) increase in GHG concentrations" [7] and "continued emission of GHG will contribute to further warming and long-lasting changes in all components of the climate system" [8]. The dramatic increase in CO_2 emissions has thickened this atmospheric blanket excessively, causing a significant rise in global temperatures. Fig. 8.3 provides a graph of the global temperature change from an established baseline (between 1881 and 1910), shown on the

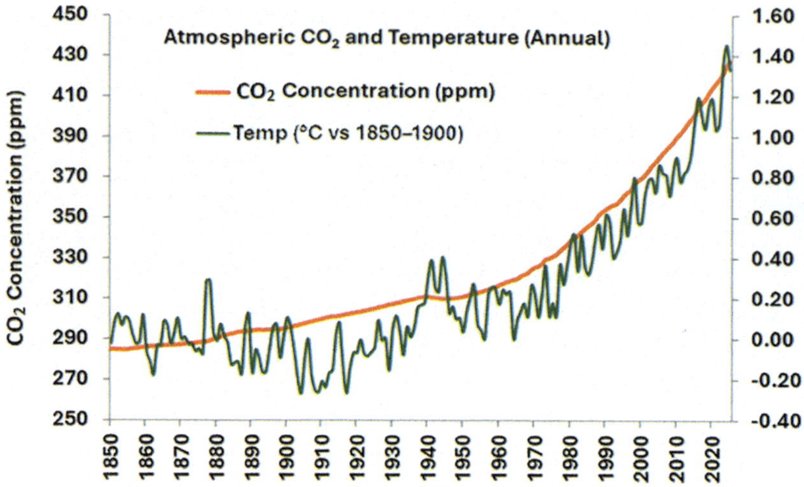

FIGURE 8.3 Correlation between CO_2 emission and global temperature rise [9]. (Note: Global temperature anomalies averaged and adjusted to the early industrial baseline (1881–1910).

left, versus the increase in average CO_2 concentration in the atmosphere, shown on the right, between 1880 and 2021. In the absence of any other variables, this graph is compelling evidence of the rise in temperature with the increase in CO_2 concentration.

During this vast stretch of time, Earth's average surface temperature fluctuated between lows of around 52 degrees Fahrenheit (11°C) and highs of 97 degrees Fahrenheit (36°C). [10] One notable "hothouse" period occurred 56 million years ago when palm trees and alligators flourished in the Arctic Circle. Following that, temperatures gradually decreased until the industrial era, when human-generated GHG emissions began a sharp upward trend. Today, the average global temperature sits around 59 degrees Fahrenheit (15°C) and continues to rise (Box 8.3).

BOX 8.3 Tech Note: CO_2 emission and global warming

Over the past 500 million years, Earth has cycled between extreme climates, shifting from hot to cold and back again. Oceans have risen and fallen, ice caps have melted and reformed, and complex life has emerged. Continents drifted apart, while geological processes released carbon dioxide into the atmosphere, then drew it down again, triggering climate shifts between two distinct states: icehouse—characterized by unipolar or bipolar continental ice sheets and steep temperature gradients from pole to equator—and greenhouse, marked by the absence of widespread continental ice sheets and milder temperature gradients across latitudes.

Scientists have long sought to reconstruct Earth's climate history to improve predictions of future climate behavior, relying on proxy data to compensate for gaps in direct measurements. Temperature records from thermometers date back only to the mid-1700s, so for earlier periods, researchers use climate proxies [11] that correlate with temperature and can infer past conditions with some uncertainty. Common proxies include tree rings, ice cores, fossil pollen, ocean sediments, and corals 1990s, a team of researchers drilled through Greenland's ice sheet, extracting thousands of cylindrical ice cores that have since provided crucial temperature data for the region. Many scientists have used these ice cores to reconstruct past temperatures by analyzing the oxygen-18 isotope ratios within the ice [12], which indicate the temperature when the ice formed. However, such interpretations are limited to data collected up to 1980, as it takes time for new snow layers to compact into solid ice [13].

Another significant temperature record source is satellite measurements, which have monitored global temperatures since 1979. In addition, fossilized sea life provides insights into historical ocean chemistry, though such fossils often yield fragmented snapshots of past global conditions. Combining these various data sources, Dr. Judd and colleagues [10] ran simulation models, refining their estimates with fossil data to reconstruct historical climates. Their findings highlighted a strong relationship between global mean surface temperature and atmospheric CO_2 concentrations, suggesting that CO_2 is a dominant factor in long-term climate change. This relationship is especially striking given that solar luminosity also influences climate across these vast timescales.

While the study revealed broad agreement between model results and the historical records, certain discrepancies, such as those from the Mesozoic era, suggest that other factors, like changes in solar luminosity, planetary albedo, or trace GHG, may contribute to long-term climate dynamics. Interestingly, during that era, the correlation between CO_2 and temperature was notably stable. This suggests that, given known CO_2 concentrations, Earth's temperature over multimillion-year scales is relatively predictable. Yet, our understanding of how these long-term climate drivers interact remains incomplete.

Greenland and Antarctica have lost about 6.4 trillion tons of ice since 1990s, with the resulting meltwater raising the global sea level by roughly 17.8 millimeters. Greenland alone is losing about 266 billion tons of ice per year contributing to ongoing sea level rise. The climate system contains numerous amplifying feedback and potential tipping points, heightening the risks associated with each additional increment of warming—though the exact impact remains uncertain. An amplifying, or positive, feedback on global warming is a process whereby an initial change that causes warming brings about another change that results in even more warming (decrease in albedo). Cooling tends to increase albedo, reinforcing cooling and less absorbed solar energy (Box 8.4).

TABLE 8.1 Feedback loops (see Fig. 8.4).

	Primary change	Effect
A	Sea ice melting or not forming	Decreasing albedo
B	Increasing thawing and decomposition	Increasing CO_2 and CH_4 emissions and loss of sequestration
C	Drying and increasing vulnerability to fire, smoldering, and decreasing soil organic carbon	Increasing release of CO_2 into the atmosphere and decreasing carbon sequestration
D	Increasing fire frequency and/or severity	Increasing CO_2 emissions, loss of sequestration, and a change in albedo
E	Changing cloud distribution and optical properties	Altered cloud albedo and greenhouse effect
F	Increasing chronic aridification and hotter drought stress extremes leading to expanding deserts	Decreasing CO_2 sequestration, and increasing albedo
G	Dieback of Amazon, boreal, and other forests	Loss of sequestration and change in albedo, and decreasing evapotranspiration
H	Changing insect distributions and abundances and decreased host tree defense	Loss of sequestration and change in albedo
I	Decreasing snow cover	decreasing albedo

BOX 8.4 Tech Note – Feedback loops

Scientists have identified 27 positive (or intensifying) feedback loops (e.g., wildfires increasing CO_2, which causes more warming, etc.) and only seven negative (or dampening) ones (e.g., Planck growth reduces CO_2 concentration, which reduces the warming intensity) [14]. These are shown in Fig. 8.4. Additional explanations are provided in Table 8.1.

The scientists warn that these feedback loops could interact, potentially triggering a series of climate tipping points that could lead to cascading climate impacts [15].

The current and projected global emissions (based on InternationalEnergy Agency [IEA] forecast) [16] are shown for two scenarios in Fig. 8.5. The Current Policies Scenario refers to existing plans without any obligations to reduce emissions further. The Stated Policies Scenario (STEPS) refers to countries' promises to reach net zero by a certain year. The reduced emissions from different sources are also indicated on this graph.

Recent data (IPCC report) suggests that the atmosphere is warming at a rate of approximately 0.17°C per decade. The Paris Agreement, adopted in 2015,

FIGURE 8.4 Map of feedback loops (A–I images show example locations where select positive feedback loops are likely operating (see Table 8.2) [15].

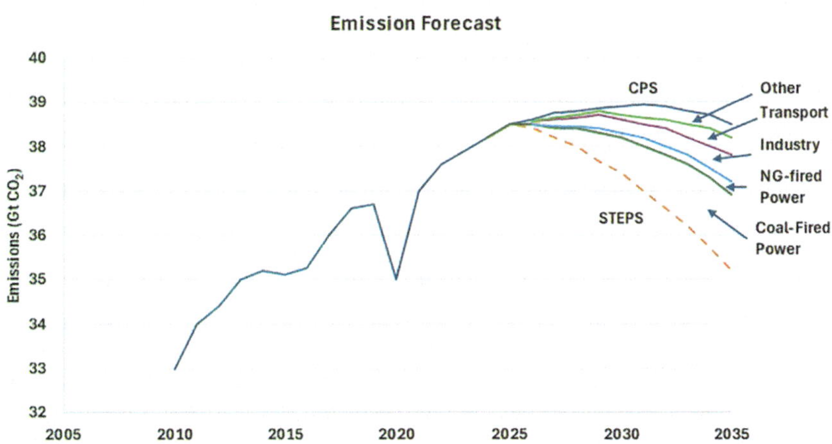

FIGURE 8.5 Forecast of CO_2 emission [16].

established a landmark legal framework for global climate action, with nearly 200 nations pledging to hold the increase in global average temperature to "well below 2°C" and to "pursue efforts" to limit it to 1.5°C above preindustrial levels.

The baseline for temperature increase is relative to a period in the mid-to-late 19th century, coinciding with the start of the Industrial Revolution. The Industrial Revolution marked a significant shift in global energy consumption, with a sharp rise in fossil fuel use. Many scientists pinpoint this period as the origin of the current climate crisis.

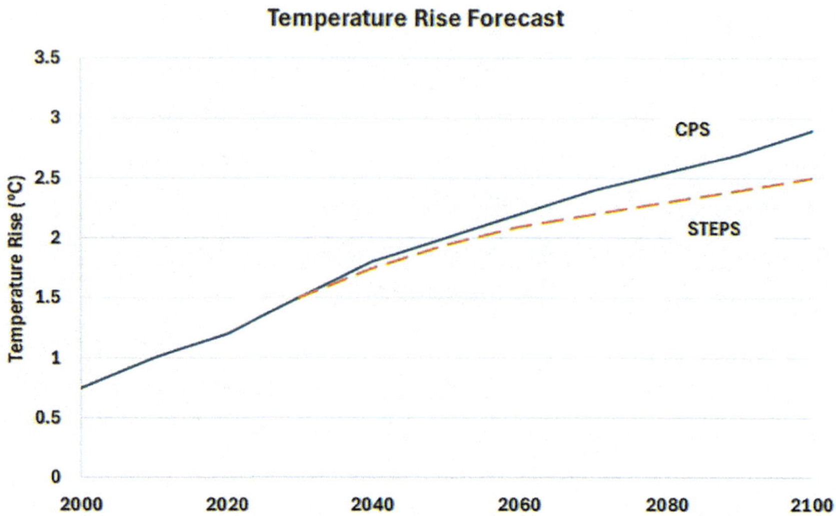

FIGURE 8.6 Forecast of temperature rise due to GHG emissions [16].

Only a decade after the Paris Agreement, temperatures reached the proposed limit and are likely to continue to rise. The corresponding temperature rise for these two scenarios is shown in Fig. 8.6. According to the published data, 2025 was the hottest year on record, and the first in which the temperature rise from preindustrial levels exceeded 1.5°C. The IEA report [16] suggests that, based on today's policies, far from limiting global warming to 1.5°C or well below 2°C, we are currently heading toward outcomes in the range of 2.5°C–3°C, with severe implications for lives and livelihoods around the world.

Particularly alarming is the rapid warming observed in the near-polar regions, like Russia, where temperatures are rising at a rate 2.5 times faster than the global average, and the Arctic more generally, which is experiencing even more accelerated warming. Despite Moscow's narrative portraying Russia as a beneficiary of climate change, with anticipated advantages such as a navigable Arctic shipping route, extended growing seasons, and expanded farmland due to thawing areas, the broader global implications of such environmental shifts demand urgent action and comprehensive global cooperation [17].

Ocean warming

Another concerning aspect of climate change is the warming of oceans. The world's oceans not only absorb around a quarter of the carbon dioxide that humans produce, but they also soak up around 90% of the excess heat. However, during 2023, the oceans displayed the most concerning evidence yet that they were struggling to cope, with the sea surface particularly feeling the heat.

Fig. 8.7 displays the daily average sea surface temperature between 1979 and 2024. The data represents the average temperature extending a hundred feet or

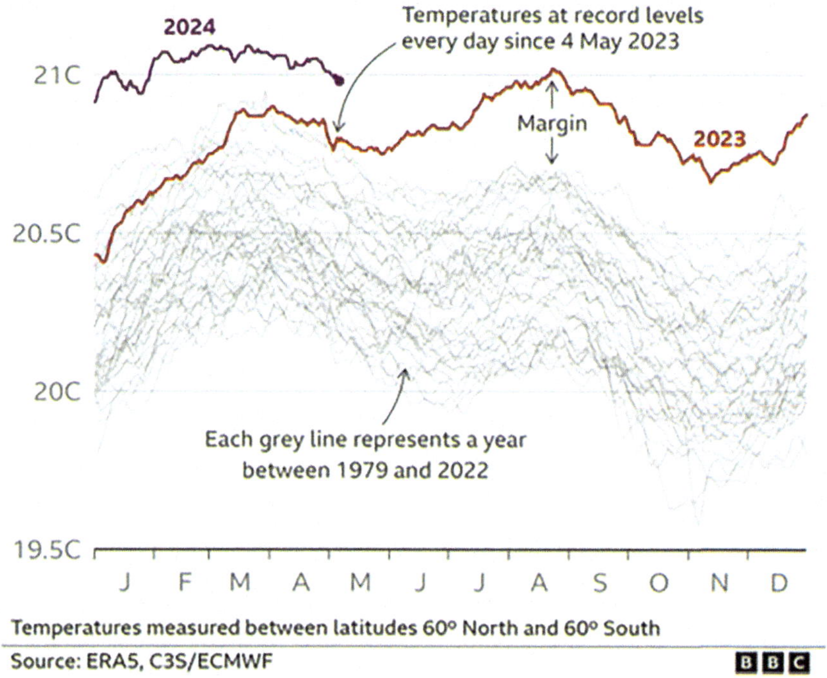

FIGURE 8.7 Daily average sea surface temperature, 1979–2024 [18].

more below the surface. They are also measured between latitudes 60 degrees north and 60 degrees south of the equator. The margin from previous temperature trends has widened. It was around 0.25°C in 2023 and widened even further in 2024.

Across the Gulf of Mexico, water temperatures are averaging 2 degrees Fahrenheit above normal. This overheated water is the fuel that allows for what hurricane watchers call "rapid intensification" [1]. Warmer oceans cause short-term impacts as well as long-term consequences. For example, ice-sheet melting and deep-ocean warming are likely to continue to fuel sea-level rise in the centuries to come. This human-driven ocean warming is having considerable impacts on global sea life and may even be shifting the seasonal cycle of sea temperatures [19, 20].

Ocean role in our planet

In the oceans, life is found everywhere, even at depths much greater than 130 ft, which is the depth that a recreational diver can go. About three-quarters of the ocean's depth is deeper than 1 km. Ocean biomass, in the form of microbes, lives at depths that sunlight cannot penetrate. Such microbes make at least half, if not 80%, of the oxygen we breathe [21]. As you go deeper, the temperature drops, and the density increases. Within these depths, there are discontinuous layers that do not mix. The deeper you go, the slower the water moves. What keeps these distinct layers of water from mixing is the fact that "there is not a mechanism that can stir up the whole ocean quickly enough to make it all the same" [22]. Manned or robotic underwater vehicles, called submersibles, are now providing more and more detail—yet still very little given the scale of what is being explored. The great depths of the sea remain a mystery, including how it could be impacted by global warming.

Global commitments to greenhouse gases reduction

For the past 30 years, over 200 countries have held annual conventions, known as "Conference of the Parties," or COP. During a preceding COP21 gathering in Paris in 2015, they pledged to transition toward renewable energy sources. The stated target for many countries has been to be "net carbon zero" by 2050. Net carbon zero, also referred to as net-zero emissions or carbon neutrality, describes a state where human-caused emissions of GHG are balanced by their removal from the atmosphere. However, there has been a slow uptake of their commitment to reducing their emissions. Also, geopolitical events such as the invasion of Ukraine have limited or derailed progress for numerous countries.

Net zero requires that any CO_2 released into the atmosphere from human activity be balanced by an equivalent amount being removed, either through nature-based solutions (including afforestation, reforestation, and other changes in land use) or technological solutions that permanently store CO_2 captured (directly or indirectly) from the atmosphere. The sooner net zero emissions are achieved, the greater the chances of meeting the most ambitious climate goals.

Four types of key actions have been suggested for achieving net zero by 2050 [23]:

- Clean electricity via renewable energy sources
- Fuel switching to electricity and the use of low-emission fuels such as biofuels
- Improving energy productivity
- Carbon management via carbon capture and storage

We need to add lowering consumption and energy conservation as other key activities. In the following section, we will explore technologies designed to reduce CO_2 emissions in the long term. It is crucial to recognize that while such processes are vital for long-term GHG mitigation, they do not replace the urgent need for immediate emissions reduction measures such as efficiency improvements and the use of renewable energies. We will address short-term actions in a separate chapter.

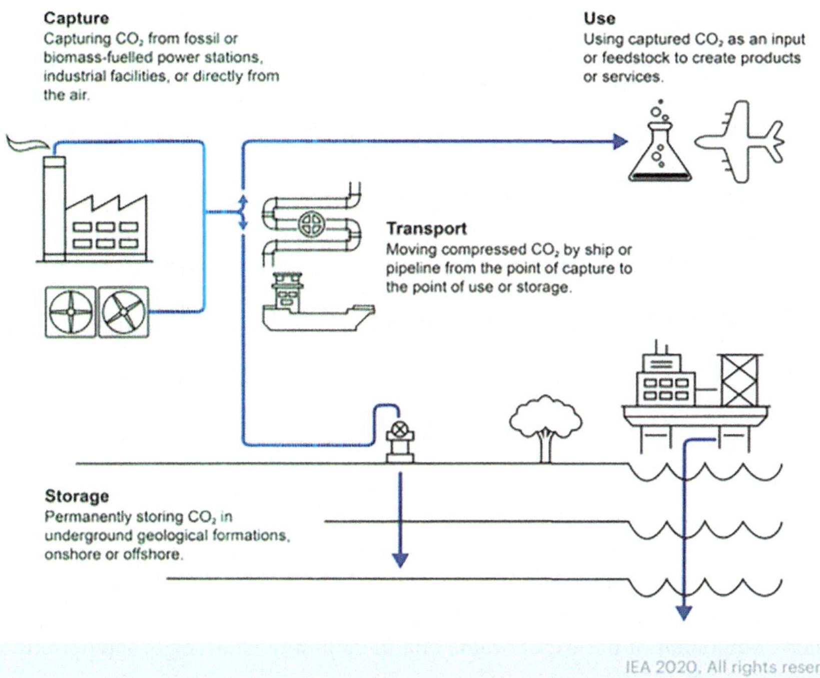

FIGURE 8.8 Typical CCUS schematic [24]. *CCUS,* carbon capture, use, and storage.

Carbon capture, use, and storage

Most world energy demand predictions identify fossil fuels as a key ingredient of the near- and medium-term mix of energy sources. However, the continued burning of these fuels will result in more CO_2 emissions that need to be captured and sequestered so that they do not, in effect, flow into the atmosphere if we are to have a chance to meet global climate goals (net carbon zero by 2050).

Carbon capture, use (sometimes utilization), and storage (CCUS) represents a set of technologies that can be employed to reduce the emissions of CO_2 to the atmosphere from the combustion of hydrocarbon (fossil) fuels like oil, natural gas, and coal. CCUS consists of:

- separation of CO_2 from industrial and energy-related sources (capture)
- transportation of CO_2 to a storage or use location
- long-term isolation of CO_2 from the atmosphere (via use and/or storage).

A schematic of the CCUS process is shown in Fig. 8.8.

CCUS is an option in the portfolio of mitigation actions for the stabilization of atmospheric GHG concentrations. The IEA projects a 20% reduction in CO_2 emissions, which may be achieved with the help of CCUS over the next 30 years.

Fulfilling this estimate may require capturing and storing up to 20 gigatons (Gt) per year of anthropogenic CO_2 from the atmosphere starting from 2050 onward.

CCUS can facilitate the transition to net-zero CO_2 emissions by:

- capturing CO_2 emissions from the flue gases of existing facilities (such as power plants, refineries, and LNG)
- providing a way to address emissions from some of the most energy-intensive and challenging industrial sectors (like cement and steel manufacturing)
- providing a cost-effective pathway to scale up hydrogen production as a potential future energy source
- allowing for CO_2 removal from the atmosphere through direct air capture (DAC).

In the Net Zero Emissions by 2050 scenario (NZE) envisioned by the IEA [25], policies are envisioned to facilitate the establishment of markets for CCUS investments. These policies also aim to promote the utilization of shared CO_2 transport and storage infrastructure among stakeholders involved in hydrogen and biofuel production, industrial hub operations, and the retrofitting of existing coal-fired power plants. The potential of CCUS as a mitigation technology could be substantial. The upcoming sections will delve into detailed discussions on each component of the CCUS value chain, including challenges and opportunities.

CO_2 capture

Various technologies, including transport, contribute to CO_2 capture within the framework of CCUS. Four primary CO_2 capture technologies include absorption (into a solvent), adsorption (within a porous material), membranes (for separating the gas stream), and cryogenic processes (where separation is done at higher pressures and cooler temperatures). Among these, amine absorption stands out as the most utilized and mature technology. In amine absorption, a gas mixture containing CO_2 flows upward through an absorption tower, where it meets an aqueous amine solution flowing downward. The amine absorbs the CO_2 from the gas, resulting in CO_2-free gas exiting the tower's top. The absorbed CO_2 is subsequently released from the amine through heating, after which the amine is recycled for further use. In order to work properly, some carbon capture processes require that other pollutants (such as sulfur dioxide, and some forms of nitrogen oxide) be removed from the exhaust gas of power plants before the solvent that captures the CO_2 is introduced.

While amine-based capture technology is mature, its deployment at a large scale is hindered by its substantial cost [26]. High water usage has also caused some concerns, and new efforts are underway to use non-aqueous solvents. Ongoing research explores alternative solvents for more effective CO_2 capture, as well as configurations with lower energy consumption, such as cryogenics.

Additionally, technologies like membranes are being investigated for specialized applications involving higher CO_2 content or lower mass flow rates. The separation of gases with a membrane is dependent on the permeability and the selectivity for the target component in the gas to be removed. The current membranes are limited in these characteristics, such that a single-stage separator cannot attain high removal and purity objectives, such as achieving 90% capture and 95% purity. Hybrid combinations of solvent or sorbent with the membrane process have been emphasized in the latest capture R&D in order to enhance performance or improve energy efficiency.

In the cryogenic process, phase change can be used to separate components from a gas stream. This is typically accomplished by cooling the gas stream until one or more of the components change phase to either a dense liquid or solid phase that can be physically separated from the noncondensing species.

Recognizing the imperative of improved technology and widespread deployment, the scientists acknowledge the necessity of advancing CO_2 capture across all industries to facilitate decarbonization efforts. Carbon capture has undergone several phases of technological evolution throughout the last century. The first phase commenced in the 1930s with the invention of the amine process, which facilitated the removal of CO_2 from natural gas to improve the gas quality. The second phase gained momentum in the 1970s with the injection of CO_2 in oil reservoirs to enhance oil recovery (EOR). Subsequently, the third phase commenced in the 1990s, focusing on the storage of CO_2 in geological formations for climate reasons.

As of early 2025, there was just over 50 million tons (Mt) of CO_2 capture and storage capacity in operation. By 2030, this could reach 270 (WEF) [27] to 430 (IEA) [28] Mt CO_2 per year.

Of the total CO_2 captured in 2050, approximately 95% is earmarked for permanent geological storage, while the remaining 5% is designated for synthetic fuel production.

The main factors that affect the cost of CO_2 capture are the concentration of CO_2 in the feed stream, the purity of the captured CO_2 stream (level of contaminants), and the scale or volume of CO_2 that needs to be captured. Increasing dilution with other gases (lower CO_2 concentration) causes more difficulty in capturing CO_2. This is expected because the percentage of unwanted or inert gas (mostly nitrogen) that needs to be processed is significantly higher, and more energy is required to separate them.

CO_2 transportation

CO_2 transport refers to the transfer of CO_2 from one location to another or between a CO_2 source and a geologic sink or point of use. CO_2 transport networks necessitate the development and construction of various infrastructure components, encompassing onshore and offshore pipelines, compressor stations, platforms, injection wells, and subsea manifolds. Fig. 8.9 illustrates the existing

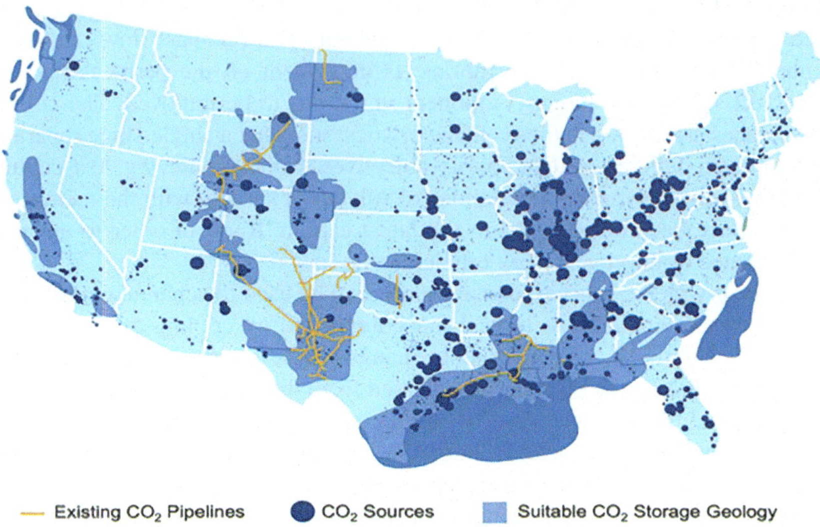

FIGURE 8.9 US CO_2 infrastructure, industrial emitters, and suitable storage sites [29].

CO_2 pipelines dispersed throughout the US oil fields, with approximately 5300 miles of CO_2 pipelines currently in operation. They convey over 50 Mtpa of CO_2 from natural and industrial sources for enhanced oil recovery in oil fields in the West, southwest, and Gulf Coast regions of the United States [29]. This network represents approximately 85% of the current global CO_2 pipeline infrastructure. In June 2020, the Alberta Carbon Trunk Line in Canada came online with a pipeline capacity of 14.6 Mt CO_2, with significant excess capacity (some 90%) to accommodate CO_2 from future CCUS facilities.

Pipeline transportation of CO_2 over longer distances is most efficient and economical when the CO_2 is in the dense phase: dense like a liquid but highly compressible and low in viscosity, like a gas. This means that the pressure in the pipeline must be kept at 1080 psi and above. Since most carbon capture technologies are limited in the purity of the captured CO_2 (around 85%), the impact of such impurities on the pipeline needs to be studied. CO_2 must also be dehydrated to avoid corrosion of the pipeline. Dehydrating the CO_2 entails removing the water from the gas mixture stream. Pipeline transport is the lowest-cost option for large volumes of carbon dioxide or where pipeline infrastructure already exists. This is the solution planned for use by the East Coast Cluster in the United Kingdom and Porthos in the Netherlands, for example, and for the bulk of planned US hubs. Where storage or pipelines are not readily available, or volumes are smaller, carbon dioxide can be chilled into a liquid state to go on ships and/or tanker trucks for transport to the injection location. The transport cost per ton is higher, but the solution offers greater flexibility for collecting multiple sources, requires less capital expenditure upfront, and has lower risk in built-up areas.

There is considerable potential to reuse or repurpose the large existing oil and gas pipelines for the transport of captured CO_2 in many parts of the world. Also, repurposing could potentially help to avoid the substantial costs of decommissioning pipelines. However, in many cases, natural gas pipelines' pressure rating is below the CO_2 transportation pressure limit (2200 psi). The existing pipelines would require a large number of pump stations, which would not be operationally practical for long distances [30]. It is recommended that the conversion of each pipeline's potential be studied based on the project-specific conditions and according to the Pipeline and Hazardous Materials Safety Administration (PHMSA) Part 195 regulations.

In populated areas, pipeline safety is a key concern due to potential leaks. The occurrence of some incidents has already alarmed the public against CO_2 pipelines. On Wednesday, April 3, 2024, near the town of Sulphur, Louisiana, a pipeline operated by Exxon-owned Denbury carrying highly pressurized CO_2 experienced a "leak" that caused a release of CO_2 that prompted a shelter-in-place order to be issued. In another incident, just after 7 pm on February 22, 2020, a CO_2 pipeline ruptured in Satartia, Mississippi. Shortly after a greenish cloud settled into the valley surrounding the little town, the presence of hydrogen sulfide in the gas mix worsened the impact of leakage. At least 45 people were hospitalized, cars stopped working, and a hobbling emergency response. The key learning from these incidents was an urgent need for an adequate emergency warning system wherever CO_2 leakage could occur.

By 2050, with large quantities of CO_2 captured globally, extensive pipeline and shipping infrastructure will be required to link the capture facilities with storage sites. A Princeton University study indicates that over 69,000 miles of CO_2 pipelines will be necessary by 2050 [31]. In the foreseeable future, an industrial CCS cluster will aggregate CO_2 from various industrial plants into a storage hub. CO_2 emissions from diverse industrial sources could be gathered and transported to a temporary terminal for subsequent offshore conveyance via new pipelines for injection into saline aquifers. The shared usage of this infrastructure by various industries within these clusters can alleviate the upfront investment burden [2].

CO_2 use

Using CO_2 as a raw ingredient for making other products inherently shifts CCUS from a fee-based waste disposal business model into a self-financing recycling one. CO_2 use represents a new generation of innovative technologies that convert CO_2 into products such as fuels, chemicals, and materials via chemical reactions or biological conversions, as discussed further [32].

- Oil recovery: Traditionally, CO_2 has been used by the petroleum industry for Enhanced Oil Recovery (EOR) [33]. The trapping of CO_2 during the EOR process is incidental to the primary purpose of producing oil. In this process, CO_2 is injected into the reservoir rock to liberate the oil within the porous rock and move it toward the production well. This exchange results in the CO_2 becoming trapped by capillary pressure within this same pore space, dissolving in the residual fluids present in the pore space, or eventually becoming mineralized (after a long time). The primary concern regarding CO_2 injection for EOR has centered on the use of natural CO_2 from reservoirs with high CO_2 concentrations rather than utilizing anthropogenic CO_2. Consequently, operators were releasing naturally trapped CO_2 from its geological confines. Although a portion of this released CO_2 would return to solution upon injection into hydrocarbon reservoirs, where it could be re-trapped, it was not feasible to recapture all of the released CO_2. The sustainable path forward for this technology hinges on employing anthropogenic CO_2 as the source of injection, thereby ensuring that the operation achieves net carbon neutrality.

  Also, it was customary to continue CO_2 injection until it was no longer economical to produce additional oil. The current thinking revolves around stopping the injection once the trapped CO_2 reaches its maximum. Generally, this happens earlier than when the injection reaches its economic limit. Finally, it is important to look at the full lifecycle analysis of CO_2 in this process in order to ensure that the produced oil will not generate more emissions.
- Chemical conversion: As a substitute, CO_2 could be used for fossil-derived inputs to a number of everyday materials, including graphene, graphite, plastic, fiber, and synthetic rubber. Climate benefits would endure as long as the material does not degrade, ensuring that the carbon dioxide remains sequestered in the end product.
- Mineralization: Examples of carbon mineralization in nature include the underground formation of carbonate rocks such as limestone. The carbonation of alkaline minerals mimics the natural alteration of calcium- and magnesium-rich rocks as they react with atmospheric CO_2 over geological timescales. As a substitute, CO_2 can be used for various inputs at numerous stages of the production of different building materials, such as cement and concrete materials, thereby sequestering it away indefinitely.
- Biological use: Photosynthetic CO_2 conversion mimics natural plant growth that occurs in sunlight. CO_2 could be used in greenhouses to enhance plant growth. Photosynthesis can use natural and artificial light to supercharge the growth of living organisms like algae and cyanobacteria to produce bioenergy. Microorganisms can capture and convert CO_2 into chemicals, chemical intermediates, or fuels. Green algae have an amino acid profile that compares favorably with common food proteins and could be an attractive source of dietary protein for animal feed and for human consumption. Climate benefits

could be substantial, especially if the carbon dioxide released while burning the biofuel is captured by itself. Multiple reports describe the state of the science and research challenges for algal biofuels [34].

- Fuel products: CO_2 can be converted into higher-energy synthetic fuels such as methanol via processes including hydrogenation. Using synthetic fuels could also reduce GHS emissions from marine vessels. However, CO_2 requires a relatively large amount of energy to convert it into products that contain higher energy. Unfortunately, most of these higher-energy products require hydrogen as a co-reactant, and this hydrogen should be produced via a more sustainable process if the carbon intensity of the final product is to be reduced. Since hydrogen is starting to be in high demand as the fuel of the future, its use in such processes may not be an option [25].

For example, CO_2 serves as a fundamental ingredient in the manufacturing of bicarbonate of soda, essential for treating kidney disease, water purification, and various industrial processes such as glassmaking, detergent production, pharmaceuticals, and food and animal feed manufacturing. Additionally, innovative processes involve transforming CO_2.

In 2022, China saw the inauguration of the world's premier CO_2-to-methanol plant, marking a significant milestone in carbon use. Operating based on the Emissions-to-Liquids (ETL) technology pioneered by Carbon Recycling International (CRI) and initially showcased in Iceland, this groundbreaking facility boasts the capability to capture 160,000 tons of carbon dioxide emissions annually. The process involves reacting to the captured carbon dioxide with recovered hydrogen in CRI's proprietary ETL reactor system, resulting in the production of 110,000 tons of methanol per year [35].

Most of these technologies will work best in a hybrid state, i.e., the use of electricity from renewable resources or net-zero fuels or the use of hydrogen that is generated via electrolysis. Also, it is important to remember that using CO_2 in products does not necessarily reduce emissions. Quantifying the potential climate benefits is complex and challenging, requiring a lifecycle approach. The climate benefits associated with CO_2 use primarily arise from displacing a product with one that has higher lifecycle CO_2 emissions. CO_2 use is the least mature component in the supply chain of CCUS technologies.

The importance of accurate and rigorous lifecycle analyses on CO_2 use cannot be overstated. There is considerable difficulty in capturing CO_2 and then converting this CO_2 into products using significant energy input. Given that the primary driver of CO_2 use is to reduce or eliminate net carbon emissions to the atmosphere, it is necessary to approach all CCUS applications with lifecycle analyses in mind to determine and quantify a realistic reduction in CO_2 emissions for a product or process [25].

Active research continues to pursue multiple technological pathways for converting CO_2 into products. Given the relative immaturity of these technologies, CO_2 use is likely to remain an outlet for only a small fraction of captured CO_2

over the next 10–20 years. However, these technologies hold promise for future emissions reduction. Knowledge of how to use and transform this molecule both efficiently from a chemical perspective and effectively from a broader systems perspective will be essential to the sustainability and resiliency of a closed anthropogenic carbon cycle.

CO_2 storage

In order to permanently put away the captured CO_2, deep underground reservoirs are required with the following characteristics [36]:

1. The formation rock must have sufficient pore space (porosity) in which CO_2 can be contained for storage and pathways connecting the pore space (permeability) so the CO_2 can be injected into and move within the formation. About 73% of the rocks on the Earth's surface meet these criteria. Although sandstones and carbonates appear to be solid rock, they are filled with small void spaces called pores. When these types of rock are about a mile or more below the surface, under normal conditions, the pores are filled with salty water, which is why they are called saline formations. Pushing the water out of the way and filling the pore spaces with CO_2 instead enables the storage of large volumes of CO_2.
2. It should be deep enough so that the natural pressure and temperature can maintain the CO_2 as a dense fluid, also called a supercritical fluid or state. Typically, the minimum depth required for this temperature and pressure is greater than or equal to about 3000 feet (about 1 km or 0.56 miles), depending on the geothermal gradient. To protect underground drinking water aquifers, CO_2 storage is only permitted in saline formations that are saltier than 10,000 ppm Total Dissolved Solids per EPA Class VI UIC regulations.
3. A prospective storage reservoir must have a geological seal above it. The sedimentary rock of a geologic seal, also referred to as caprock, must have a very low permeability that prevents CO_2 from leaving the storage formation. Seals are often made up of clay (shale), salt, or carbonate rocks with pores that are too small to enable the CO_2 to enter or pass through them.

The goal of underground storage is to safely contain CO_2 over long periods to combat climate change. Once CO_2 is captured from flue gas using pre-, post-, or oxyfuel combustion methods, it is transported to a storage site. Then, an appropriate underground geological formation is selected for CO_2 storage. According to the National Petroleum Council (NPC) report [36], optimal CO_2 storage site options include:

- Depleted oil fields offer favorable conditions and present economic incentives by enhancing short-term oil recovery upon CO_2 injection. They boast substantial storage capacities and benefit from existing infrastructure courtesy of oil and gas extraction practices. The key challenges with using

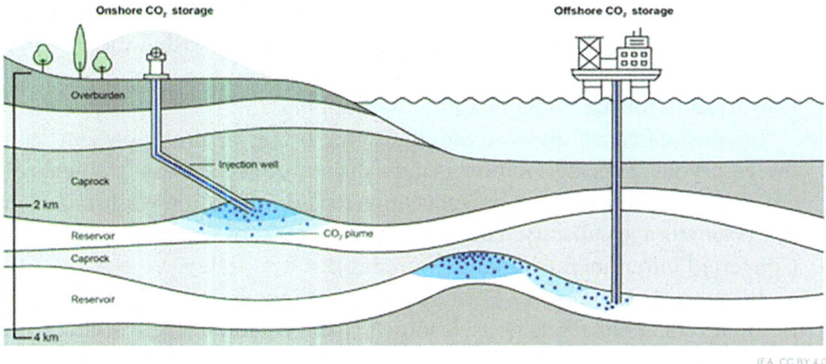

FIGURE 8.10 Onshore and offshore CO_2 underground storage [38].

depleted oil and gas fields are availability, location, and leakage risks from existing well penetrations. Ownership of the pore space is another challenge. We may need a suite of new legal frameworks to resolve the challenges facing CO_2 storage in depleted hydrocarbon reservoirs.

- Salt formations, like salt domes and caverns, boast excellent sealing properties and provide highly efficient storage capabilities. However, they are geographically constrained. Coal beds could also offer good storage for CO_2 due to their high adsorption capability.
- Saline aquifers, as previously discussed, represent abundant and spacious storage sites, offering the largest storage volume. Characterizing such aquifers, however, will require additional data acquisition since, unlike oil and gas fields, not much information is available about their storage properties [37].
- Basalt formations also show promise as long-term storage sites, as CO_2 can form stable carbonate minerals through high-temperature reactions. The high porosity and permeability of basalt increase its reactivity with CO_2, making it an ideal medium for CO_2 injection and storage.

The proximity between CO_2 sources and potential storage sites, coupled with the availability of existing CO_2 pipelines, enhances the feasibility of CCUS projects. Pipeline crossing through populated areas makes the permitting process more complicated. A typical storage site is presented in Fig. 8.10.

Upon reaching the reservoir, CO_2 is securely sequestered through four trapping mechanisms.

1. Sealed by a Caprock (structural trapping):

   A nonporous, impermeable caprock acts as a lid, isolating injected CO_2 beneath it. This thick rock layer prevents CO_2 from escaping and securely traps it.
2. Locked in the Pores (residual trapping):

Reservoirs are not cavernous but consist of porous rock strata resembling a sponge. During residual trapping, CO_2 is contained within the tiny pores of the rock, immobilizing it effectively.

3. Dissolved in formation fluid (solubility trapping):

   Injected CO_2 can dissolve into the saline brines naturally present within the reservoir. This dissolution occurs due to concentration gradients, increasing brine density, and prevents upward CO_2 migration while enhancing reservoir storage capacity. This process takes 100s of years to complete.

4. Converted into minerals (mineral trapping):

   In mineral trapping, CO_2 interacts with minerals in the rock formation through geochemical reactions, leading to the formation of new, solid carbonate minerals such as calcite, magnesite, and dolomite. This process integrates CO_2 into the rock structure, permanently sequestering it. In sedimentary rocks, this process takes thousands of years to complete. However, at high temperatures in basalt, this process takes only months [39].

Storage capacity

Potential storage for sedimentary basins around the world is estimated conservatively at around 8200 Gt CO_2 and between 20,000 and 35,000 Gt CO_2 in high estimates. However, many factors could reduce these estimates, including the maximum achievable reservoir pressure, well injectivity, and reservoir architecture. A reasonable CO_2 footprint for sequestration in sedimentary reservoirs lies between 0.5 and 5.0 tons CO_2/m^2 and depends on the architecture of the reservoir and the caprock, the petrophysical properties of the rocks, the pressure and the temperature in the reservoir, and the extent of secondary trapping mechanisms. Comparing these values with annual global CO_2 emissions highlights the potential these storage options have of significantly reducing CO_2 emissions to the atmosphere. The United States Geological Society (USGS) has published its recommended methodology for assessing CO_2 storage capacity [40].

CUS risk mitigation

The key CCUS project risks are volume, impurities, project development, permitting, and public opposition. Injections for storage purposes can lead to very large, source-dependent CO_2 plumes. The operator will potentially need to garner storage rights, develop monitoring plans, ensure containment, characterize the storage, and forecast its development via numerical models.

A Risk Management Plan is a crucial component of any CCUS project, requiring proponents to furnish evidence that the selected site is suitable for CO_2 sequestration. Geologic sequestration involves the use of highly engineered Class VI wells approved by the EPA, namely [41]:

- Extensive site characterization requirements.

- Injection well construction requirements for materials that are compatible with and can withstand contact with CO_2 over the life of the project.
- An injection pressure management system is used to maintain the reservoir pressure below the fracture pressure. The operator may need to extract brine from saline aquifers in order to allow for more CO_2 injection. Significant challenges remain, and more research is needed to assess the viability of brine extraction as a means of pressure management.
- Comprehensive monitoring requirements that address all aspects of well integrity, CO_2 injection and storage, CO_2 leakage, and groundwater quality during the injection operation and the post-injection site care period.
- Financial responsibility requirements ensure the availability of funds during the required period following the termination of injection.
- Reporting and recordkeeping requirements that provide project-specific information to evaluate Class VI operations and confirm safety operations continually.

These wells are designed to safeguard underground drinking water sources and ensure secure long-term CO_2 storage. Furthermore, stringent regulations mandate continuous monitoring of storage sites for many years post-CO_2 injection.

Early-phase site characterization and ongoing monitoring mitigate environmental risks. The EPA-regulated Monitoring, Reporting, and Verification (MRV) program includes meticulous injection well construction and operation standards, comprehensive monitoring of well integrity and CO_2 storage, groundwater quality surveillance, and sufficient funding for the project lifespan. Simulation models inform monitoring program design, guiding decisions like well placement and sampling frequency.

Areas of concern for CO_2 sequestration encompass fault assessment, induced seismicity, groundwater flow alterations, impacts on subsurface pore space users, non-saline groundwater effects, and public safety in release events. EPA guidelines mandate oversight of the CO_2 stream, injection pressures, well integrity, groundwater quality, and post-injection monitoring to ensure safety and prevent endangerment to deep waters.

Proper risk management, regulatory oversight, and robust monitoring protocols are essential for the safe and effective implementation of CCUS projects, addressing environmental concerns and ensuring the long-term security of CO_2 storage complexes.

Norway offers an excellent model of the potential of CCUS. The Norwegian Petroleum Directorate highlighted Norway's extensive experience in injecting and storing CO_2, both onshore and offshore, over several decades. This expertise is instrumental in Norway's efforts to achieve its climate targets and decarbonize its industries. Bolstered by its accumulated knowledge and successful track record in CCUS operations, Norway is confidently expanding its initiatives in this field. Through meticulous mapping and evaluation, they have determined that offshore Norway has the potential to store as much as 80 billion tons of CO_2, underscoring the nation's pivotal role in global carbon mitigation efforts.

CCUS in other industries

As previously mentioned, the steel and cement industries rank as the second and third most polluting sectors, trailing closely behind the fossil fuel industry. In the subsequent discussion, we will provide a concise overview of CCUS activities within these sectors.

Steel industry

Processing iron and manufacturing steel are profoundly high-temperature, emissions-intensive processes responsible for 11% of GHG emissions. Traditionally reliant on coal to generate the high temperatures necessary for steel production and as a raw material for the process itself, the industry faces significant environmental challenges. CCUS present a promising avenue for mitigating emissions across the steel value chain. According to estimates by the IEA, CCUS could contribute up to 15% of the reductions required for the industry to achieve net-zero targets, amounting to a total of 10 gigatons by 2060.

Retrofitting an existing plant or fitting a new one is a costly endeavor. Including CCS when building a new power plant will increase the plant's capital costs by between roughly 40% and almost 100% per kilowatt of plant capacity, compared with the capital outlay for a representative conventional power plant [42]. That additional cost varies depending on whether the plant is fueled by coal or natural gas and whether the CO_2 is captured before or after combustion.

Also, space constraints in existing plants can make it harder to install CCS equipment and can require longer pipes and ductwork to connect the equipment to existing machinery. In addition, at power plants that were built before the current, more stringent emission standards were put in place, scrubbers would need to be installed to remove pollutants (such as sulfur dioxide, nitrogen oxide, and mercury) from the plant's exhaust stream so those gases did not impede the solvent from capturing CO_2. As part of achieving their carbon-neutral goals, many steel producers in Europe, Japan, and Southeast Asia are seeking carbon sequestration opportunities.

Cement industry

The cement production industry stands as the third most polluting sector globally (7% of global GHG emissions). Cement serves as the cornerstone of concrete, the single most utilized human-made material on the planet, crucial for housing, infrastructure, and various construction endeavors. The annual production of over 4 billion tons of cement to meet worldwide concrete demands is responsible for approximately 7%–8% of annual CO_2 emissions. In the United States, the cement industry alone contributes roughly 1%–2% of the nation's CO_2

emissions, totaling around 70 million metric tons per year [43]. The imperative for the US cement sector to accelerate its decarbonization efforts cannot be overstated, aligning with broader industry aspirations for net-zero emissions.

As an example, at the heart of Portland cement lies Clinker, a binding agent created by heating limestone and aluminosilicate materials such as clay at high temperatures. Clinker production constitutes a staggering 85% of emissions within the cement production process. While various strategies for emissions reduction are emerging, their readiness for technological adoption varies. These approaches include clinker substitution, utilization of alternative fuels, and CCUS.

In the cement industry, governmental and consumer pressures are prompting a shift toward the adoption of green cement. Major buyers, especially government agencies and prominent private-sector entities, are beginning to commit to procuring low-carbon materials on a large scale. They can enforce stringent requirements for their purchases of these materials, such as concretes utilizing low-carbon cement, thereby compelling suppliers to invest in innovative approaches. The US Department of Energy's Industrial Decarbonization Roadmap projects that the US cement industry may need to achieve reductions of approximately 10% in emissions by 2030, 35% by 2035, and 60% by 2040 to stay on track for net zero. Correspondingly aggressive standards for low-carbon procurement could be implemented, with the federal government already initiating some requirements in this regard.

The scalability of green cement production is paramount for the United States to achieve overall net-zero emissions, positioning the nation at the forefront of global efforts to decarbonize the cement industry, including the widespread implementation of domestically developed technologies.

In emerging markets and developing economies, where coal-fired power plants have been recently constructed in large numbers, retrofits hold significant promise, especially when combined with opportunities for carbon storage. Conversely, in advanced economies, gas-fired plants with CCUS technology play a more prominent role, offering dispatchable electricity at relatively low costs, particularly in regions with abundant natural gas resources and existing infrastructure. It is projected that by 2050, approximately 7.6 gigatons of CO_2 will be captured annually from a diverse array of emission sources.

CCUS projects

As of 2023, there were 21 CCUS facilities around the world with the capacity to capture up to 40 MtPa CO_2, with a total capacity of 420 MtPa in 2024. Some of these facilities had been operating since the 1970s and 1980s, when natural gas processing plants in the Val Verde area of Texas began capturing CO_2 and supplying it to local oil producers for EOR operations. The first industrial-scale CO_2 injection project started in 1996 at the Sleipner field, located about 240 kilometers off the coast of Norway in the North Sea. This pioneering initiative, operated by Equinor (previously Statoil) and the Sleipner partners, aimed to mitigate GHG emissions by injecting CO_2 extracted from natural gas into the Utsira Sand, a significant saline aquifer of the late Cenozoic age. Commencing in October 1996, the project injected CO_2 at a rate of 1 million tons per year, requiring an incremental investment of approximately $80 million. With a CO_2 tax of $50 per ton, the project's payback period was estimated to be a mere 1.6 years. The CO_2 tax is an example of an incentive that is crucial to drive the expansion of CCUS (Please refer to Chapter 11 for more discussion on carbon tax).

While CCUS technologies hold promises for combating global climate change, their implementation has presented challenges. Such projects require stable revenue sources such as subsidies, carbon tax credits, green premiums, and lower capital costs. There has been considerable progress in the design of government support mechanisms, market instruments, and legal and regulatory approaches that could help the private sector manage some key risks and enable viable business models that are necessary for scaling up CCUS. The Infrastructure Investment and Jobs Act, passed in 2021, was a landmark legislation in the United States. It provided over US$12 billion for CCUS and related activities. US$2.5 billion for carbon storage validation, US$8 billion for hydrogen hubs, including blue hydrogen, and over US$200 million for CCUS technology development. Unfortunately, this act was stopped during the following administration.

This raises fundamental questions about the characteristics of CCUS projects that make them difficult to finance and scale, despite the technology being in use for decades, particularly in the oil and gas industry. In the power sector, some notable projects illustrate its application. The Boundary Dam project in Estevan, Saskatchewan, Canada, captures post-combustion CO_2 from a lignite coal power plant, with a reservoir size capable of storing 300 years' worth of supply. Similarly, the Petra Nova project in Texas, United States, captures CO_2 from a post-combustion pulverized coal power plant, injecting 1.6 million tons of CO_2 annually for enhanced oil recovery.

Despite their significance, these projects have encountered operational challenges. Notably, the Boundary Dam facility in Saskatchewan, Canada, the world's longest-running carbon-capture facility attached to a coal-fired power plant, failed to meet its emissions capture goal, with most captured CO_2 directed toward enhanced oil recovery [44]. A 2016 Parliamentary Budget Office report found that CCUS doubled the cost of Boundary Dam's coal-derived electricity [45]. In 2021, technical problems reduced Boundary Dam's CO_2 capture rate by 43% and caused it to go offline altogether for months at a time.

The existing and the future CO_2 infrastructure under development are presented in Fig. 8.11[46]. The existing global capacity is 51 MtPa. This is expected to increase to 450 MtPa by 2035. However, any expansion requires climate finance (financial resources and instruments that are used to support action on climate change) because of the large-scale investments that are needed to transition to a low-carbon global economy. The rich nations need to help the

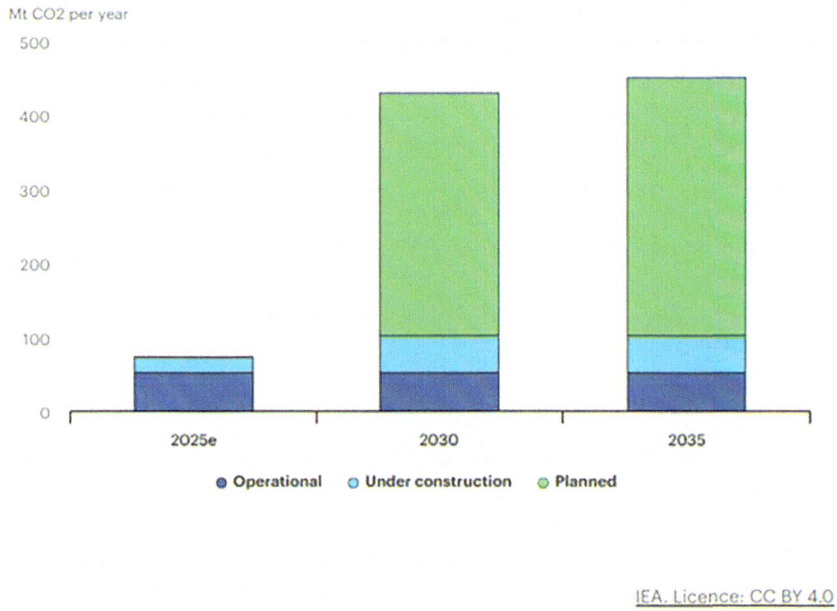

FIGURE 8.11 CO_2 projects (operational and planned capacity) [46].

rest of the world on their quest to become net carbon zero. Emerging countries are particularly vulnerable to the impacts of climate change, and they need significant financial support to adapt. Current estimates on addressing adaptation financing needs sit at up to $387 billion per year until 2030, and some mechanisms need to be established to start delivering finance flows to where they are needed most [47].

Carbon dioxide removal

Negative emissions technologies (NETs) that remove carbon dioxide from the atmosphere consist of Carbon Dioxide Removal (CDR) methods that are designed to extract CO_2 directly from the atmosphere [48]. This can be achieved through either (1) bolstering natural carbon sinks or (2) employing chemical engineering to capture CO_2, all with the aim of diminishing atmospheric CO_2 concentrations. CDR methods involve interventions in the ocean, on land, and through technical systems, including strategies such as iron fertilization, large-scale afforestation, and direct capture of CO_2 from the atmosphere using engineered chemical processes. While certain CDR methods may be classified as geoengineering, this categorization might not apply universally, as it hinges on factors such as the magnitude, scale, and repercussions of the specific CDR endeavors. The boundary between CDR and mitigation is not always distinct, as there may be some overlap between the two concepts based on prevailing definitions.

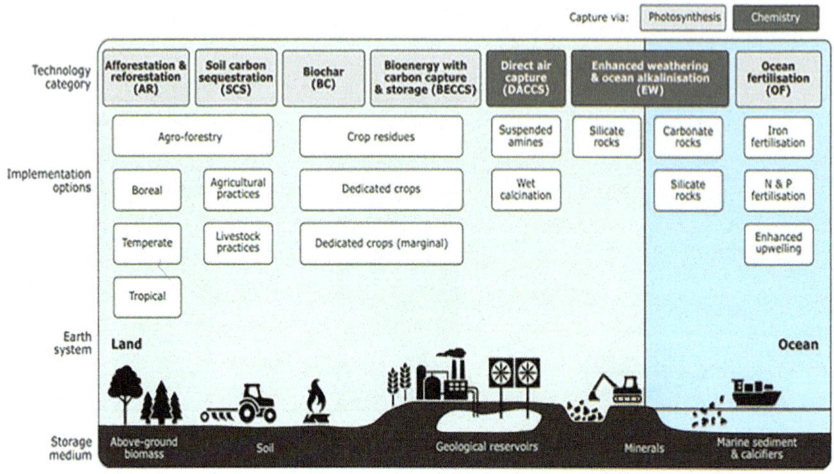

FIGURE 8.12 A taxonomy of negative emissions technologies (NETs) [46].

Fig. 8.12 shows a proposed framework for different categories of CDR. Definitions of these categories are provided in Appendix A. Afforestation and reforestation, Biochar, Bioenergy (plants and trees), Soil carbon sequestration (soil), and Ocean fertilization use photosynthesis to capture CO_2 emissions above ground. Diract Air Capture (DAC) and Enhanced weathering use chemicals to capture carbon and store it underground. We will discuss some of the more significant technologies below.

Direct air capture

Plans are underway to construct numerous DAC plants across various locations. These facilities consist of colossal fans housed within steel structures, resembling oversized air conditioning units expanded to remarkable proportions. Their primary function will be to extract carbon dioxide from the ambient air—an endeavor that carries a poignant irony. In the very region renowned for its prolific extraction of billions of barrels of fossil fuels from underground reservoirs throughout the 20th century, these DAC plants will now be engaged in the reverse process: reclaiming CO_2 from the atmosphere and injecting it back into depleted reservoirs.

Capturing CO_2 from the atmosphere poses greater energy intensity and, consequently, higher costs compared to capturing it from point sources. This heightened energy demand stems from the fact that atmospheric CO_2 is far more diluted (0.04%) than the concentrations found, for instance, in the flue gas emitted by power stations or cement plants (around 15%). DAC represents a relatively straightforward technology that employs large suction fans to capture CO_2 directly from the air. An Artist's rendering of a DAC facility is shown in Fig. 8.13.

FIGURE8.13 Typical direct air capture (DAC) units [49].

In a typical DAC unit, the air is drawn through a filter soaked in potassium hydroxide solution, which effectively absorbs CO_2 from the air. The liquid is directed to a secondary chamber where it interacts with calcium hydroxide, which captures the dissolved CO_2, resulting in the formation of small calcium carbonate flakes. These flakes are separated and introduced into a third chamber, termed a calciner, where they undergo decomposition upon heating. This process liberates pure CO_2, which is subsequently captured and stored. At each stage, the residual chemical byproducts are recycled back into the system, forming a closed-loop reaction that perpetuates endlessly without generating waste materials.

Two primary technological approaches [50] are typically employed for capturing CO_2 from the air: solid and liquid DAC (S-DAC and L-DAC). S-DAC relies on solid adsorbents operating at ambient to low pressure and moderate temperature conditions (typically between 80°C and 120°C). On the other hand, L-DAC utilizes an aqueous basic solution (such as potassium hydroxide) to release captured CO_2 through a series of units operating at elevated temperatures ranging from 300°C to 900°C.

In theory, DAC can be deployed virtually anywhere to extract carbon emissions from the atmosphere. It has been hailed as an efficient process, boasting land efficiency up to 100 times greater than reforestation and the capability to sequester carbon for centuries. Key advantages of DAC include scalability, permanence, and verifiability. However, DAC technology is still in its nascent stages of deployment. In 2023, according to the IEA, [51] there were 130 DAC plants under development worldwide, with 27 facilities commissioned in Europe, North America, Japan, and the Middle East, and 18 completed. Despite these numbers, all existing facilities remain relatively small-scale, collectively capable of removing approximately 11,000 tons of CO_2 annually (Box 8.5).

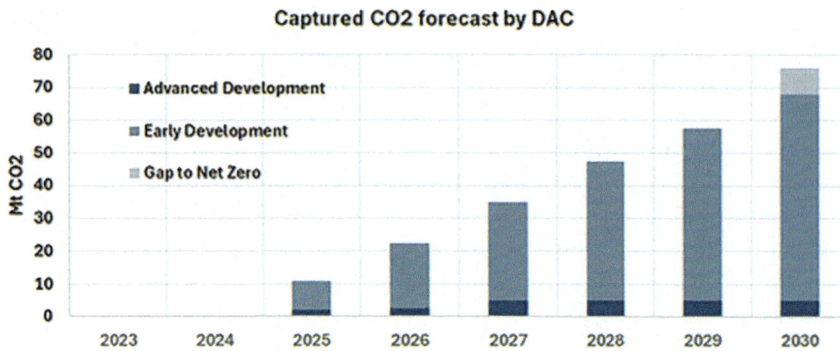

FIGURE 8.14 Forecast of CO_2 capture by direct air capture (DAC) [52].

The announced Inflation Reduction Act in 2022 by the United States established new funding for reaching the net zero target, including: increasing the 45Q tax credit to USD 180/t CO_2 captured for storage via DAC, with a capture threshold of as little as 1 kt CO_2 /year, and allocating $3.5 billion to establish four commercial-scale DAC hubs. Fig. 8.14 shows the projected CO_2 capture by DAC, based on planned projects and in the Net Zero Emissions by 2050 Scenario, 2020–30 (Box 8.5).

BOX 8.5 Tech Note: Emerging DAC technologies

To mitigate the energy intensity associated with the separation process in direct air capture (DAC), numerous innovators are actively developing new technologies, as highlighted in an IEA report [48]:

- Electro Swing Adsorption-DAC utilizes an electrochemical cell wherein a solid electrode absorbs CO_2 when negatively charged and releases it upon application of a positive charge. This promising technology is currently undergoing development in both the United States and the United Kingdom.
- Zeolites, renowned for their porous structure conducive to CO_2 adsorption, are now being harnessed for DAC applications. The inaugural operational DAC plant leveraging zeolites commenced operations in Norway in 2022, with ambitious plans to scale up the technology to a capacity of 2000 metric tons of CO_2 per year by 2025.
- Passive DAC represents an innovative approach that capitalizes on expediting the natural process whereby calcium hydroxide and atmospheric CO_2 transform into limestone. A pioneering company in the United States is engineering this process, employing renewably powered kilns to facilitate the separation of CO_2 from limestone.

These advancements underscore the dynamic landscape of DAC technology, with researchers and industry stakeholders striving to enhance efficiency, scalability, and sustainability in carbon capture efforts.

Environmental concerns about direct air capture

Numerous environmental organizations express apprehension toward DAC technology. The energy demand for capturing CO_2 from the air, which constitutes approximately 420 parts per million or 0.04% of the atmosphere, results in costs ranging from \$180 to \$2750 per ton of CO_2, according to a Reuters report [53]. They contend that the extensive energy requirements for capturing, purifying, and sequestering CO_2 render DAC one of the most exorbitant and inefficient methods of carbon sequestration.

Utilizing renewable energy to power a DAC plant can significantly enhance its overall carbon footprint. However, the success of DAC operations hinges crucially upon access to suitable sites for storing the captured CO_2. Establishing and operationalizing a CO_2 storage site typically entails a lengthy process, spanning approximately four to ten years from the initial project conception to the commencement of CO_2 injection. The prolonged timeline for developing such storage sites could potentially impede the widespread deployment of DAC technology, underscoring the urgent need for accelerated efforts in identifying and developing CO_2 storage locations [54].

Several European environmental scientists have suggested that to merely match the current global CO_2 emissions, standing at 36 gigatons annually, we will need the establishment of approximately 30,000 large-scale DAC plants—an endeavor demanding more than three times the number of existing coal-fired power stations worldwide. With each plant commanding a construction cost of up to \$500 million, the cumulative expense could soar to a staggering \$15 trillion. Moreover, each facility must be equipped with an ample supply of solvent for CO_2 absorption, translating to a demand for around four million tons of potassium hydroxide annually—an amount surpassing the entire global supply of this chemical by one and a half times. Furthermore, powering these thousands of DAC plants poses another formidable challenge. Operating a global industry capable of capturing 10 gigatons of CO_2 annually would consume approximately 100 exajoules of energy, equivalent to roughly one-sixth of the world's total energy output. Consequently, each DAC plant would necessitate a gas furnace, alongside a steady supply of gas, to fulfill its energy needs.

CCUS economic analysis

According to a 2020 report, the implementation of the Clean Air Act legislation of the 1980s still produces 3.8 trillion dollars in annual economic benefits and saves nearly four hundred thousand American lives each year. The climate writer David Roberts notes: "The air quality benefits alone are enough to pay for the energy transition." And slowing climate change may also be the most effective mitigant of wildfires, which thrive in hotter, drier conditions [1]. More recent legislation, such as the USA infrastructure bill of 2023, could provide similar

TABLE 8.2 Levelized cost of CO_2 capture by sector and initial CO_2 concentration [48, 53].

Sector	CO_2 concentration	Levelized cost (USD/ton CO_2)	Status/Difficulty
Natural Gas Processing	High (>90%)	\$15–\$30	Most mature capture is often part of the standard process.
Ethanol/Bio-refineries	High (>95%)	\$20–\$35	"Low-hanging fruit"; high purity allows for cheap capture.
Ammonia Production	High (>95%)	\$25–\$40	Similar to ethanol; process produces a nearly pure CO_2 stream.
Hydrogen (Blue)	Moderate (30%–90%)	\$45–\$75	Growing rapidly; cost depends on the reformer type (SMR vs. ATR).
Steel (Blast Furnace)	Moderate (20%–30%)	\$60–\$100	High thermal energy requirement; complex integration.
Cement	Moderate (15%–30%)	\$60–\$120	Large "process emissions" make this a priority for 2030 targets.
Power (Coal)	Low (12%–15%)	\$50–\$100	Huge volumes of flue gas require massive, expensive scrubbers.
Power (Natural Gas)	Very Low (3%–5%)	\$80–\$130	Difficult due to high nitrogen content; requires more energy.
Direct Air Capture (DAC)	Extreme Low (0.04%)	\$450–\$1000+	Most expensive; involves processing massive amounts of ambient air.

benefits to most people. It is expected that other nations will follow this act and pass similar legislation.

Economic screening plays a crucial role in the selection of potential storage sites for CO_2. While lateral compartmentalization aids in retaining CO_2 within the desired storage site, it can also impede CO_2 injectivity, necessitating elevated injection pressures and increasing costs. To justify the investment, which encompasses the expenses associated with CO_2 capture, transportation, and injection, and to lower the cost per ton of sequestered CO_2, priority should be given to sites with greater storage capacity and longer sequestration lifespans. Hence reservoirs with individual CO_2 sequestration capacities exceeding 1 Mt each are deemed favorable candidates for initial consideration.

Capture cost

The typical cost range for CO_2 capture in different industries is shown in Table 8.2. For most sectors, CO_2 capture is by far the most expensive part of the process, with the exact cost depending largely on the mixture of gases it has to deal with. If there is a high proportion of carbon dioxide at high pressure and on a large scale, such as in power plants, it is relatively easy to capture, making costs lower than for dilute or low-pressure exhaust gases, such as in cement or steel plants. For industries with high-purity carbon dioxide streams, such as those making fertilizers or ethanol, capture costs varied from \$19 to \$40 per ton of carbon dioxide. Notice that DAC was by far the most expensive process [55].

Table 8.2 covers capture only. For a full "Levelized Cost of CCS," you must typically add Transportation (\$5–\$15 via pipeline) and Storage (\$5–\$20 via injection) per ton.

As of 2024, the developers of technologies to capture CO_2 from low-concentration gas streams have generally aimed for capture rates of 85%–90%. These rates result in the lowest cost per ton of CO_2 captured. In a net-zero emissions energy system, the residual emissions from fossil fuel-based power plants and industrial facilities would have to be mitigated as well, either by raising CO_2 capture rates, by co-firing biomass, or by integrating other carbon removal options. There are no technical barriers to increasing capture rates well beyond 90% for most mature capture technologies. This can be achieved by modifying the CO_2 separation process, such as using larger equipment and multiple process steps. These modifications add higher energy consumption per ton of CO_2 captured and increase the unit costs.

Transport and storage costs

Transport and storage costs depend on the distance to cover and the infrastructure available. In 2017, the Global CCS Institute calculated the cost of pipeline transport and storage to be \$7–\$12 per ton for onshore and \$16–\$37 per ton for offshore. In Europe, where pipelines are less prevalent, the Clean Air Task Force (CATF) [56] estimates transport and storage costs to fall from a range of \$75–\$270 per ton, depending on location, to less than \$65 everywhere as new storage sites are identified. CATF tools can be used to estimate CCUS costs in many locations.

In 2021, the researchers at MIT suggested [57] that the widely used assumption of \$10/tCO_2 for the transport and storage of CO_2 is reasonable in some regions (i.e., the United States, Middle East, Russia, Canada, and Mexico) but not in others (i.e., Europe, Japan, China, Africa, and India).

Today's market models, unfortunately, make it more favorable for industrial companies to emit carbon dioxide (even where carbon has a cost) than to invest in carbon capture and have the carbon dioxide stored. As carbon prices rise, incentives develop, and mandates kick in, CCUS will become cheaper than emitting. To incentivize and reduce the burden on CCUS development, the US Internal Revenue Service started offering a tax credit for carbon dioxide sequestration, commonly known as the 45Q tax credit. This is calculated per metric ton of qualified carbon dioxide captured and sequestered. While primarily involving CO_2, the injection and sequestration of other carbon oxides, such as carbon monoxide, are also feasible. The credit amounts to $31 per ton of carbon stored underground and approximately $20 per ton if utilized for enhanced oil recovery [58]. The 2022 inflation reduction bill raised the tax credit to $85 for each metric ton of carbon dioxide captured and stored [59]. The value of the credit for DAC was increased to $180 per metric ton of CO_2 for geologic sequestration. The 45Q credit is valid only for 12 years. It usually takes about 3 years for permitting and another 3 years for the project Front End Engineering Design and Execution. Hence, the remaining 6 years do not seem to be sufficient for project financing. Liability insurance for decades for the post-injection phase is a major hurdle. The timing of the liability phase needs to be changed. Granting relief after a few years of safe operation could help the operators (Box 8.6).

BOX 8.6 Tech Note: A notional cost analysis

A notional cost analysis for capturing 1.5 MTPA CO_2 with coke/coal as fuel is illustrated in the following Fig. 8.15 for both sequestration and utilization.

Typical costs for carbon capture, transport, and storage can be estimated using a cost assessment tool that was developed for public use by the NPC in 2020 [61].

According to a report by the United States Government Accountability Office, falling natural gas prices and uncertainty regarding markets for carbon credits have adversely impacted the economic viability of coal plants equipped with carbon capture technologies. Additionally, integrating CCUS into power plants escalates the cost of electricity production. Overall, CCUS costs are expected to fall rapidly as the industry grows and more cost-effective capture technologies mature. The capture costs above are already substantially lower than earlier cost calculations, reflecting closer work on actual facilities. CCUS hubs are expected to accelerate this process through economies of scale on transport and storage, as well as standardization effects as new industries deploy and develop new capture technologies.

Oxford Energy [62] suggested that in designing frameworks to support the investment and scaling up of CCUS. To mitigate these risks and create a sustainable revenue stream that makes CCUS projects more attractive for private sector investment, several key elements are essential. A stable and supportive legal and regulatory framework provides the necessary certainty for long-term investments. Mechanisms that enable revenue stacking across the supply chain ensure financial viability for operators at different stages of the process. Additionally, varying levels of government participation in the CCS supply chain help facilitate risk-sharing and mitigation, making these projects more feasible and financially sustainable.

It is also becoming clear that no one CCUS business model fits all projects or sectors. Instead, various business models have emerged with different degrees

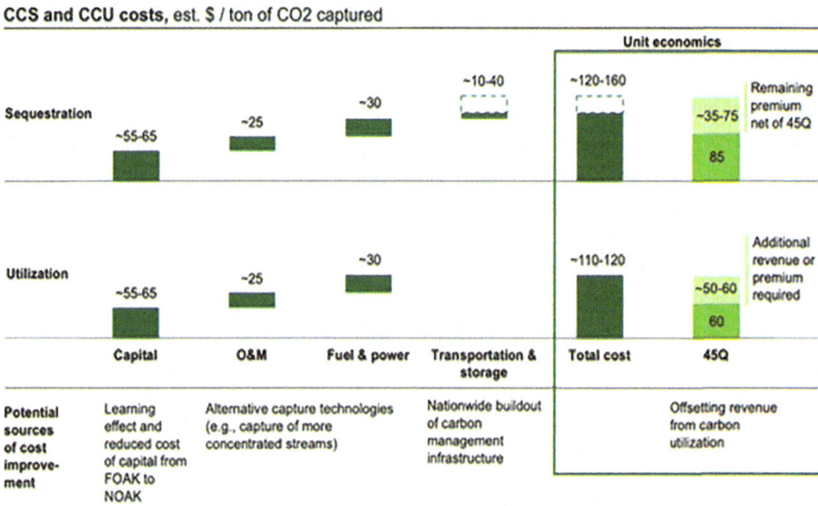

FIGURE 8.15 Illustrative economics for carbon capture, utilization, and storage (CCUS) deployment [60].

of integration in the supply chain, including varying degrees of government support, multiple support mechanisms, and various mixes of ownership structures and public–private partnerships.

The role of agriculture and trees in CO_2 sequestration

Forests, grasslands, wetlands, and coastal ecosystems are the most important natural CO_2 sinks, with "blue carbon" systems (mangroves, seagrasses, salt marshes) being especially powerful. This is done through a process called photosynthesis. During photosynthesis, trees absorb CO_2 from the atmosphere through small pores in their leaves called stomata. The CO_2 molecules are then used, along with water and sunlight, to produce glucose (a type of sugar) and oxygen.

The glucose produced during photosynthesis serves as food for the tree, providing energy for growth and maintenance. Some of the glucose is used immediately by the tree for its metabolic processes, while the rest is stored in various parts of the tree, including its roots, trunk, branches, and leaves.

As trees grow, they continue to absorb CO_2 from the atmosphere and store it in their biomass (trunks, branches, leaves, and roots) through the process of carbon sequestration. This stored carbon remains locked up in the tree's biomass until the tree dies or decomposes. Forests play a crucial role in absorbing carbon dioxide from the atmosphere, underscoring the imperative to preserve the natural world by curbing deforestation. Unlike Rocky Mountain pine forests, which rely on regular wildfires for regeneration, the Pacific temperate rainforest has evolved with minimal large-scale disturbances. Windstorms and landslides may occasionally topple patches of trees, but widespread disruptions are rare. Coupled with moderate temperatures and abundant rainfall, this stability enables these forests to amass an extraordinary amount of carbon.

The Tongass National Forest, spanning 16.8 million acres and covering nearly 80% of Southeast Alaska, harbors a staggering 2.8 billion tons of carbon—almost half of the total carbon stored in the US national forests. In contrast, the aboveground carbon density in the woods of the Colorado Rockies averages a mere 8 metric tons per acre. Even the most ambitious tree-planting initiatives pale in comparison to the carbon storage capacity of the Tongass [63].

Every year, substantial amounts of fallen needles, roots, moss, and soil accumulate in forests, each serving as a repository for carbon. However, when forests are cleared using heavy machinery, approximately two-thirds of this carbon is released into the atmosphere. Between 2000 and 2017, activities such as logging, agriculture, droughts, and wildfires led to the reduction of global tree cover by nearly a billion acres, resulting in the emission of approximately 4 billion tons of CO_2 annually—an alarming 10 percent of the roughly 40 billion tons of carbon pollution generated by human activities each year. To stabilize the climate, it is imperative to reduce and ultimately eliminate industrial and agricultural emissions while also devising strategies to capture the remaining emissions. Planting an extensive network of trees could potentially sequester over 200 billion tons of carbon. Some experts have even proposed establishing a strategic national carbon reserve akin to the strategic petroleum reserve, aimed at safeguarding the carbon sequestered in old forests and large trees on federal lands nationwide.

In the United States, the management of forests and other land (e.g., cropland and grasslands) has acted as a net sink of CO_2, which means that more CO_2 is removed from the atmosphere and stored in plants and trees than is emitted. This carbon sink offset about 13% of total emissions in 2021. Thus, trees play a crucial role in countries' efforts to achieve net-zero emissions targets. However, trees need to be planted in suitable locations for maximum efficacy and wider ecosystem benefits. Needless to say, this process is slow and would necessitate vast tracts of valuable land, potentially equivalent to foresting an area the size of the United States. Additionally, it is important to recognize the limitations of trees in carbon removal, as they will eventually decompose and release stored carbon unless managed within a closed system. During the wildfires, this carbon sink is also released to the atmosphere.

Bioenergy with Carbon Capture and Storage (BECCS) offers another avenue for carbon sequestration, wherein plants sequester carbon through photosynthesis, and the captured carbon is stored after biomass combustion. While theoretically carbon-negative, BECCS requires substantial land resources. IPCC scenarios involving BECCS drawdown would necessitate approximately 12% of the world's current cropland—an allocation that could impede global food production. Despite its potential, the world's largest BECCS operation at an Archer Daniels Midland facility in Illinois has encountered challenges, storing only half the initially promised carbon emissions [64]. In addition, after ADM's BECCS component went online, carbon emissions from the facility rose by 200,000 tons a year [65].

CO_2 storage in the oceans

The ocean harbors approximately 40 times more carbon than the atmosphere. A significant contributor to ocean carbon storage is the biological carbon pump, which transports organic carbon from the surface ocean to deeper waters through sinking particles, vertically migrating organisms, and physical circulation [66]. The marine processes that contribute to the ocean's biological pump begin with phytoplankton, microscopic organisms that soak up carbon dioxide from the atmosphere as they grow. When they die, phytoplankton collectively sink through the water column as "marine snow," carrying that carbon with them. At various depths, the particles are consumed by microbes, which convert the particles' organic carbon and respire it into the deep ocean in an inorganic, mineral form, in a process known as remineralization. This raises the question: could these mechanisms be harnessed for carbon sequestration?

The history of ocean-based CDR has seen periods of optimism, coinciding with advancements in research and pivotal climate agreements such as the Kyoto Protocol in 1997 and the Paris Agreement in 2015. However, it has also been marked by controversy and disillusionment as doubts arose regarding the efficacy of these approaches and the motivations behind their advocacy. Interestingly, the recent resurgence of interest has not been driven by technological breakthroughs or a reduction in scientific uncertainties but rather by new socio-technical configurations, alliances, and narratives [67].

A study by MIT scientists suggests that the amount of carbon that the ocean is able to draw down and sequester from the atmosphere varies widely, depending on which mathematical description for the biological pump they used. The results suggest that the ocean's pumping strength and the processes that govern how fast marine snowfalls are still an open question [68].

Some scientists also proposed injecting CO_2 into deeper ocean depths. In the shallow depths, we are dealing with the "Dissolution Type," i.e., the CO_2 plume is rapidly dissolving in the ocean water. This could potentially impact the ocean ecosystem due to changes in the water's pH or acidity. However, if CO_2 is injected in the deeper zones (below 3 km), because of its higher density, the CO_2 plume will sink to the ocean floor and create "Lake-type" storage. However, because of dangers to deep marine biology, this idea was abandoned.

Methane emissions

Methane is responsible for around 30% of the rise in global temperatures since the Industrial Revolution, and making rapid and sustained reductions in methane

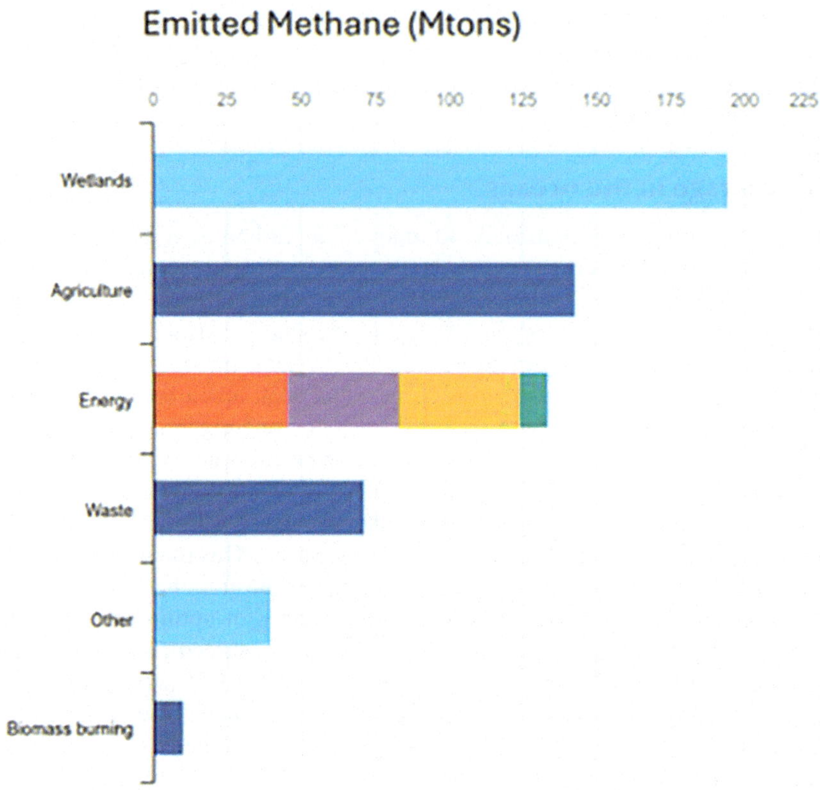

FIGURE 8.16 Emitted methane (IEA, 2023) [70].

emissions is critical for improving the climate. Methane, the primary component of natural gas, stems from various human activities in the world. In terms of climate impact, methane is more potent than carbon dioxide, particularly in the first decades after its release into the atmosphere, although its lifespan is shorter (12 years). Methane is 80 times more effective at trapping radiation than CO_2 and has an impact factor 28 times greater than CO_2 over 100 years [69]. Consequently, reducing methane emissions presents one of the most immediate and cost-effective strategies to combat climate change.

The anthropogenic sources of CH_4 include fossil fuels (flares, natural gas, coal mining), enteric fermentation, part of domestic ruminant animals' digestive processes, landfills, domestic livestock manure management, and flooded land. The global methane emissions in 2025 were around 610 million tons (Mt/yr), and the main emitters were: China, the United States, and Russia.

The amount and sources of global methane emissions are presented in Fig. 8.16. The global energy sector was responsible for nearly 135 million tons of methane emissions.

Historically, gas flaring and uncontrolled leaky equipment in the oil industry have been significant contributors to methane emissions, a factor often overlooked in discussions about climate change. Globally, according to the World Bank [71], about 148 billion cubic meters of gas were flared in 2023, resulting in 381 million tons of CO_2 (equivalent). New studies also show that flaring may have a bigger impact on the climate than previously thought. Governments and industry have historically assumed that flares combust methane at the 98% efficiency rate. But scientists using airborne sampling across three US fossil fuel basins found flares effectively only destroyed 91.1% of methane [72]. There is growing recognition among scientists and policymakers of the critical importance of reducing gas flaring.

According to the IEA's Global Methane Tracker 2025 [73], methane concentration in the atmosphere from human activities has more than doubled since preindustrial levels, contributing approximately 30% to current global warming. The energy sector, encompassing oil, natural gas, coal, and bioenergy, accounts for nearly 40% of human-induced methane emissions, with agriculture and landfills also contributing significantly. Cutting methane emissions from the energy sector by 75% by 2030 is one of the least costly opportunities to limit global warming in the near term.

CH_4 emission in the oil and gas industry

Estimates from the IEA indicate that methane emissions from the oil and gas industry amount to around 2.4 gigatons of carbon dioxide equivalent, nearly half of the industry's total Scope 1 and 2 emissions. Routine and nonroutine flaring have come under scrutiny as key sources of methane emission due to the inefficiency of flares, leading some states to enact bans on this practice.

Previously, methane emissions in the oil and gas industry were often assessed using standard emission factors derived from global data and specific basin data. However, this approach did not always capture site-specific emissions accurately and often overlooked key sources such as leaks, venting, and flaring. Consequently, estimates of methane emissions varied widely. In recent years, there has been a concerted effort to enhance methane emissions measurement in the oil and gas sector through robust MRV practices. The International Association of Gas Producers has published a framework and guidelines to help the oil and gas operators with the selection and deployment of methane emissions detection and quantification technologies. These are tailored to their sites and objectives and are accompanied by an online technology filtering tool, detailed technology data sheets covering over fifty technologies, and decision trees to guide deployment [74].

National oil companies (NOCs) account for roughly half of global oil and gas production, giving them a decisive influence on methane abatement efforts. More than 30 NOCs have joined the Oil and Gas Decarbonization Charter and are actively participating in initiatives to reduce methane emissions and flaring. This creates a significant opportunity for NOCs seeking to implement best-practice methane management to learn from peer experience and adopt strategies tailored to their specific operating contexts. Key best practices include minimizing flaring and venting, implementing robust leak detection and repair programs, and deploying low-emissions equipment [75].

IEA estimates that methane emissions from fossil fuels should fall by around 75% by 2030, and by around 85% by 2035 compared with current levels (125 Mt). Most of this decline stems from the explicit deployment of methane abatement technologies and mitigation policy enforcement.

CH_4 emission in the coal industry

Coal operations represent a significant source of methane emissions, posing challenges for reduction compared to oil and gas operations. Methane, known as coal mine methane (CMM), is naturally present in coal seams and can be released during mining operations or afterward. Emissions originate from various sources, including seepage from exposed coal seams in surface or open-pit mines, drainage systems used to lower methane levels in coal deposits, and ventilation systems, where methane is extracted and vented from underground mines for safety. Post-mining activities such as processing, storage, and transport also contribute to emissions, along with abandoned mines. In 2025, the IEA estimated that over 40 million metric tons of CMM were released globally, accounting for more than 10% of total methane emissions from human activity. China, the world's largest coal producer, is responsible for roughly half of global CMM emissions.

While there are numerous opportunities to decrease these emissions, achieving wholesale reductions is generally more complex than in the oil and gas sector. One of the most effective approaches to mitigating CMM is likely to involve reducing coal consumption itself. However, prioritizing the widespread deployment of abatement measures remains crucial, particularly considering the persistent demand for coal in the foreseeable future [76].

To address CMM emissions, the EIA has proposed various strategies, including mitigating methane from ventilation systems. This is considered one of the most critical measures the coal industry can undertake to reduce emissions, potentially decreasing CMM on a global scale by almost 30%. Targeting high-concentration methane sources through degasification systems implemented prior to mining operations can also significantly reduce emissions. Methods such as degasification wells and drainage boreholes capture methane from coal deposits, thereby reducing emissions during production.

Landfills: A significant source of methane emissions

Consumerism, or the excessive consumption of goods and services, is a major driver of landfill issues. Many products are designed to become obsolete quickly, encouraging consumers to replace them frequently. Also, single-use items like plastic bags, bottles, and utensils contribute significantly to landfill waste. More importantly, constant advertising and marketing pressure consumers to buy more, even if they do not need it. As consumption increases, so does the volume of waste that ends up in landfills.

Landfills, the final resting place for much of our waste, play a significant role in contributing to GHG emissions, particularly methane. As organic waste decomposes within a landfill in the absence of oxygen, it produces methane. This methane is released into the atmosphere, contributing to climate change and air pollution. Landfills also emit hazardous air pollutants that are precursors to ozone and produce PM that causes odor nuisance compounds. Leachate, a toxic liquid generated from decomposing waste, can contaminate groundwater and soil. Unfortunately, communities that are adjacent to such landfills bear the brunt of these impacts.

Some of the world's largest landfills and their approximate sizes are:

- Apex Regional Landfill, Las Vegas, Nevada, USA: 2200 acres. It holds 50 million tons of waste and receives 300 tons of garbage every hour.
- Bordo Poniente, Mexico City, Mexico: 927 acres. This site was closed in 2011.
- Laogang, Shanghai, China: 830 acres.
- Malagrotta, Rome, Italy: 680 acres. This site was closed in 2012.
- Puente Hills, Los Angeles, California, USA: 630 acres. This site was closed in 2013.
- Sudokwon, Incheon, South Korea: 570 acres
- Delhi Landfills, New Delhi, India: 500 acres

Although some of these sites are shut down, they are still generating GHG. Note that landfill sizes can change over time as waste is added and management practices evolve. Additionally, many landfills are not officially tracked or reported, making it difficult to determine the exact size and number of landfills worldwide.

Unfortunately, there is no precise figure for the total methane generated by all major landfills worldwide, but some estimates provide context. On a global scale, the World Bank estimated that in 2020, the world generated 2.2 billion tons of waste, with 40% deposited in urban landfills. This translates to approximately 0.79 kg of waste per person per day. Assuming a conservative estimate of 0.05 tons of methane generated per ton of landfilled waste, global methane emissions from landfills would amount to roughly 44 million tons.

In the United States, landfills are a major source of methane emissions. According to the EPA, US landfills released around 3.7 million metric tons (295 $MMTCO_2e$) of methane in 2023. This is comparable to the annual emissions from 66 million cars or 79 coal-fired power plants.

The actual methane emissions from landfills vary based on factors such as waste composition, landfill management practices, and climate conditions.

Additionally, many landfills, particularly in developing countries, may lack accurate data on their methane emissions.

Reducing landfill emissions

By implementing a combination of source reduction, waste diversion, landfill gas capture and utilization, advanced waste treatment technologies, and improved landfill design and management, we can significantly reduce landfill emissions and contribute to a more sustainable future.

As will be discussed in more detail in Chapter 12, source reduction and waste minimization can be achieved by changing our consumption habits—purchasing only what is necessary, opting for reusable products, and avoiding single-use items. Recycling and composting help divert recyclable and compostable materials from landfills. Composting facilities prevent methane generation by breaking down organic waste aerobically with oxygen, while anaerobic digesters generate methane in an enclosed tank that limits fugitive methane emissions.

Landfill gas capture and utilization involve installing methane-capturing systems in landfills to collect the generated methane, which can then be used to generate electricity or heat. This process provides a valuable energy source while reducing GHG emissions. For instance, the captured gas at the Apex landfill generates 11 megawatts, enough to power 6000 homes. However, the generated CO_2 in these landfills, which can be as high as 50%, is generally not captured, contributing to further pollution.

Advanced waste treatment technologies, such as anaerobic digestion, break down organic waste in the absence of oxygen, producing biogas that can be used for energy generation or upgraded to biomethane for transportation fuel. Thermal treatment technologies, including incineration and gasification, convert waste into energy or valuable materials, significantly reducing the amount of waste sent to landfills.

The key to any GHG reduction in landfills is comprehensive monitoring of the emitted gases. Implementing best practices in landfill design, including proper waste layering, gas collection systems, effective cover practices to control surface methane, and leachate management, can minimize methane emissions and environmental impacts.

Methane emission mitigation

Eliminating methane emissions presents a win-win scenario, contributing to GHG reduction while enhancing system efficiency. Furthermore, captured methane can serve as a valuable resource for producing other environmentally friendly products, such as methanol and hydrogen, fostering sustainability and innovation in energy practices.

In the oil and gas sector, implementing well-known measures such as leak detection and repair programs and upgrading leaky equipment can reduce methane

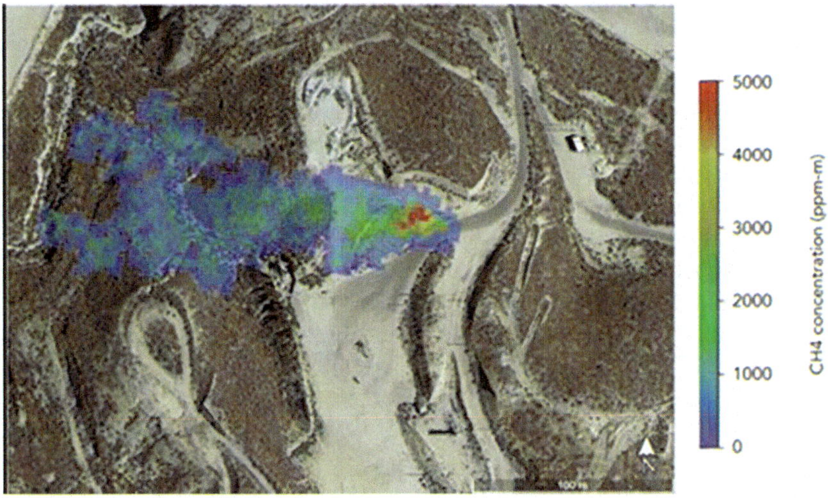

FIGURE 8.17 Satellite image of CH_4 leakage from a pipeline [78].

emissions. In the coal sector, more than half of methane emissions could be cut by making the most of CMM utilization or by flaring or oxidation technologies when energy recovery is not viable [75]. Under the Global Methane Pledge, over 150 countries have committed to reducing methane emissions by at least 30% from 2020 levels by 2030. This initiative aims to eliminate 0.2°C of warming by 2050 [77]. Thanks to advancements in technologies for detecting and mitigating leaks and a concerted effort to avoid nonemergency flaring and venting, methane emissions in the oil and gas industry can now be addressed in many locations at relatively low or minimal cost.

Methane monitoring

In 2014, recognizing the need for improved methane management, the United Nations Environment Program (UNEP) established the Oil & Gas Methane Partnership. Building on this initiative, UNEP expanded the program in 2020, now known as OGMP 2.0, to enhance its scope and ambition. OGMP 2.0 aims to advance the accuracy and transparency of methane emissions reporting, facilitating the identification and prioritization of mitigation measures within the sector. Notably, OGMP 2.0 emphasizes the importance of data-driven approaches to methane management.

The impact of these efforts is exemplified by a methane plume detected by NASA's AVIRIS-NG in the summer of 2020, originating from a leaking gas line in an oil field in California, as shown in Fig. 8.17. This image underscores the significance of advanced monitoring technologies in identifying methane leaks promptly. Following detection, the operator confirmed and promptly repaired the leak, highlighting the practical implications of improved methane emissions measurement and management practices.

In 2022, the United Nations launched its Methane Alert and Response System, empowering the UNEP to validate emissions reported by companies and track changes over time. Concurrently, some companies are intensifying leak detection and repair campaigns, aiming to mitigate fugitive emissions by enhancing flare efficiency and leveraging artificial intelligence to prevent emergency venting of methane. The Global Methane Pledge [79] is an organization that was launched at COP26 by the European Union and the United States to help with mitigating CH_4 emissions. Participants joining the Pledge agree to take voluntary actions to contribute to a collective effort to reduce global methane emissions by at least 30% from 2020 levels by 2030. This is an international, not a national, reduction target.

Methane conversion to methanol

Methanol is primarily used as a feedstock for producing chemicals such as formaldehyde, and it is a building block in thousands of products from car parts to paint to construction materials. Roughly 98 million metric tons of methanol are produced annually, according to the Methanol Institute [80] (51% derivative chemicals, 25% petrochemical feedstocks, 14% gasoline blending, 6% alternative fuels, 4% others). However, the compound could also play a significant role in the energy transition because when combusted as fuel, it generates less CO_2 and PM than many other fossil fuel-derived products. That means it could help curb the climate impact of sectors like marine shipping. Although about 20% of global supplies are currently blended into gasoline and used in alternative fuels, the chemical compound still has some challenges in reaching widespread adoption because it uses less energy per liter when compared to conventional fuels like diesel.

The majority of the world's methanol is made from natural gas or coal at large, centralized facilities where the fossil fuel is converted into syngas by heating it with steam in the presence of a catalyst. That process, done at scale, is cost-effective, even after paying for natural gas. As a fuel, methanol derived from flare gas is better than methanol produced from fossil fuels [81].

Methane mitigation technologies

According to the IEA, the global oil and gas industry has the potential to reduce methane emissions by 70% using existing technologies, with approximately two-thirds achievable at no net cost. The advent of new technologies, including satellite monitoring, drones, and advanced sensors, has revolutionized methane emission detection and quantification. These innovations enable the industry to tackle methane emissions more effectively, emphasizing the importance of regulatory support and enforcement. Flaring and methane emissions represent a significant waste of natural resources, particularly amid tight natural gas markets. According to IEA [82], implementing robust policies and effective on-the-ground measures for flaring and methane emissions reduction could potentially conserve up to 200 billion cubic meters of additional gas globally.

Adopting emission reduction technologies

Unfortunately, many countries are deferring significant emissions reductions until closer to their net-zero target year. This delay underscores the critical importance of breakthroughs, commercialization, and incentives related to CCUS and CDR technologies in achieving their goals. Leading R&D efforts and pilot projects in this domain are being spearheaded by Australia, China, the EU, Japan, the United Kingdom, and the United States, according to the Global Carbon Capture Sequestration Institute. However, achieving deployment levels sufficient to meet the objectives of the Paris Agreement hinges on either technological advancements that substantially reduce costs or governmental intervention through subsidies and taxes that escalate the costs associated with fossil fuels. Presently, there is no significant market demand for CO_2 on a large scale.

European hydrocarbon-producing nations, particularly Norway and the United Kingdom, are likely the best positioned for large-scale CCUS deployment in the next decade. This advantage stems from supportive government policies and regulatory frameworks, including carbon-pricing initiatives. Meanwhile, the United States boasts several strengths that position it as a potential leader in CCUS. US companies are making substantial investments and possess experience in utilizing CCUS to enhance oil and gas yields. According to the USGS, nearly half of the world's operational carbon capture facilities are located in the United States, which also boasts significant geologic storage capacity. Moreover, it is expected that more countries will invest in and advocate for CCUS as a crucial strategy for offsetting emissions while prolonging fossil fuel production and consumption. Oil and gas companies are intensifying their research and development efforts in CCUS for similar reasons.

Public perception of carbon capture, utilization, and storage

CCUS stands at the forefront of decarbonizing fossil fuel-burning industries such as steel and cement, and even in emerging industries such as hydrogen fuel generation. However, many in the public perceive such endeavors as merely prolonging the use of fossil fuels despite acknowledging their ongoing necessity. They criticize the government tax credit as compensating companies for the carbon dioxide they capture and store rather than for the CO_2 they avoid.

In a signed letter to the Canadian Deputy Prime Minister & Minister of Finance [83], more than 100 scientists voiced their opposition to a proposed CCUS investment tax credit. They commented that "CCUS is neither economically sound nor proven at scale, with a terrible track record and limited potential to deliver significant, cost-effective emissions reductions." The high cost, difficulty in scaling up, and risks were mentioned as the main reasons that CCUS was not a feasible solution for emission reduction. They recommended that tax credits be contingent on the development of stringent requirements for MRV, and this tax should be available only for those sectors where there are no decarbonization options. They proposed moving forward with proven climate solutions that will contribute the most to emissions reductions. Although their concerns are valid, CCUS projects go through very stringent review and monitoring requirements and a long permitting process. The application of new technologies should reduce the cost of carbon capture and scale up.

Public concerns extend beyond environmental implications to encompass potential health and safety hazards associated with CO_2 transport and storage infrastructure. Addressing these apprehensions is paramount for the successful execution of CCUS projects, emphasizing the need for robust engagement with local communities and the public at large.

It is important to recognize both the benefits and drawbacks of CCUS. On the negative side, poorly designed CCUS projects can present several risks. One major concern is groundwater pollution, as migrating CO_2 can dissolve and alter minerals in rocks and soils, potentially contaminating freshwater supplies. Surface leaks of CO_2 could also accumulate in low-lying areas and enclosed spaces, creating hazards for humans and wildlife. Additionally, CO_2 leakage may disrupt ecosystem biodiversity.

Another risk is the potential for induced earthquakes. Similar to wastewater injection, CO_2 injection can trigger seismic activity depending on local stress conditions. Although no documented cases of CO_2 injection-induced seismicity existed until recently, new studies suggest otherwise. For example, research at the Decatur CCS site identified induced seismic events linked to CO_2 injection. Despite careful injection practices, these events occurred, indicating that stress transfers from CO_2 injection reactivated preexisting fractures in the critically stressed basement, leading to seismic activity.

Furthermore, while CCUS has been successfully deployed at numerous sites worldwide, concerns remain about the long-term containment of CO_2, especially in large-scale storage scenarios. Uncertainty about the permanent underground storage of CO_2 continues to hinder the widespread adoption of CCUS technology.

As a reminder, this process of underground gas storage mirrors the longstanding practice of storing natural gas underground, which has been conducted for years, even in densely populated areas, without any incidents. Storing natural gas in subsurface storage sites during periods of low demand and withdrawing it during peak demand is a proven strategy for efficiently delivering energy to the market on a seasonal basis. According to information from the Energy Infrastructure Councilwebsite [84], the United States hosts approximately 400 active natural gas storage facilities across 30 different states. Underground storage facilities play a crucial role in the US energy supply, providing roughly

205 of the natural gas consumed during the 5-month winter season each year. There are three main types of underground storage sites used in the country: depleted natural gas or oil fields, which account for 80%; aquifers, at 10%; and salt formations, which are also 10%. Between 2002 and 2014, the capacity of operational underground natural gas storage in the United States increased by 18.2%, helping to ensure a reliable supply of natural gas when demand is highest. These storage facilities can collectively store or withdraw approximately 4 trillion cubic feet (Tcf) of natural gas to meet consumer demand.

The Earth's naturally occurring geology provides the oldest proof that large quantities of CO_2 can be safely and securely trapped underground for millions of years. CO_2 injection in the oil and gas reservoirs has been in operation for more than fifty years. Also, large-scale geologic storage in saline formations is understood and has been practiced since 1996, when CO_2 injection started in Norway at the Sleipner Field. The United States has been active in research in this area. It has been storing CO_2 in saline formations at various scales of injection since the early 2000s, notably via the US Department of Energy regional carbon sequestration projects.

Currently, there are stringent regulations and permitting processes for any new carbon storage projects. In 2010, the EPA developed UIC Class VI rules for wells used to inject CO_2 specifically for long-term geologic storage. These Class VI rules cover items such as CO_2 injection and site characterization, well permitting, well construction and operational standards, testing, plugging, recordkeeping, corrective action, emergency and remedial response, closure, and post-closure care, and associated financial assurance requirements [85].

The Congressional Budget Office report [86] reported that, given the uncertainty inherent in technological change, funding R&D programs to develop both CCUS and other clean energy technologies may be an appropriate diversification of public investment in ways to mitigate climate change. The widespread and timely deployment of CCUS on a gigaton scale is pivotal for achieving the emissions reduction targets outlined in the Paris Agreement and combating climate change. Moreover, emerging technologies like the green hydrogen industry will require CO_2 sequestration, further underlining the importance of CCUS.

It is important to assure the public that such events are less likely to happen under the current intense and lengthy permitting process. For example, in a recent study published in 2023, environmental researchers from Spain and various US universities evaluated the potential for basin-wide carbon dioxide leakage in aquifers with caprocks, where billions of tons of CO_2 are injected underground. Their findings indicate that even under worst-case scenarios involving rocks with numerous fractures, carbon dioxide would remain contained deep within the subsurface for millions of years [87].

Community engagement is essential for all CCUS projects to address safety concerns related to various aspects of implementation. One key concern is the location of carbon capture sites, especially when they are situated near residential areas. Additionally, the construction of CO_2 pipelines raises issues when they cross both public and private lands. Storing CO_2 on private property is another major consideration, affecting both pore space owners and mineral rights holders who may have concerns about long-term impacts and property rights.

Most people are not willing to sacrifice their convenience for the global good. The public needs to be aware of why CCUS is important in order to be comfortable with its implementation. Only then might they pay premium prices on decarbonization efforts as they do when buying organic foods. Transparent communication efforts can facilitate informed decision-making and garner public support for CCUS initiatives. The ultimate outcome for winning the public is building win-win strategies by providing near-term economic benefits to individual businesses, municipalities, or countries, and at the same time contributing toward meeting climate mitigation, adaptation, and other sustainability goals [88].

Public actions on emission reduction

Some of the best practices by the public on ways of reducing GHG emissions are discussed in Chapter 12. These include plastic recycling, waste reduction, and conscious consumerism.

Summary

CCUS and methane mitigation are increasingly important for meeting climate goals because they address emissions that are difficult or costly to eliminate through electrification or renewable energy alone. Heavy industries such as cement, steel, chemicals, and refining produce large volumes of process-related CO_2 that cannot be avoided without fundamentally changing production methods. CCUS enables these sectors to continue operating while significantly reducing their carbon footprint. It also supports energy security by allowing low-carbon use of existing infrastructure and fuels during the transition, and it can deliver negative emissions when combined with bioenergy or DAC. As a result, most credible net-zero pathways see CCUS as a necessary complement to renewables, efficiency, and electrification rather than a substitute for them. Unfortunately, cost and permitting hurdles are impeding the progress of such projects. Advanced technologies are needed to reduce cost, and financial incentives are needed to promote their development.

Key takeaways

- For a carbon-neutral future, CCUS technology is necessary for the continuous consumption of fossil fuels.
- Methane leak mitigation is as important as CCUS and is more cost-effective.
- Landfills need to have methane mitigation facilities.

- New advancements (technology, policies, etc.) are needed to make CCUS more economical.
- Community engagement is essential for all emission reduction projects (including CCUS) to address safety concerns related to various aspects of implementation.
- To win the public, operators need to build win-win strategies by providing near-term economic incentives to individuals, businesses, municipalities, or countries, and at the same time advance toward meeting climate mitigation, adaptation, and other sustainability goals.

References

[1] E Kolbert, The hazy days of summer; The New Yorker, 2023. https://www.newyorker.com/magazine/2023/07/03/the-hazy-days-of-summer (Accessed 20 January, 2026).

[2] U.S. Energy Information Administration. *Energy and the Environment Explained: Greenhouse Gases*; U.S. EIA: Washington, DC. https://www.eia.gov/energyexplained/energy-and-the-environment/greenhouse-gases.php (Accessed June 2024).

[3] U.S. Environmental Protection Agency. GHG reporting program (GHGRP): greenhouse gas emissions. https://www.epa.gov/ghgreporting/ghgrp-emissions-ghg (Accessed January 20, 2026).

[4] Secretariat of the Convention on Biological Diversity. *Convention on Biological Diversity*; United Nations: Montreal, Canada. https://www.cbd.int/doc/legal/cbd-en.pdf (Accessed January 20, 2026).

[5] Based on North Carolina State University, College of Natural Resources. Carbon footprint measurement and management: case study of the school forest enterprise. https://bioresources.cnr.ncsu.edu/resources/carbon-footprint-measurement-and-management-case-study-of-the-school-forest-enterprise/ (Accessed January 20, 2026).

[6] U.S. Environmental Protection Agency. Basics of climate change. https://www.epa.gov/climatechange-science/basics-climate-change (Accessed January 20, 2026).

[7] Intergovernmental Panel on Climate Change*Climate Change 2013: The Physical Science Basis*; Contribution of Working Group I to the Fifth Assessment Report of the Intergovernmental Panel on Climate Change, Cambridge University Press, Cambridge, U.K. and New York, NY, 2013 https://www.ipcc.ch/report/ar5/wg1/ (Accessed Jan 20, 2026).

[8] Intergovernmental Panel on Climate ChangeClimate Change 2014: Synthesis Report, IPCC, Geneva, Switzerland, 2014 https://www.ipcc.ch/report/ar5/syr/ (Accessed January 20, 2026).

[9] https://www.ncei.noaa.gov/access/monitoring/global-temperature-anomalies (Accessed March 8, 2025).

[10] E.J. Judd, J.E. Tierney, et al., A 485-million-year history of Earth's surface temperature, Science 385 (2024) eadk3705. https://doi.org/10.1126/science.adk3705.

[11] National Centers for Environmental Information (NCEI), NOAA. What are proxy data? https://www.ncdc.noaa.gov/news/what-are-proxy-data (Accessed January 20, 2026).

[12] Wikipedia. Oxygen isotope ratio cycle. https://en.wikipedia.org/wiki/Oxygen_isotope_ratio_cycle (Accessed January 20, 2026).

[13] Carbon Brief. Factcheck: what Greenland ice cores say about past and present climate change. https://www.carbonbrief.org/factcheck-what-greenland-ice-cores-say-about-past-and-present-climate-change/ (Accessed January 20, 2026).

[14] W.J. Ripple, C. Wolf, T.M. Lenton, J.W. Gregg, S.M. Natali, P.B. Duffy, J. Rockström, H.J. Schellnhuber, Many risky feedback loops amplify the need for climate action, One Earth 6 (2) (2023) 86–91. https://doi.org/10.1016/j.oneear.2023.01.004.
[15] E. Kolbert When the arctic melts: what the fate of greenland means for the rest of the Earth. The New Yorker, 2024. https://www.newyorker.com/magazine/2024/10/14/when-the-arctic-melts (Accessed January 20, 2026).
[16] Data from International Energy AgencyWorld Energy Outlook 2025, International Energy Agency, Paris, France, 2025 https://www.iea.org/reports/world-energy-outlook-2025 (Accessed January20, 2026).
[17] S. Pinkham A hotter Russia: review of Thane Gustafson's Klimat: Russia in the age of climate change; The New York Review of Books, 2022. https://www.nybooks.com/articles/2022/06/23/a-hotter-russia-klimat-thane-gustafson/(Accessed January 20, 2026).
[18] M. McGrath, M. Poynting, J. Rowlatt Climate change: World's oceans suffer from record-breaking year of heat. *BBC News—Climate & Science*, 2024. https://www.bbc.com/news/science-environment-68921215 (Accessed January 20, 2026).
[19] B. McKibben Hurricane Idalia's explosive power comes from abnormally hot oceans. The New Yorker 2023. https://www.newyorker.com/news/annals-of-a-warming-planet/hurricane-idalias-explosive-power-comes-from-abnormally-hot-oceans (Accessed January 20, 2026).
[20] J.R. Shi, B.D. Santer, Y.O. Kwon, et al., The Emerging Human Influence on the Seasonal Cycle of Sea Surface Temperature, Nat. Clim. Change 14 (2024) 364–372. https://doi.org/10.1038/s41558-024-01958-8.
[21] J.C. See Venter, D.E. Duncan, in: The Voyage of Sorcerer II: The Expedition That Unlocked the Secrets of the Ocean's Microbiome, Belknap Press of Harvard University Press, Cambridge, MA, 2023, p. 44.
[22] H Czerski, in: The Blue Machine: How the Ocean Works, W. W. Norton & Company: New York, NY, 2023, p. 464.
[23] International Energy Agency*Net Zero Roadmap: A Global Pathway to Keep the 1.5° C Goal in Reach* (2023 Update), International Energy Agency, Paris, 2023 https://www.iea.org/reports/net-zero-roadmap-a-global-pathway-to-keep-the-15-c-goal-in-reach (Accessed January 20, 2026).
[24] International Energy AgencyEnergy Technology Perspectives 2020, International Energy Agency, Paris, 2020 https://www.iea.org/reports/energy-technology-perspectives-2020 (Accessed January 20, 2026).
[25] International Energy AgencyNet Zero by 2050: A Roadmap for the Global Energy Sector, International Energy Agency, Paris, 2021 https://www.iea.org/reports/net-zero-by-2050 (Accessed January 20, 2026).
[26] National Petroleum Council. *Meeting the Dual Challenge: A Roadmap to At-Scale Deployment of Carbon Capture, Use, and Storage*; Chapter 6: CO_2 Capture (updated Mar 12, 2021). https://www.energy.gov/sites/default/files/2022-10/CCUS-Chap_6-030521.pdf (Accessed January 20, 2026).
[27] World Economic Forum. Carbon capture and storage is at a turning point. Here's Why. 2025. https://www.weforum.org/stories/2025/08/carbon-capture-storage-decarbonization-pivotal-moment/ (Accessed January 20, 2026).
[28] M. Fajardy, C. Greenfield, et al., CCUS Projects around the World Are Reaching New Milestones, *International Energy Agency* Commentary, 2025 https://www.iea.org/commentaries/ccus-projects-around-the-world-are-reaching-new-milestones (Accessed January 20, 2026).

[29] U.S. Department of Energy, National Energy Technology Laboratory. NATCARB/ATLAS. https://www.netl.doe.gov/coal/carbon-storage/strategic-program-support/natcarb-atlas (Accessed January 20, 2026).

[30] National Petroleum CouncilMeeting the Dual Challenge: A Roadmap to At-Scale Deployment of Carbon Capture, Use, and Storage; Chapter Six—CO2 Transport, National Petroleum Council, Washington, DC, 2019 https://www.energy.gov/sites/default/files/2022-10/CCUS-Chap_6-030521.pdf (Accessed January 20, 2026).

[31] E. Luna-Ortiz, Reusing existing infrastructure for CO2 transport: risks and opportunities, Presented at Offshore Technology Conference Asia (Virtual and Kuala Lumpur, Malaysia), 2022 Paper OTC-31457-MS https://doi.org/10.4043/31457-MS (Accessed January 20, 2026).

[32] National Petroleum Council, Chapter Nine—CO2 useMeeting the Dual Challenge: A Roadmap to At-Scale Deployment of Carbon Capture, Use, and Storage, National Petroleum Council, Washington, DC, 2019 https://www.energy.gov/sites/default/files/2022-10/CCUS-Chap_9-030521.pdf (Accessed January 20, 2026).

[33] National Petroleum CouncilMeeting the Dual Challenge: A Roadmap to At-Scale Deployment of Carbon Capture, Use, and Storage; Chapter Eight—CO2 Enhanced Oil Recovery, National Petroleum Council, Washington, DC, 2019 https://www.energy.gov/sites/default/files/2022-10/CCUS-Chap_8-030521.pdf (Accessed January 20, 2026).

[34] U.S. Department of Energy, Bioenergy Technologies Office. Rewiring Carbon Reduction; 2017. https://biomassboard.gov/sites/default/files/pdfs/tac_2017_q3_rowe.pdf (Accessed January 20, 2026).

[35] Hydrocarbon Processing. World's largest CO2-to-methanol plant starts production. *Hydrocarbon Processing* (online) 2022. https://www.hydrocarbonprocessing.com/news/2022/10/worlds-largest-co2-to-methanol-plant-starts-production (Accessed January 20, 2026).

[36] National Petroleum Council, Chapter Seven—CO2 geologic storage;Meeting the Dual Challenge: A Roadmap to At-Scale Deployment of Carbon Capture, Use, and Storage, National Petroleum Council, Washington, DC, 2019 https://www.energy.gov/sites/default/files/2022-10/CCUS-Chap_7-030521.pdf (Accessed January 20, 2026).

[37] M.R. Fassihi, J.-P. Blangy, Integrated Aquifer Characterization and Modeling for Energy Sustainability: Key Lessons from the Petroleum Industry, CRC Press, Boca Raton, FL, 2023 https://doi.org/10.1201/9781003272809.

[38] International Energy Agency (IEA)CO2 Storage Resources and Their Development: An IEA CCUS Handbook, IEA, Paris, 2022 https://iea.blob.core.windows.net/assets/42d294af-ce07-44c7-9c96-166f855088e8/CO2storageresourcesandtheirdevelopment-AnIEACCUSHandbook.pdf (Accessed January 20, 2026).

[39] P. Kelemen, S.M. Benson, H. Pilorgé, P. Psarras, J. Wilcox, An overview of the status and challenges of CO2 storage in minerals and geological formations, Front. Clim. 1 (9) (2019). https://doi.org/10.3389/fclim.2019.00009. (Accessed January 20, 2026).

[40] S.T. Brennan, R.C. Burruss, M.D. Merrill, P.A. Freeman, L.F. Ruppert, A Probabilistic Assessment Methodology for the Evaluation of Geologic Carbon Dioxide Storage; U.S. Geological Survey Open-File Report 2010–1127, U.S. Geological Survey, Reston, VA, 2010 https://pubs.usgs.gov/of/2010/1127/ofr2010-1127.pdf (Accessed January 20, 2026).

[41] U.S. Environmental Protection AgencyHow EPA's UIC Class VI Regulations Work to Ensure the Protection of Groundwater Resources, EPA, Washington, DC, 2023 https://www.epa.gov/system/files/documents/2023-04/EPA%20Pamphlet_How%20Class%20VI%20Regulations%20Ensure%20Groundwater%20Protection.pdf (Accessed January 20, 2026).

[42] Canary Media. Chart: heavy industry is the next big climate problem to tackle. Canary Media (online), 2023. https://www.canarymedia.com/articles/emissions-reduction/chart-heavy-industry-is-the-next-big-climate-problem (Accessed January 20, 2026).

[43] U.S. Department of EnergyPathways to Commercial Liftoff: Low-Carbon Cement, U.S. Department of Energy, Washington, DC, 2023 https://liftoff.energy.gov/wp-content/uploads/2023/12/20230921-Pathways-to-Commercial-Liftoff-Cement.pdf (Accessed January 20, 2026).

[44] Institute for Energy Economics and Financial Analysis (IEEFA)Boundary Dam 3 Coal Plant Achieves CO_2 Capture Goal Two Years Late, IEEFA, 2021 http://ieefa.org/wp-content/uploads/2021/04/Boundary-Dam-3-Coal-Plant-Achieves-CO2-Capture-Goal-Two-Years-Late_April-2021.pdf (Accessed January 20, 2026).

[45] Office of the Parliamentary Budget Officer (PBO)Climate Change: Implications for Canada's Economy, Government of Canada, Ottawa, ON, 2016 https://www.pbodpb.gc.ca/web/default/files/Documents/Reports/2016/ClimateChange/PBO_Climate_Change_EN.pdf (Accessed January 20, 2026).

[46] International Energy Agency (IEA). CCUS projects explorer. https://www.iea.org/data-and-statistics/data-tools/ccus-projects-explorer (Accessed January 20, 2026).

[47] The Climate Reality Project. 4 Big Climate Stories to Watch in 2024. Climate Reality Project (online), 2024. https://www.climaterealityproject.org/blog/4-big-climate-stories-watch-2024/ (Accessed January 20, 2026).

[48] J.C. Minx, W.F. Lamb, M.W. Callaghan, S. Fuss, J. Hilaire, F. Creutzig, T. Amann, T. Beringer, W. de Oliveira Garcia, J. Hartmann, T. Khanna, D. Lenzi, G. Luderer, G.F. Nemet, J. Rogelj, P. Smith, J.L. Vicente Vicente, J. Wilcox, M. Del Mar Zamora Domínguez, Negative emissions—Part 1: research landscape and synthesis, Environ. Res. Lett. 13 (2018) 063001. https://doi.org/10.1088/1748-9326/aabf9b.

[49] Office of Clean Energy Demonstrations, U.S. Department of Energy. Regional direct air capture hubs. https://www.energy.gov/oced/DACHubs (Accessed January 20, 2026).

[50] G. Realmonte, L. Drouet, A. Gambhir, J. Glynn, A. Hawkes, A.C. Köberle, M. Tavoni, An inter-model assessment of the role of direct air capture in deep mitigation pathways, Nat. Commun. 10 (2019) 3277. https://doi.org/10.1038/s41467-019-10842-5.

[51] International Energy Agency (IEA)Direct Air Capture 2022, IEA, Paris, 2022 https://www.iea.org/reports/direct-air-capture-2022 (Accessed January 20, 2026).

[52] Adapted from International Energy Agency (IEA)Direct Air Capture: A Key Technology for Net Zero, IEA, Paris, 2022 https://iea.blob.core.windows.net/assets/78633715-15c0-44e1-81df-41123c556d57/DirectAirCapture_Akeytechnologyfornetzero.pdf (Accessed January 20, 2026).

[53] M. Scott, T. Slavin Fossil-fuel industry embrace raises alarm bells over direct air capture. Reuters (online), 2023 (updated Oct 12, 2023). https://www.reuters.com/sustainability/climate-energy/fossil-fuel-industry-embrace-raises-alarm-bells-over-direct-air-capture-2023-10-10/ (Accessed January 20, 2026).

[54] *Science* (ScienceInsider). U.S. unveils plans for large facilities to capture carbon directly from air. 2023. https://www.science.org/content/article/us-unveils-plans-for-large-facilities-to-capture-carbon-directly-from-air (Accessed January 20, 2026).

[55] J. Valentine, A. Zoelle, S. Homsy, H. Mantripragada, M. Woods, N. Roy, A. Kilstofte, M. Sturdivan, M. Steutermann, T. Fout, Direct Air Capture Case Studies: Sorbent System, National Energy Technology Laboratory (NETL), Pittsburgh, PA; Morgantown, WV; Albany, OR, 2022 https://netl.doe.gov/projects/files/DirectAirCaptureCaseStudiesSorbentSystem_070822.pdf (Accessed January 20, 2026).

[56] Clean Air Task Force2022 Annual Report, Clean Air Task Force, Boston, MA, 2023 https://cdn.catf.us/wp-content/uploads/2023/10/03100558/2022-annual-report.pdf (Accessed January 20, 2026).

[57] E. Smith, J. Morris, H. Kheshgi, G. Teletzke, H. Herzog, S. Paltsev, The cost of CO_2 transport and storage in global integrated assessment modeling, Int. J. Greenh. Gas Control 109 (2021) 103367. https://doi.org/10.1016/j.ijggc.2021.103367.

[58] Energy Improvement and Extension Act of 2008, Pub. L. No. 110-343, div. B, § 115, 122 Stat. 3765, 3829 (2008).

[59] Inflation Reduction Act of 2022, H.R. 5376, 117th Cong. § 50161(g)(3) (2022).

[60] U.S. Department of EnergyPathways to Commercial Liftoff: Low-Carbon Cement, U.S. Department of Energy, Washington, DC, 2023 https://liftoff.energy.gov/wp-content/uploads/2023/09/20230921-Pathways-to-Commercial-Liftoff-Cement.pdf (Accessed January 20, 2026).

[61] National Petroleum CouncilGaffney-Cline CCS Cost Assessment Tool User Manual: Carbon Capture, Transport, and Storage Cost Assessment Tool, National Petroleum Council, Washington, DC, 2020 (User manual accompanying NPC CCUS cost assessment tool).

[62] IEACapture Carbon Capture Value, Oxford Institute for Energy Studies, Oxford, UK, 2024 https://www.oxfordenergy.org/wpcms/wp-content/uploads/2024/02/CM08-Capture-Carbon-Capture-Value_Final.pdf (Accessed January 20, 2026).

[63] K. Langlois, The Giving Trees: A Journey through Alaska's Tongass National Forest, the Greatest U.S. Natural Carbon Reserve, Sierra Club, 2022 *Sierra* (Summer https://www.sierraclub.org/sierra/2022-2-summer/feature/giving-trees-alaska-tongass-national-forest (Accessed January 20, 2026).

[64] U.S. Environmental Protection AgencyGreenhouse Gas Reporting Program (GHGRP) Data for Facility ID 1005661, EPA, Washington, DC, 2019 https://ghgdata.epa.gov/ghgp/service/html/2019?id=1005661&et=undefined (Accessed January 20, 2026).

[65] U.S. Environmental Protection Agency. *GHGRP Facility Detail for Facility ID 1005661, 2019 Reporting Year.* EPA: Washington, DC. https://ghgdata.epa.gov/ghgp/service/facilityDetail/2019?id=1005661&ds=E&et=&popup=true (Accessed January 20, 2026).

[66] OCEAN NETs. Ocean-based negative emission technologies.https://www.oceannets.eu/(Accessed January 20, 2026).

[67] K. De Pryck, M. Boettcher, The Rise, fall, and rebirth of ocean carbon sequestration as a climate "Solution", Glob. Environ. Change 85 (2024) 102820. https://doi.org/10.1016/j.gloenvcha.2024.102820.

[68] Massachusetts Institute of Technology (MIT) News. How much carbon does the ocean absorb? MIT News (online),2021. https://news.mit.edu/2021/how-much-carbon-ocean-absorbs-0405 (Accessed January 20, 2026).

[69] International Energy Agency (IEA)Strategies to Reduce Emissions from Coal Supply – Global Methane Tracker 2023 – Analysis, IEA, Paris, 2023 https://www.iea.org/reports/global-methane-tracker-2023 (Accessed January 20, 2026).

[70] International Energy Agency (IEA)Global Methane Tracker 2023, IEA, Paris, 2023 https://www.iea.org/reports/global-methane-tracker-2023 (Accessed January 20, 2026).

[71] World Bank. *Global Gas Flaring Data*; World Bank: Washington, DC. https://www.worldbank.org/en/programs/gasflaringreduction/global-flaring-data (Accessed January 20, 2026).

[72] G. Plant, E.A. Kort, A.R. Brandt, R. Duren, M. Omara, E. Fortner, M. Smith, S. Schwietzke, S. Hamburg, A. Townsend-Small, et al., Inefficient and unlit natural gas flares both emit

large quantities of methane, Science 377 (6614) (2022) 1566–1571. https://doi.org/10.1126/science.abq0385.

[73] International Energy Agency (IEA)Global Methane Tracker 2025, IEA, Paris, 2025 https://iea.blob.core.windows.net/assets/b83c32dd-fc1b-4917-96e9-8cd918801cbf/GlobalMethaneTracker2025.pdf (Accessed January 20, 2026).

[74] International Association of Oil & Gas Producers (IOGP)Recommended Practices for Methane Emissions Detection and Quantification Technologies – Upstream, IOGP, London, UK, 2025 https://www.ogci.com/wp-content/uploads/2025/04/IOGP_methane_practices.pdf (Accessed January 20, 2026).

[75] International Energy Agency (IEA)Advancing Methane Emissions Reductions by National Oil Companies, IEA, Paris, 2025 https://iea.blob.core.windows.net/assets/d65617f6-0c50-4fab-b255-afd86e3fea5a/AdvancingMethaneEmissionsReductionsbyNationalOilCompanies.pdf (Accessed January 20, 2026).

[76] International Energy Agency (IEA)Strategies to Reduce Emissions from Coal Supply — Global Methane Tracker 2023 — Analysis, IEA, Paris, 2023 https://www.iea.org/reports/global-methane-tracker-2023 (Accessed January 20, 2026).

[77] Climate and clean air coalition to reduce short-lived climate pollutants (CCAC). Global Methane Pledge. https://www.globalmethanepledge.org/(Accessed January 20, 2026).

[78] BBC FutureThe Search for the World's Largest Methane Sources, BBC Future, 2021 (online) https://www.bbc.co.uk/future/article/20210805-the-search-for-the-worlds-largest-methane-sources (Accessed January 20, 2026)*Image from NASA/JPL-Caltech*.

[79] International Energy Agency (IEA)Global Methane Tracker 2024, IEA, Paris, 2024 https://www.iea.org/reports/global-methane-tracker-2024 (Accessed January 20, 2026).

[80] International Renewable Energy Agency (IRENA)Innovation Outlook: Renewable Methanol 2021, IRENA, Abu Dhabi, 2021 https://www.methanol.org/wp-content/uploads/2020/04/IRENA_Innovation_Renewable_Methanol_2021.pdf (Accessed January 20, 2026).

[81] Bloomberg News. *Florida Startup Backed by Bill Gates Has Unique Methane Solution. Bloomberg* (online), Sep 29, 2023. https://www.bloomberg.com/news/articles/2023-09-29/florida-startup-backed-by-bill-gates-has-unique-methane-solution (Accessed January 20, 2026).

[82] International Energy Agency (IEA)Methane Abatement, IEA, Paris, 2025 https://www.iea.org/energy-system/fossil-fuels/methane-abatement (Accessed January 20, 2026).

[83] Environmental Defence CanadaLetter from Scientists, Academics, and Energy System Modellers on CCUS Tax Investment Credit, Environmental Defence, Toronto, ON, 2022 https://www.environmentaldefence.ca/wp-content/uploads/2024/01/Letter-from-Academics-re-CCUS-tax-investment-credit_January-2022.pdf (Accessed January 20, 2026).

[84] Energy Infrastructure. *Natural Gas Storage — Energy 101*; Energy Infrastructure: Washington, DC. https://www.energyinfrastructure.org/energy-101/natural-gas-storage (Accessed January 20, 2026).

[85] U.S. Environmental Protection Agency (EPA). *Class VI – Wells Used for Geologic Sequestration of Carbon Dioxide.* https://www.epa.gov/uic/class-vi-wells-used-geologic-sequestration-carbon-dioxide (Accessed January 20, 2026).

[86] Congressional Budget Office (CBO)Carbon Capture and Storage in the United States, CBO, Washington, DC, 2024 https://www.cbo.gov/publication/59832#footnote-013 (Accessed January 20, 2026).

[87] I.R. Kivi, R.Y. Makhnenko, C.M. Oldenburg, J. Rutqvist, V. Vilarrasa, Multi-layered systems for permanent geologic storage of CO_2 at the gigaton scale, Geophys. Res. Lett. 49 (2022) e2022GL100443. https://doi.org/10.1029/2022GL100443.
[88] J. Hinkel, et al., Transformative narratives for climate action, Climatic Change 160 (2020) 495–506. https://doi.org/10.1007/s10584-020-02761-y.

Further readings

Additional information on carbon utilization is provided in LeClerc et al., 2024, The CO2 Tree: The potential for carbon dioxide utilization pathways, https://doi.org/10.1021/acssuschemeng.4c07582

Basic information about landfill gas /US EPA, https://www.epa.gov/lmop/basic-information-about-landfill-gas#:⟨:text=Municipal%20solid%20waste%20(MSW)%20landfills,States%2C%20accounting%20for%20approximately%2014.4

NPC report, Dec. 2019, Meeting the dual challenge—A Roadmap to At-Scale Deployment of CARBON CAPTURE, USE, AND STORAGE, CHAPTER ONE – The Role of CCUS in the Future Energy Mix

Solid Waste Management – World Bank, https://www.worldbank.org/en/topic/urbandevelopment/brief/solid-waste-management#:⟨:text=In%202020%2C%20the%20world%20was,kilograms%20per%20person%20per%20day.

Chapter 9

Sustainable water, soil, and minerals

Synopsis

The story of human civilization is inseparable from the fate of Earth's finite resources. For millennia, water and soil—the fundamental elements of life—have sustained our existence. The fertile land that feeds us and the water that quenches our thirst have long been the cornerstones of human survival. Yet, as we navigate the challenges of the 21st century, the unsustainable exploitation of these vital resources threatens our future. Climate change, population growth, and increasing demand are pushing water and soil to a critical tipping point.

This chapter serves as a guide to a more sustainable future, where we balance the challenges and opportunities of managing water, soil, and mineral resources. We will explore innovative strategies for sustainable water management, from conservation techniques to the transformative potential of wastewater reclamation into a valuable, reusable asset.

Moving from water to the ground beneath our feet, we will delve into soil science—an intricate ecosystem that extends far beyond plant growth. Soil serves as a natural water filter, purifying groundwater as it moves through its layers. It is also a crucial carbon sink, storing vast amounts of carbon dioxide and playing a key role in mitigating climate change. Healthy soil supports biodiversity, forming the foundation of thriving ecosystems and sustaining the delicate web of life on our planet.

Our exploration will then shift to the realm of sustainable mineral extraction. Minerals are the building blocks of modern civilization, essential to everything

F.A.C.T.S. Flexibility, Affordability, Conservation, Transition, Sustainability: Our Path to the Future of Energy and the Environment. DOI: https://doi.org/10.1016/B978-0-443-52574-2.00008-2

from smartphones to wind turbines. However, their extraction often carries high environmental costs. We will examine sustainable mining practices and emerging technologies that aim to reduce ecological damage while ensuring responsible resource use.

Water sustainability

Water, the elixir of life, is a finite resource facing growing pressure from a multitude of factors. Scientific reports show that human-caused global warming is intensifying the hydrological cycle, increasing both drought risk in some areas and heavy precipitation (which can cause flooding) in others. The water-holding capacity of air increases by about 7% per 1°C warming, which leads to increased water vapor in the atmosphere, and this probably provides the biggest influence on precipitation [1]. Growing populations and urbanization increase water demand for domestic, industrial, and agricultural purposes. To navigate these challenges and ensure a sustainable future, we must embrace a multifaceted approach to water resource management.

The cornerstone of sustainable water management lies in conservation and efficiency. This involves implementing practices that reduce overall water demand. At the individual level, this encompasses simple actions like taking shorter showers, fixing leaky faucets, and utilizing water-efficient appliances. On a larger scale, industries can adopt water-saving processes and technologies in manufacturing. Traditional irrigation practices, responsible for a significant portion of global water consumption, can be optimized through precision agriculture techniques. These techniques utilize sensors and data analytics to deliver water exactly where and when crops need it, minimizing waste. Public awareness campaigns that educate consumers about the importance of water conservation and promote water-saving habits are also crucial [2–4].

A critical appraisal of freshwater resources

Freshwater, while constituting a seemingly vast expanse across the Earth's surface (encompassing roughly 71%), represents a mere 2.5% of the total global water volume [5]. Furthermore, a substantial portion of this freshwater is locked away in glaciers and polar ice caps, leaving a critically limited amount readily accessible for human consumption, agriculture, and industrial processes (Fig. 9.1) [6].

Global water withdrawals total approximately 4300 cubic kilometers (km^3) per year. Though this may seem vast, it obscures a deeper crisis, uneven distribution, and increasing demand that outpaced supply in many regions.

Agriculture accounts for about 70% of global withdrawals, making it the largest water-consuming sector. Irrigation, essential for food production in arid and semi-arid areas, is particularly water-intensive. Improving water use

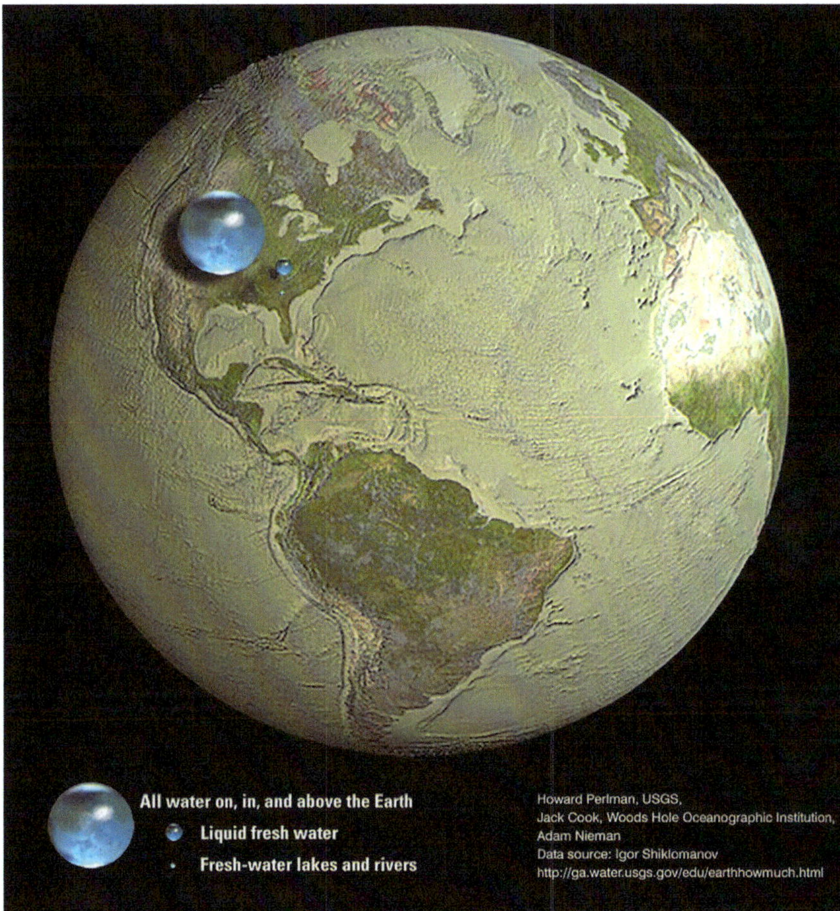

FIGURE 9.1 Water distribution in the world [6].

efficiency in agriculture is critical to reducing withdrawals while maintaining food security.

Industry, including manufacturing, energy production, and mining, is responsible for roughly 20% of global water use. Water is essential for cooling, production, and waste disposal. Reducing industrial freshwater consumption through conservation measures and alternative processes is key to fostering more sustainable operations.

Municipal use, covering drinking water, sanitation, and hygiene, represents around 10% of global withdrawals. Consumption varies based on climate and economic development, but ensuring universal access to clean water remains a major challenge for billions of people worldwide.

Freshwater challenges

Freshwater scarcity has significantly affected various industries, including agriculture. Fortunately, established and emerging technologies provide a diverse range of solutions to address this challenge (Box 9.1).

BOX 9.1 Tech note – Technologies for producing fresh water from brines

There are several common technologies for generating fresh water from brines, each with its own advantages and applications. Here are some common ones:

Reverse osmosis: Reverse osmosis (RO) is a widely used technology for desalination, including the treatment of brackish water and seawater. In RO, pressure is applied to the brine side of a semi-permeable membrane, causing water molecules to pass through the membrane while leaving behind dissolved salts and other contaminants. RO systems can produce high-quality fresh water suitable for various purposes, including drinking water, agriculture, and industrial processes.

Multi-stage flash distillation: Multi-stage flash distillation (MSF) is a thermal desalination process where brine is heated and successively flashed into steam at decreasing pressures, separating freshwater vapor from concentrated brine. It has been widely used in large-scale seawater desalination plants. However, due to its high energy consumption and significant waste production, MSF has gradually been replaced by RO, a trend that continues today.

Multi-effect distillation: Multi-effect distillation (MED) involves passing brine through multiple stages of heat exchangers, each at a lower pressure and temperature than the previous stage. As the brine moves through each stage, it releases water vapor, which is then condensed to produce fresh water. MED is a thermal desalination process known for its energy efficiency and is often used in combination with power plants or industrial processes that generate waste heat.

Electrodialysis: Electrodialysis (ED) utilizes an electric field to selectively transport ions through ion-exchange membranes, separating freshwater from the brine. In ED, positively charged ions migrate toward a negatively charged electrode (cathode), while negatively charged ions migrate toward a positively charged electrode (anode), leaving behind fresh water in between. ED is often used for brackish water desalination and can be combined with other processes for higher salinity brines.

Forward osmosis: Forward osmosis (FO) involves the natural diffusion of water through a semi-permeable membrane from a solution of higher osmotic pressure (brine) to one of lower osmotic pressure (a more concentrated solution or draw solution). In FO, a drawing solution with high osmotic pressure is used to draw water from the brine across the membrane, leaving behind a concentrated drawing solution and fresh water. FO is gaining interest as a desalination technology due to its potential for lower energy consumption and reduced fouling.

Membrane distillation: Membrane distillation (MD) utilizes a hydrophobic membrane to separate water vapor from a heated brine solution. In MD, the heated brine is brought into contact with one side of the membrane, while the other side is maintained at a lower pressure to allow water vapor to pass through. MD is particularly suitable for treating highly saline brines and can be powered by waste heat or renewable energy sources.

Desalination is a key technology for freshwater production, removing dissolved salts from seawater to make it suitable for consumption, agriculture, and industry. Although energy-intensive, advancements in efficiency and the integration of renewable energy sources like solar and wind power offer a more sustainable path forward. However, managing brine discharge remains critical to minimizing environmental impacts.

Water treatment facilities are essential for purifying freshwater sources such as rivers and lakes. Using filtration, coagulation, and disinfection, these facilities ensure water quality meets safety standards. Expanding and upgrading treatment infrastructure will be necessary, particularly in regions with aging systems or increasing demand.

Water reuse has become an effective way to reduce dependence on pristine freshwater. Advanced wastewater treatment technologies enable the safe reuse of treated water for irrigation, toilet flushing, and industrial applications. By enhancing treatment processes and encouraging responsible reuse practices, communities can establish a more sustainable water cycle and reduce pressure on freshwater resources. Some countries have divided their water supply into potable and non-potable water. The latter is used for outdoor irrigation and agriculture.

Atmospheric water generators provide a decentralized method of water production by capturing moisture from the air, even in arid environments. Though energy-intensive, improvements in efficiency and the integration of renewable energy sources could enhance their viability, especially in remote areas or disaster-stricken regions where conventional water sources are scarce.

The optimal technology for freshwater production depends on various factors, including location, water quality, energy availability, and cost. A one-size-fits-all approach may not be the most effective solution. For instance, desalination might be ideal for coastal areas with limited freshwater sources, while rainwater harvesting could be a suitable solution for regions with high rainfall. Incorporating a combination of established and emerging technologies may be the most effective approach in many regions. This necessitates careful planning, feasibility studies, and ongoing evaluation to ensure the chosen technologies align with specific needs and environmental considerations see Box 9.1 on the previous page.

All of the technologies mentioned previously offer a range of options for generating fresh water from brines, catering to different salinity levels, resource availability, and energy requirements. The choice of technology depends on factors such as the quality of the source water, the desired output water quality, energy availability, and economic considerations.

Water challenges in energy production

Water plays a crucial role in energy production, but its use presents significant challenges across various sectors. Thermal power plants, which dominate global electricity generation, depend heavily on water for cooling. Inefficient water use in these facilities depletes freshwater resources and contributes to thermal pollution. Optimizing water usage and adopting alternative cooling technologies are essential for achieving net zero emissions.

Hydropower, though a renewable energy source, can disrupt natural water flows and ecosystems. Sustainable water management is necessary to balance power generation with environmental protection. Minimizing dam footprints and implementing measures to maintain natural river flows are key considerations (please see the Klamath River Dam case study at the end of this chapter).

Bioenergy production, particularly the cultivation of biofuels, can be water-intensive, potentially worsening water scarcity in certain regions. Sustainable bioenergy practices involve using drought-resistant crops and wastewater for irrigation. Comprehensive life cycle assessments (LCAs) are critical to ensure that bioenergy production delivers net environmental benefits.

In the petroleum industry, water is an essential resource throughout its operational lifecycle. Drilling operations require drilling mud, which relies on water for cooling, lubricating, and maintaining wellbore stability. Depending on availability and geological compatibility, fresh, brackish, or saltwater can be used.

Hydraulic fracturing, or fracking, depends on fracking fluids—a mixture of water, sand, and specialized chemicals—pumped at high pressure to create fractures in rock formations, enhancing oil and gas flow. Water usage in fracking varies based on geological and well-specific factors but remains a significant component of the process.

Produced water in the oil industry: A challenge and potential opportunity

The petroleum industry generates vast amounts of produced water, with approximately 100 million m^3 surfacing daily alongside oil and gas in 2024. As the second largest global producer of wastewater after desalination plants, the industry faces a significant challenge due to the high salinity and complex mix of dissolved elements and contaminants in produced water, making it unsuitable for most immediate uses.

Currently, much of this wastewater remains unused, raising environmental and sustainability concerns. However, it also presents opportunities for resource recovery and reuse. Developing cost-effective and efficient treatment technologies could enable the recycling of produced water for use in other industries. Treated water could support irrigation in drought-prone regions or be utilized in industrial processes that can accommodate its unique composition, helping to reduce freshwater demand.

Beyond reuse, produced water holds potential as a resource for valuable mineral extraction. Advanced brine mining techniques could recover elements such as lithium, bromine, and magnesium, transforming wastewater into a source of critical materials. By shifting the perception of produced water from a waste product to a valuable resource, these approaches could contribute to both environmental sustainability and economic benefit.

More broadly, the search for unconventional source of critical minerals is expanding across a wide range of potential sources. These include deep-ocean mineral deposits, and even interplanetary resources such as asteroids. While interplanetary and deep-sea solutions may still be in their infancy, there's a potential goldmine that has been under our noses for decades: Oil and gas produced water. For every barrel of crude oil extracted, between 6 and 10 barrels of water are produced, amounting to a staggering 600–1000 million barrels of water daily, that is an incredible 160 billion liters every single day. This is not ordinary water. Over millions of years, this water has been in intimate contact with diverse minerals in the reservoir. As a result, it is mineral-rich, with total dissolved solids levels reaching a saline 37%, making it ten times saltier than seawater. Moreover, the periodic table finds a near-complete representation in this water; nearly every element can be found dissolved within, albeit in varying concentrations.

Surprisingly, this water, ripe with mineral potential, has been historically overlooked. Due to technological constraints and perhaps a lack of foresight, this resource has largely been disposed of back into reservoirs without extraction or treatment.

The advantages of tapping into this source are multifold. For one, the sheer volume of this water, available in copious amounts across many nations, ensures a substantial and steady supply. Furthermore, redirecting our focus from the far reaches of space or the deep ocean to this terrestrial resource could alleviate many environmental and economic concerns (solving the issue versus escaping from it).

However, realizing this vision requires concerted efforts. Scientists must innovate and develop novel extraction techniques suitable for this unique medium. Concurrently, federal agencies and the private sector need to champion and fund research in this direction, recognizing the potential impact on global mineral supply chains.

In summary, while the allure of outer space or the ocean's depths is undeniable, the path to securing our critical mineral future might very well lie in

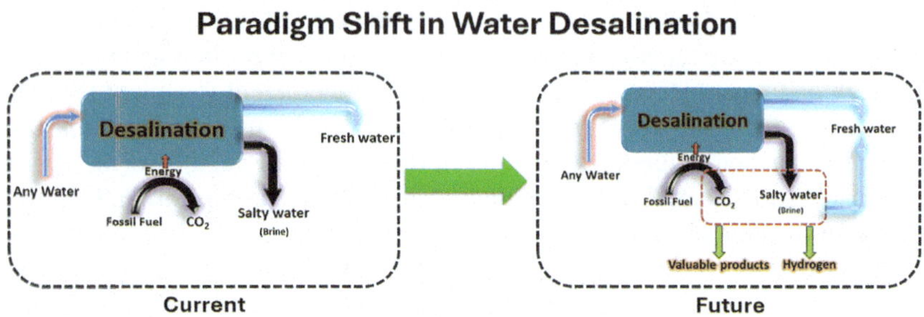

FIGURE 9.2 Current and desired future for water desalination.

the byproducts of our present. Before we reach for the stars or plunge into the ocean's abyss, it is time we tapped into the untapped oil and gas produced water.

Paradigm shift in water desalination

Water desalination holds the key to achieving equitable and sustainable water access, offering a promising solution to the global challenge of water scarcity. However, the path to establishing new desalination facilities is hindered by two critical challenges: (1) *the environmental impact of produced brine and* (2) *the emission of CO_2 from fossil fuel-based energy sources*. The approach taken to address these challenges is pivotal—recognizing them as untapped resources rather than mere waste holds the potential for transformative change through resource recovery. We can overcome economic, environmental, and social barriers through a convergence approach that addresses both brine disposal and CO_2 emissions and transforms them into value-added products, including critical minerals. This holistic strategy has the power to enhance sustainability, circular economy, and inclusivity, ultimately ensuring an ample supply of fresh water accessible to society as a whole (Fig. 9.2).

The concentrated byproduct of desalination, known as brine, frequently finds its way into oceans and water bodies, presenting a potential hazard to aquatic ecosystems due to its elevated salinity and modified chemical composition. Yet, a shift in perspective that views brine as a valuable resource, rather than mere waste, introduces a realm of innovative possibilities. One such avenue is brine mining, a concept that has the potential to address the environmental concerns tied to brine disposal. By harnessing methods that allow for the extraction of valuable minerals from brine, not only can its detrimental impact be mitigated, but economic opportunities can also be unlocked through the creation of marketable products. These products, often possessing a value surpassing that of water itself, promise to eventually enhance the accessibility and affordability of water resources.

A major obstacle to utilizing these f desalination procedures is that they are predominantly dependent on energy sources derived from fossil fuels, which results in the emission of CO_2 into the atmosphere, thereby amplifying the issue of climate change. While an immediate transition to sustainable energy alternatives like solar and wind power might not be feasible, converting CO_2 into value-added commodities is an alternative approach. This approach not only underscores the commitment of the desalination industry to responsible climate change mitigation but also leads to the production of significant, valuable salts.

The crux of this paradigm shift lies in harnessing the potential synergy between brine management and CO_2 mitigation. As we discussed in Part 1, the application of the 4Z Index in all aspects of energy and environment could help with optimization of the processes. By merging the challenges of brine disposal (zero liquid discharge) and CO_2 emissions (zero gas discharge), innovative processes like mineral extraction from brine (zero mineral discharge), CO_2 mineralization, and carbon capture and utilization can emerge. These synergistic approaches can yield value-added products, such as minerals for various industries and solid carbonates that sequester CO_2. This dual resource recovery not only enhances the economic viability of desalination but also bolsters its environmental and social acceptance. Moreover, more fresh water will be produced.

The proposed approach will address key economic, environmental, and social barriers to water sustainability. By recovering valuable minerals and other resources, it will generate new revenue streams that help offset the costs of desalination. Environmentally, the project will transform waste products into beneficial resources, reducing the harmful impact of brine discharge and mitigating CO_2 emissions. Socially, integrating resource recovery and carbon mitigation into desalination efforts aligns withenhancing public perception and acceptance of such initiatives.

Case example: Jubail seawater desalination plant, Saudi Arabia

To understand the concept of "circular" desalination, we can study Jubail 3A, a modern large-scale plant in Saudi Arabia. This facility produces about 600,000 m^3 of potable water per day, with reported energy performance around 2.8 kWh per m^3 of water. Using a representative grid emissions intensity, this electricity demand corresponds to roughly 0.3–0.4 million tons of CO_2 per year associated with power use.

On the brine side, seawater RO typically recovers on the order of 35%–50% of the feed as product water. At that range, a 600,000 m^3/day plant implies a concentrated stream on the order of 0.6–1.1 million m^3/day; using a mid-range recovery assumption, this yields about 900,000 m^3/day of brine returning to the sea.

If the concentrate stream is treated not as a disposal liability but as a feedstock for resource recovery and CO_2 utilization, the carbon story can change dramatically. Depending on the chosen chemistry and recovery pathways, CO_2 can be converted into stable solid products while also producing marketable salts

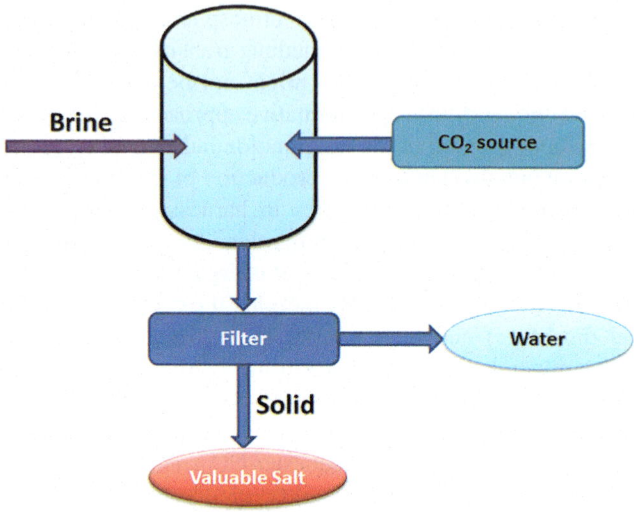

FIGURE 9.3 A general scheme for CO_2 desalination.

and minerals. Under an aggressive, integrated scheme, the CO_2 utilization can reach a multi-million-ton-per-year scale. For example, in a best-case scenario, it can approach about 17 million tons of CO_2 per year, meaning the site could, in principle, remove substantially more CO_2 than is associated with its own electricity use. Also, the system could produce on the order of 60 million tons per year of high-purity mineral products, recover roughly 300 million m^3 per year of additional fresh water, and generate annual mineral sales on the order of USD 20 billion, depending on product mix, achieved purity, and market prices.

These figures are presented at scale to show what is possible when brine management and CO_2 mitigation are merged together. Of course, the outcome depends on local seawater chemistry, recovery targets, energy sourcing, process configuration, permitting constraints, and commodity markets.

Brine refining or CO_2-driven desalination is indeed a novel, promising approach that addresses multiple challenges concurrently (Fig. 9.3).

CO_2-driven desalination presents a range of benefits that address water scarcity, global warming, critical mineral shortages, energy efficiency, and economic opportunities. By refining brine from sources such as seawater, brackish water, desalination reject, and produced water from oil, gas, or geothermal operations, this process provides fresh water from otherwise unusable sources. This is particularly valuable in arid regions where access to freshwater is limited (African Sahara).

Case study: Water rights

Water is a public resource essential for life, sustained by vast networks of dams and aqueducts. Water rights ownership grants legal entitlement to a specific amount of water. However, modern corporate farming seems to be operating much like an extraction industry—akin to oil drilling, mining, or fracking. While climate change plays a role in water shortages, an often-overlooked factor is water allocation. As droughts become more frequent worldwide, water rights grow increasingly valuable. Powerful agribusinesses control these rights, diverting water to non-essential crops at the expense of broader public needs. In the following, we discuss two cases where the sale of acquired water rights by one billionaire helped with access to water by locals. In another case, the water right acquisition by a monopoly appears to have resulted in extensive damage to the environment.

Texas case: T. Boone Pickens, the billionaire oil investor, foresaw the growing importance of water as a critical resource. Through his company, Mesa Water, he spent over a decade acquiring water rights in the Texas Panhandle, amassing control over 200,000 acre-feet of land atop the Ogallala Aquifer.

The Ogallala Aquifer is one of the world's largest, spanning approximately 174,000 square miles (450,000 km^2) across eight U.S. states—South Dakota, Nebraska, Wyoming, Colorado, Kansas, Oklahoma, New Mexico, and Texas (Fig. 9.4). It consists of a shallow water table surrounded by sand, silt, clay, and gravel, with a saturated thickness ranging from just a few feet to over 1000 feet, reaching depths of up to 1200 feet (300 m) [7]. Mesa Water controlled approximately 65 billion gallons of water annually—enough to supply the yearly needs of millions. In fact, Pickens effectively owned more water than any other individual in the United States.

By 2011, the Texas Panhandle and South Plains regions were experiencing their worst drought in decades, with nearly all areas at the highest levels of dryness. After months of negotiations, Pickens sold his water rights to a regional water supplier, securing water access for approximately 443,000 acres in the Texas Panhandle. This acquisition ensured a stable water supply for industrial and municipal use across 11 communities, projected to last well beyond the year 2100 [9].

California case: The Wonderful Company owns 175,000 acres of farmland in California, which, according to Forbes, uses around 150 billion gallons of water annually [10]. The company has a 57.7% stake in Kern Water Bank, a privately owned groundwater storage reservoir in the San Joaquin Valley, which can store up to 1.5 million acre-feet (489 billion gallons) of water [11]. California uses a long-term average of 40 million acre-feet of water each year. In 2025, there was a massive wildfire in Los Angeles. Some critics claimed (although unproven) that while firefighters struggled to get water from street hydrants, the company did not release their water to the city and hindered access to it. They also mentioned that many small farmers in the San Joaquin Valley are suffering because of a lack of access to the Kern Water Bank.

The 2025 documentary *Pistachio Wars*, directed by Rowan Wernham and Yasha Levine, exposed how the aforementioned company that monopolizes pistachio production also dominates water access, leaving others without. The film reveals how corporate agriculture exacerbates droughts in neighboring regions

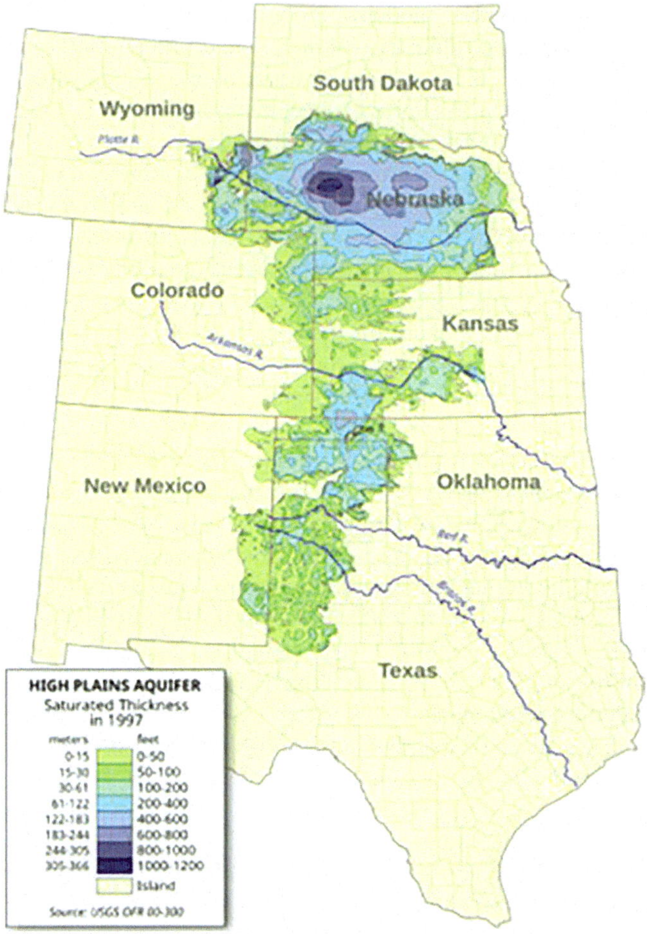

FIGURE 9.4 Saturated thickness of the Ogallala Aquifer in 1997 [8].

through its unchecked control of water sources. Beyond this, *Pistachio Wars* explores the intersection of climate change, overdevelopment, and corporate opportunism, ultimately shedding light on the complex ecological, economic, and social consequences of industrial farming in California. The Wonderful Company has denied the documentary allegations.

Ensuring fair and responsible water rights is essential for a stable future. Governments and local authorities must implement policies that protect public access, prevent large water monopolies, prioritize essential needs, and prevent overexploitation. Sustainable water management will require innovation, investment in infrastructure, and cooperation among stakeholders to secure this precious resource for future generations.

Sustainable wastewater management

Wastewater, a byproduct of human activity, poses a significant environmental challenge when not properly managed. Globally, approximately 1.3 trillion m^3 are produced each year, a figure expected to rise with population growth and urbanization. A large portion remains untreated, particularly in developing regions, threatening both ecosystems and public health.

Sources of wastewater include domestic activities such as household waste from showers, toilets, sinks, and laundry; industrial processes generating effluents requiring specialized treatment; and agricultural runoff carrying pollutants like fertilizers and pesticides. If left untreated, wastewater can contaminate freshwater supplies, harm aquatic life, and spread diseases.

Treatment typically involves multiple stages to improve water quality. Primary treatment removes large solids through screening and sedimentation. Secondary treatment relies on microorganisms to break down organic pollutants, while tertiary treatment further enhances water quality through filtration, disinfection, and nutrient removal, making it suitable for various applications.

Managing wastewater effectively requires a coordinated effort among governments, industries, and individuals. Minimizing waste at the source by modifying industrial processes and promoting water conservation reduces treatment demands and environmental impact. Optimizing plant operations and using renewable energy sources like solar or biogas can cut energy consumption and greenhouse gas emissions.

Resource recovery offers additional benefits by extracting valuable materials such as phosphorus and minerals from treated wastewater, supporting circular economy principles. Reusing treated water for irrigation and industrial purposes decreases reliance on freshwater, ensuring its availability for essential needs. By adopting these strategies, wastewater management can shift from an environmental burden to a sustainable resource solution.

Soil health and responsible mineral extraction

The pursuit of net zero emissions extends beyond energy production to encompass the very foundation of life: soil. Achieving sustainability requires attention to soil health and responsible mineral extraction, both of which play critical roles in mitigating climate change and ensuring long-term environmental stability.

Healthy soils serve as a vast carbon sink, absorbing atmospheric CO_2 and storing it as organic matter. However, this sequestration capacity has been severely compromised by intensive agriculture, deforestation, and soil degradation. Disrupted carbon cycles have led to increased CO_2 emissions, with degraded soils worldwide releasing billions of tons annually. Addressing this challenge requires a shift toward regenerative agricultural practices that restore soil health while enhancing productivity. Techniques such as cover cropping, reduced tillage, and compost application improve carbon storage, soil fertility, and water retention. By fostering diverse microbial ecosystems, these methods reduce reliance on synthetic fertilizers, cutting emissions associated with their production and transport. Integrating livestock into farming systems further enriches soil through manure-based nutrient recycling, yet widespread adoption demands dietary shifts, particularly a reduction in beef consumption. Given that the U.S. beef industry oversees over 93.6 million cows and calves, transitioning to regenerative grazing would require more than double the current farmland allocated to animal agriculture.

Restoring peatlands is another effective carbon sequestration strategy. For example, a project at the Pocosin Lakes National Wildlife Refuge in North Carolina has successfully stored 70 million pounds of carbon and 2.2 million pounds of nitrogen annually by restoring degraded peatlands.

Bioenergy can contribute to the net zero transition when paired with sustainable soil management. Techniques such as no-till farming (in which crops are grown without plowing or turning the soil between planting seasons, and instead of tilling, seeds are placed directly into undisturbed soil, often through the residue of the previous crop) and responsible use of crop residues help preserve soil health while maintaining its carbon storage potential. Additionally, prioritizing bioenergy feedstocks with low carbon footprints—such as fast-growing perennial grasses or energy crops cultivated on marginal lands—further reduces environmental impact.

The shift to clean energy also requires an increased extraction of copper, nickel, lithium, and rare Earth elements, all essential for batteries, wind turbines, and solar panels. However, irresponsible mining practices threaten soil health, contribute to deforestation, and pollute water sources. To ensure a sustainable transition, the approach to mineral extraction must evolve, incorporating responsible and environmentally conscious practices that minimize ecological damage while securing the materials needed for a renewable future.

Mining tailings dams

The mining industry produces enormous volumes of waste. The amount of waste produced is several thousand million tons per year [12]. The chief waste stream is tailings, which are often stored in impoundments behind dams (see Fig. 9.5). These tailings consist of finely ground rock particles and water left over after the valuable minerals or metals have been extracted from the ore. The primary purpose of tailing dams is to safely store the waste materials, known as tailings, produced during mining operations. This prevents the tailings from being released into the environment in an uncontrolled manner. Mining processes often use chemicals like cyanide, sulfuric acid, and others, which can contaminate the tailings. Tailings dams help isolate these hazardous materials, preventing them from polluting soil, groundwater, and surface water. They also store water used during the mineral processing stage. This water can sometimes be recycled back into the mining operation, reducing the need for fresh water.

FIGURE 9.5 Mining area at Berkeley, Montana, USA. Gray tailings are stored in the tailings pond shown to the left of the dark-colored Berkeley Pit [13].

Tailings dams are often built to last for decades, providing a long-term solution for the management of mining waste even after mining operations have ceased.

Globally, it is estimated that there are more than 18,000 tailings dams, with hundreds built annually as mining operations expand. The size and scale of these dams vary significantly, with some being as large as several square kilometers and towering hundreds of meters high.

The volume of water and slurry stored in these dams can be staggering. For instance, tailings facilities in large-scale operations often hold billions of cubic meters of water mixed with fine sediment. This water is typically contaminated with heavy metals, chemicals used in mineral processing (e.g., cyanide and sulfuric acid), and other harmful substances. The increasing scale of modern mining operations has led to larger dams, raising concerns about their structural stability and environmental impact. If improperly managed or in the event of a dam failure, these pollutants can seep into the surrounding groundwater or nearby rivers and lakes, causing long-term ecological damage. This could contaminate the soil, making the surrounding land unsuitable for agriculture or vegetation growth. Acid mine drainage—a process where sulfide minerals in tailings react with air and water to produce sulfuric acid—further exacerbates soil and water pollution.

Tailings dams are commonly constructed from readily available local materials, rather than the concrete used in water retention dams. Although the initial dike is commonly made of locally derived soil, waste rock, and the tailings themselves are often used in construction. As mining activities continue, intermediate retaining embankments are constructed and then raised. Thus, these dams could become unstable under excessive loading. Addressing the risks posed by the tailings dams requires a combination of proactive management, innovative technology, and robust regulation.

Tailings dam failures are among the most catastrophic industrial accidents [14]. Failures release enormous amounts of slurry into surrounding areas, polluting water systems and destroying ecosystems. Since 1915, a total of 257 failures have been recorded with circa 2650 fatalities and 250 million m^3 of contaminated residues released to the environment. Noticeable incidents include a major tailings dam failure that occurred at the Fundao dam (Samarco, Brazil, 2015), where overloading and poor structural management caused a dam failure, which killed 20 people and released 43 million m^3 of tailings and contaminated 700 km of the Doce River, reaching the Atlantic Ocean. The estimated fines were $47 billion for clean-up and reparations for several deaths.

Three years after the Samarco tragedy, another mining dam consisting of iron ore tailings failed catastrophically in Brumadinho, Brazil [15]. More than 11 million m^3 of toxic waste—equivalent to a 10-m high wave of mud that spread 10 km downhill – reached the Paraopeba River and caused the death of 270 people.

The frequent occurrence of such disasters calls for a major revision in the construction of future tailings dams. Modern tailings dams incorporate advanced engineering techniques to enhance stability. Continuous monitoring systems, including sensors and drones, are used to detect early signs of structural weakness or leakage. Instead of storing slurry, dry-stack tailings involve dewatering the waste to a solid form before storage. This method significantly reduces the risk of dam failure and water contamination. Decommissioned tailing dams can be capped and covered with soil to prevent erosion and water infiltration. Vegetation can be introduced to stabilize the area and restore ecosystems. Governments and international organizations are implementing stricter regulations for tailing dam construction, monitoring, and maintenance. For example, the Global Industry Standard on Tailings Management was introduced in 2020 to establish best practices for managing tailings facilities.

Mining companies are increasingly being held accountable for the social and environmental impacts of their operations. Engaging local communities in monitoring and decision-making processes helps ensure transparency and builds trust.

Responsible mining

Conducting comprehensive LCAs is essential for evaluating the environmental impact of mineral extraction across the entire supply chain. This process accounts for energy consumption, water usage, and waste generation associated with exploration, mining, processing, transportation, and end-of-life management. By fostering transparency, LCAs enable stakeholders to make informed decisions and prioritize mining operations with minimal environmental footprints.

Minimizing the environmental impact of mineral extraction requires the adoption of innovative techniques that reduce land disturbance, water consumption, and energy use. Precision drilling, the utilization of recycled water in processing operations, and the integration of renewable energy sources into mining activities play a crucial role in sustainable extraction. Advances in ore processing can further lower water and chemical requirements, reducing overall ecological harm.

Reclamation and restoration efforts are vital to rehabilitating post-mining landscapes and ensuring they return to a productive state. Effective restoration plans should focus on rebuilding soil health, re-establishing native vegetation, and promoting biodiversity recovery. By implementing these strategies, long-term environmental damage can be minimized, and ecosystems can regenerate, providing essential services such as water purification, flood control, and habitat creation.

Establishing ethical sourcing practices is fundamental to ensuring that mining operations uphold human rights, fair labor conditions, and community engagement. Preventing child labor and safeguarding worker safety must remain top priorities. Engaging with local communities throughout the mining process—from exploration to reclamation—enhances transparency, mitigates social conflicts, and fosters mining projects that contribute positively to local development.

Achieving a sustainable future requires a holistic approach to mineral extraction that goes beyond individual mining operations. Implementing circular economy principles by promoting recycling, remanufacturing, and extending product lifespans can significantly reduce the demand for virgin materials. Developing efficient battery recycling technologies is crucial to ensuring a sustainable, clean energy future. Establishing robust recycling infrastructure and creating economic incentives for battery collection and reprocessing can minimize reliance on virgin mineral extraction while reducing the environmental footprint of the clean energy sector.

Investing in exploration for new mineral deposits and unconventional resources is essential to minimizing environmental impact. This includes identifying untapped resources outside ecologically sensitive areas and advancing techniques such as brine mining. Additionally, fostering innovation in battery chemistry and materials science can reduce dependence on specific critical minerals. Research into alternative battery chemistries that utilize more abundant and less environmentally hazardous elements holds great promise for the long-term sustainability of clean energy technologies.

Summary

The path to net zero emissions is deeply intertwined with the health of our water, soil, and mineral resources. The relationship between water and various clean energy technologies underscores the need for responsible water stewardship.

Solar energy, while praised for its low water footprint compared to fossil fuels, still relies on water throughout its lifecycle. Panel cleaning is essential for optimal performance, requiring sustainable water sourcing practices. Additionally, the extraction and processing of raw materials for solar panels can be water-intensive. Implementing closed-loop water systems in manufacturing and exploring alternative cleaning methods, such as robotics or specialized cleaning solutions, can enhance the overall sustainability of solar energy.

Geothermal energy harnesses the Earth's heat to generate electricity, offering a renewable and clean power source. However, some geothermal plants require water for steam generation, making responsible water use and reinjection practices essential. Reinjecting extracted water back into the geothermal reservoir helps conserve resources and maintain pressure for long-term sustainability. Advancements in closed-loop geothermal technologies also hold promise for further reducing water consumption.

Hydrogen fuel presents exciting opportunities for decarbonizing transportation, particularly for heavy-duty vehicles and long-distance travel. Many hydrogen production processes depend on water, with electrolysis splitting water molecules into hydrogen and oxygen. Optimizing water use throughout the production chain is crucial, including using renewable energy sources to power electrolysis and reducing environmental impacts. Alternative production methods, such as photocatalytic water splitting using sunlight, offer additional potential for minimizing hydrogen's water footprint.

By restoring and preserving soil ecosystems, advancing regenerative agricultural practices, and adopting responsible mineral extraction techniques, we can harmonize clean energy development with environmental sustainability. Achieving this vision requires a collaborative effort that brings together governments, industry leaders, scientists, and communities in a shared commitment to sustainable resource management.

Key takeaways

- Freshwater scarcity will be a major issue for many industries in the future. This will make wastewater remediation a top activity for these industries.
- In many locations, the carbon sequestration capacity of soil has been severely compromised by intensive agriculture, deforestation, and soil degradation. Practices that can rejuvenate soil health need to be encouraged.
- Tailings dams could pose major environmental risks and need to be monitored closely.
- Establishing ethical sourcing practices is fundamental to ensuring that mining operations uphold human rights, fair labor conditions, and community engagement.
- Integration of circular economy principles, ecosystem restoration, and technological advancements is necessary for balancing economic growth with ecological resilience.
- Soil and minerals are fundamental building blocks for a clean future.

References

[1] K.E. Trenberth, The impact of climate change and variability on heavy precipitation, floods, and droughts, in: M.G. Anderson (Ed.), Encyclopedia of Hydrological Sciences, John Wiley & Sons, Ltd., Chichester, U.K., 2008 https://doi.org/10.1002/0470848944.hsa211.

[2] J.A.A. Jones, Water Sustainability: A Global Perspective (2010) 464.

[3] P.D. Raskin, E. Hansen, R.M. Margolis, Water and sustainability: global patterns and long-range problems, Nat. Resour. Forum 20 (1) (1996) 1–15. https://doi.org/10.1111/j.1477-8947.1996.tb00629.x.

[4] M. Konar, T.P. Evans, M. Levy, C.A. Scott, T.J. Troy, C.J. Vörösmarty, M. Sivapalan, Water resources sustainability in a globalizing world: who uses the water? Hydrol. Process. 30 (18) (2016) 3330–3336. https://doi.org/10.1002/hyp.10843.

[5] I.A. Shiklomanov, Appraisal and assessment of world water resources, Water Int. 25 (1) (2000) 11–32. https://doi.org/10.1080/02508060008686794.

[6] U.S. Geological Survey (USGS). How much water is there on Earth? https://www.usgs.gov/special-topics/water-science-school/science/how-much-water-there-earth (Accessed 20 Jan 2026).

[7] V. L. McGuire Changes in water levels and storage in the high plains aquifer, predevelopment to 2009; fact sheet 2011–3069; U.S. Geological Survey: Reston, VA, 2011 https://pubs.usgs.gov/fs/2011/3069/pdf/FS2011-3069.pdf (Accessed 19 Jan 2026).

[8] B. C. Fischer, K. M. Kollasch, V. L. McGuire Digital map of the saturated thickness of the high plains aquifer in parts of Kansas, Nebraska, New Mexico, Oklahoma, South Dakota, Texas and Wyoming, 1996–1997; U.S. geological survey data release, 2000. https://doi.org/10.5066/P9VNUGYE (Accessed 19 Jan 2026). https://www.sciencebase.gov/catalog/item/631405cdd34e36012efa3338.

[9] The Oklahoman. T. Boone Pickens sells water rights to Texas water supplier for $103 million. 2011, https://www.oklahoman.com/story/business/2011/06/24/t-boone-pickens-sells-water-rights-to-texas-water-supplier-for-103-million/61154706007/ (Accessed 20 Jan 2026).

[10] C. Sorvino Amid drought, billionaires control a critical california water bank. Forbes 2021, https://www.forbes.com/sites/chloesorvino/2021/09/20/amid-drought-billionaires-control-a-critical-california-water-bank/ (Accessed 20 Jan 2026).

[11] C-WIN. Fact check: the resnicks do not own most of California's water. 2025, https://www.c-win.org/blog/2025/1/27/fact-check-the-resnicks-do-not-own-most-of-californias-water (Accessed 20 Jan 2026).

[12] D. Kossoff, W.E. Dubbin, M. Alfredsson, S.J. Edwards, M.G. Macklin, K.A. Hudson-Edwards, Mine tailings dams: characteristics, failure, environmental impacts, and remediation, Appl. Geochem. 51 (2014) 229–245. https://doi.org/10.1016/j.apgeochem.2014.09.010.

[13] NASA Earth Observatory. Image (courtesy of NASA Earth Observatory). (Accessed 20 Jan 2026).

[14] ailings.info. Tailings (Basics). https://www.tailings.info/basics/tailings.htm (Accessed 20 Jan 2026).

[15] L.H.S. Rotta, E. Alcântara, E. Park, R.G. Negri, Y.N. Lin, N. Bernardo, et al., The 2019 brumadinho tailings dam collapse: possible cause and impacts of the worst human and environmental disaster in Brazil, Int. J. Appl. Earth Obs. Geoinf. (2020). https://doi.org/10.1016/j.jag.2020.102119.

Part V

Solutions for reaching sustainability

Synopsis

The transition to a sustainable energy future is an urgent and complex challenge that requires a multifaceted approach. In Part V, we explore the short-term and long-term solutions that can drive this transformation, focusing on the importance of policy, innovation, and cooperation. This overview of the strategies needed to achieve a sustainable energy future will empower readers with the knowledge and tools to become active participants in this critical transformation. Chapter 10 focuses on strategies for increasing energy efficiency. Incentivizing projects that are aligned with reducing the carbon footprint is discussed in Chapter 11. Finally, in Chapter 12, we look at how members of the public can advocate for an energy transition that will reduce the impact of climate change. By setting ambitious targets for renewable energy, reducing greenhouse gas emissions, and improving energy efficiency, we can create a roadmap for a cleaner, more sustainable future.

Chapter 10

Energy efficiency and smart grids

Synopsis

This chapter delves into a range of solutions that can accelerate the journey toward a sustainable energy supply. Energy efficiency, the practice of using less energy to achieve the same level of output, is a cornerstone of a sustainable energy future. Also, with the increased use of renewable decentralized energy systems, the majority of energy provision is shifting to the distribution grids. Improving energy efficiency and embracing smart grid technologies are discussed in this chapter.

Increasing energy efficiency

It is generally believed that increasing efficiency reduces energy demand. According to the International Energy Agency (IEA) study, a 10% improvement in efficiency is equivalent to saving 5 Gt CO_2 per year. The IEA's net zero emission scenario requires raising the rate of improvement in the global energy intensity from today's 2% to 4% by 2030. To achieve a doubling of energy intensity improvement, annual investment needs to triple within a little more than five years. Decarbonizing the power and fuel sectors is crucial, but the largest emissions come from end uses by consumers, and efficiency improvements are pivotal in curbing fossil fuel demand. This section explores

F.A.C.T.S. Flexibility, Affordability, Conservation, Transition, Sustainability: Our Path to the Future of Energy and the Environment. DOI: https://doi.org/10.1016/B978-0-443-52574-2.00003-3

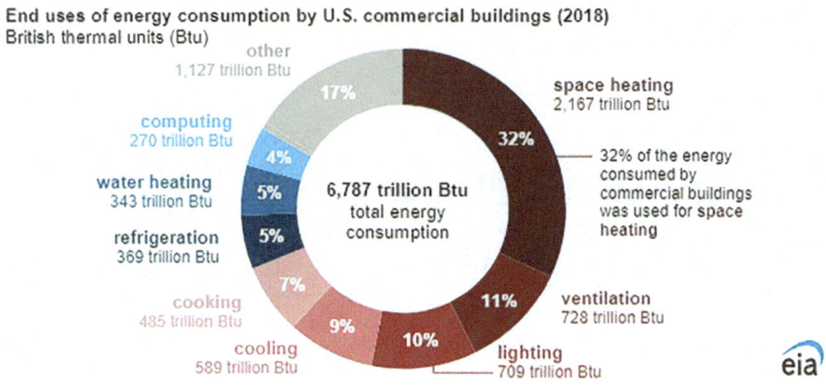

FIGURE 10.1 U.S. commercial building consumption in 2018 [1].

strategies for improving energy efficiency across various sectors, including buildings, industries, and transportation.

Building efficiency

Buildings account for a significant portion of energy consumption, making them a prime target for efficiency improvements (Fig. 10.1). Space heating takes the largest share of energy consumption, followed by ventilation and lighting.

Enacting and enforcing stringent energy codes for new construction can ensure that buildings are designed and built to minimize energy consumption. These codes can specify requirements for insulation, windows, HVAC systems, and lighting, promoting efficient practices.

Existing buildings can be retrofitted by incorporating energy-efficient features, which can significantly reduce their energy consumption. This includes measures like insulating walls and ceilings, upgrading windows, installing energy-efficient lighting and HVAC systems, and sealing air leaks.

Also, replacing older, less efficient appliances with newer, energy-star rated models can reduce energy consumption in homes and businesses. This includes refrigerators, dishwashers, washing machines, and other appliances. Of course, implementing smart home technologies can enable automated control of lighting, heating, cooling, and other systems, optimizing energy use based on occupancy and real-time conditions.

Industrial efficiency

Industries are significant energy consumers, and there is ample room for improving their energy efficiency. Primary action plans include replacing inefficient motors and drives with high-efficiency models. Identifying and eliminating energy-intensive processes or inefficiencies can lead to substantial energy savings. This can involve streamlining production lines, improving heat recovery systems, and reducing waste heat generation.

Combining heat and power generation (cogeneration) can improve overall energy efficiency. Cogeneration systems capture waste heat from industrial processes and use it to generate electricity or provide heating, reducing the overall energy demand. For example, in a Paper Pulp Mill, biomass is used as fuel. The generated electricity runs the Mill. Recovered heat is used to dry paper and heat pulp digesters. Also, implementing advanced energy management systems (EMS) (using AI) can help industries monitor and optimize their energy consumption, identify areas for improvement, and track energy savings.

Transportation efficiency

Transportation is another major energy-consuming sector. Encouraging the adoption of electric vehicles (EVs) can significantly reduce transportation-related emissions. Governments can play a crucial role in promoting EVs through incentives, infrastructure development, and raising public awareness. Global EV sales have been improving in the past few years, and the EV share of the market is around 24% (see Fig. 10.2).

Implementing and enforcing stricter fuel efficiency standards for conventional vehicles can help reduce energy consumption and emissions. Even better, promoting the use of public transportation can reduce the number of individual vehicles on the road, leading to much greater energy savings. Practices such as ride sharing, walking, cycling, and other active transportation modes can reduce reliance on motorized vehicles and improve overall public health.

Opposing sides of increasing energy efficiency

Energy efficiency plays a crucial role in reducing overall energy demand [3]. By using less energy to achieve the same level of output, we can lessen the strain on the electricity grid, reducing the need for additional power generation capacity and infrastructure [4]. This can help stabilize energy prices, improve grid reliability, and defer the need for expensive new power plants. However, according to Jevons' Paradox, improvements in resource efficiency could surprisingly lead to increased resource consumption. Historically, the Jevons Paradox was observed by William Stanley Jevons in the 19th century regarding coal use in England. He noted that the development of more efficient steam engines actually led to an increase in coal consumption, not a decrease, because the lower cost of using coal encouraged its broader adoption across various industries. When technology or processes become more efficient at utilizing a resource, the cost of using that resource for a specific purpose decreases. This lower cost makes the resource more appealing to use. For example, if cars become more fuel-efficient, the cost of driving a certain distance is reduced, potentially encouraging people to drive

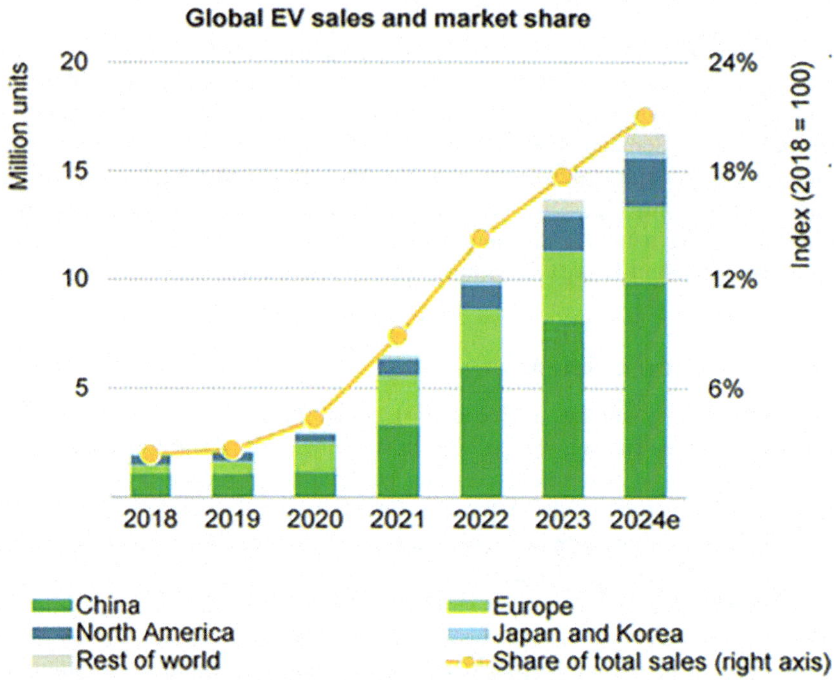

FIGURE 10.2 Global EV sales and market share [2]. *EV*, Electric vehicle.

more. This increased attractiveness can then lead to a rise in the overall demand for the resource. People might begin using it in new ways or for purposes they could not previously afford. Ultimately, this increase in demand can sometimes outweigh the initial savings from the efficiency gains, resulting in a net increase in the total consumption of the resource. For example, Fig. 10.3 demonstrates that both the global GDP and energy consumption have been increasing during 1990–2025. However, the ratio of energy growth to GDP has been decreasing.

This paradox has significant implications for sustainability efforts, suggesting that simply focusing on improving efficiency might not be sufficient to reduce resource consumption. Other measures, such as policies that promote conservation or limit overall demand, are necessary. The Jevons Paradox highlights the complex interplay between efficiency, cost, and demand, indicating that technological advancements alone may not solve resource depletion or environmental problems, and emphasizing the need for a comprehensive approach to sustainability that considers both efficiency and consumption patterns.

For businesses, improved energy efficiency can enhance competitiveness by reducing operating costs and increasing profitability [6]. This can attract investment, create new jobs [7], and stimulate economic activity. Lower energy

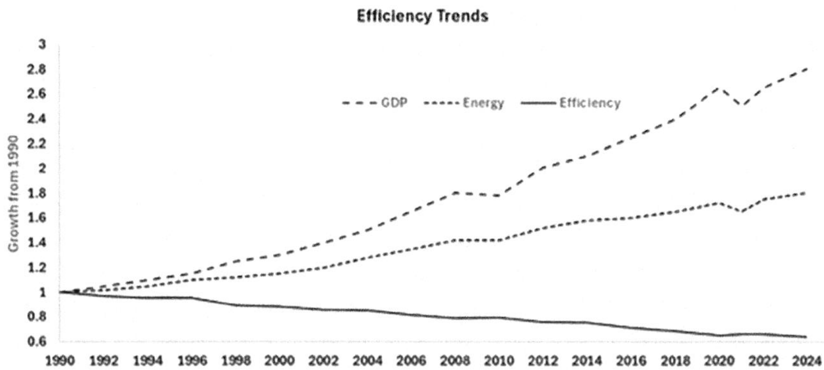

FIGURE 10.3 Energy efficiency trends [5].

demand can reduce the need for new power plants and transmission infrastructure, leading to cost savings for utilities and consumers.

By using less energy, we can reduce the amount of greenhouse gases emitted into the atmosphere, helping to mitigate climate change [8]. Energy efficiency reduces the need to extract and burn fossil fuels, conserve natural resources, and minimize environmental impacts associated with fossil fuel extraction and processing. Also, reducing energy consumption can lead to lower emissions of GHG gases and particulate matter, improving air quality and public health [9].

In summary, improved energy efficiency offers a wide range of economic and environmental benefits [3]. By reducing energy consumption, we can lower costs, create jobs, reduce greenhouse gas emissions, and preserve natural resources. Moreover, energy efficiency plays a vital role in alleviating strain on the electricity grid and contributing to a sustainable energy future.

Promoting energy efficiency

Promoting energy-conscious behavior among individuals and businesses is essential for achieving long-term energy efficiency gains. This can involve raising awareness about energy-saving practices, providing education and training, and offering incentives for energy-efficient behavior. Achieving significant energy efficiency gains requires a comprehensive approach that encompasses policies, technological advancements, and behavioral changes [10]. This section explores key strategies to promote energy-efficient practices, fostering a more sustainable and resilient energy future.

Policy and regulatory framework

Implementing and enforcing stringent energy efficiency standards for buildings, appliances, and vehicles can significantly reduce energy consumption. These standards set minimum efficiency requirements, encouraging manufacturers to develop and produce more energy-efficient products. For example, France has set ambitious national goals for reducing energy consumption and GHG emissions. They follow strict building codes requiring energy performance upgrades during renovation.

Offering tax credits, rebates, or subsidies for energy-efficient investments can incentivize individuals and businesses to adopt energy-saving measures. This can include incentives for energy-efficient appliances, electric cars, building retrofits, and renewable energy installations.

Setting renewable portfolio standards can mandate a certain percentage of electricity generation to come from renewable sources, encouraging the development and adoption of clean energy technologies and reducing the demand for fossil fuels.

Finally, implementing mandatory energy labeling for appliances and buildings can help consumers make informed choices about energy-efficient products. Energy labels provide information about the energy consumption of products, allowing consumers to compare options and select the most efficient models.

Technological advancements

Ongoing research and development in energy-efficient technologies can lead to advancements in appliances, lighting, HVAC systems, and industrial equipment. This can make it more affordable and convenient for consumers and businesses to adopt energy-efficient practices. Implementing smart grid technologies can enable real-time monitoring and control of energy consumption, allowing for demand-side management and the integration of renewable energy sources. Research and development into new building materials and technologies can improve the energy efficiency of buildings. This includes advancements in insulation materials, windows, and HVAC systems. Finally, implementing time-of-use pricing or tiered rate structures can incentivize consumers to shift their energy consumption to off-peak hours, reducing the overall strain on the electricity grid. Providing energy audits for homes and businesses can identify areas for improvement and help prioritize energy-saving investments. The audit cost can be covered by tax incentives. Finally, recognizing and promoting green building certifications can encourage the construction of energy-efficient buildings and set a standard for sustainable development. The cost can be covered by tax incentives.

Smart grid technologies

In the past, energy was provided by large, centralized power plants, and the energy flow was predominantly through a one-way street from the transmission grid to the distribution grid. However, with the increased use of renewable decentralized energy systems – primarily wind power and solar power plants – the majority of energy provision is shifting to the distribution grids.

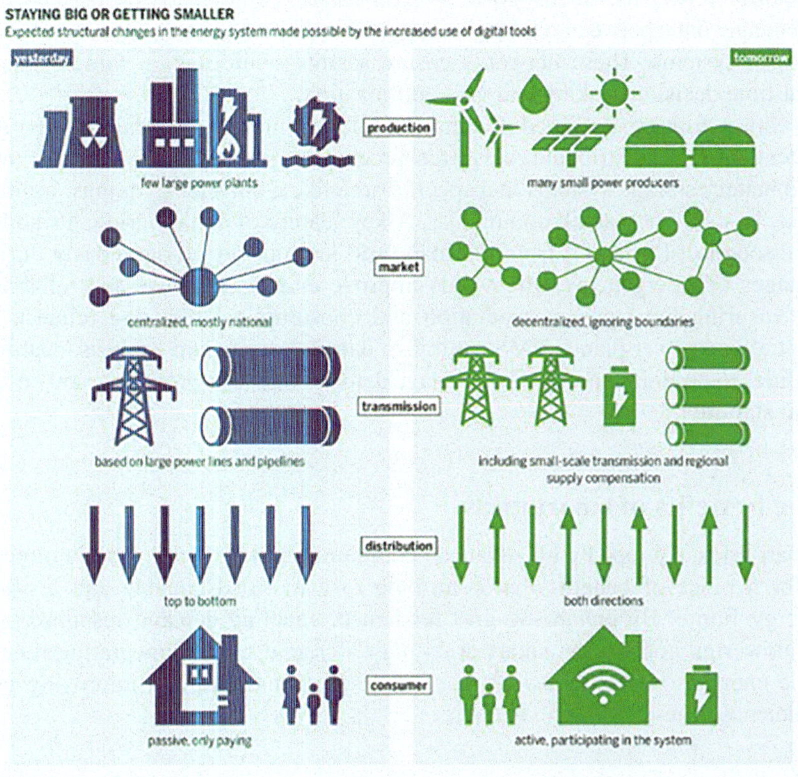

FIGURE 10.4 Characteristics of a traditional system (yesterday, *left*) versus the smart grid (tomorrow, *right*) [11].

These distribution grids were not initially built to harmonize modern renewable systems. Moreover, the conventional demand-side systems are turning into more intelligent local systems combining generation units and consumption devices. These systematic changes require the utilization of advanced technologies, the implementation of new applications and concepts, as well as the realization of more information transparency in the distribution grids. Smart grids, equipped with advanced communication and control technologies, play a pivotal role in integrating renewable energy sources and optimizing energy flow. These intelligent systems enable real-time monitoring, demand response, and enhance grid resilience, paving the way for a more sustainable and efficient energy future. Fig. 10.4 shows characteristics of a traditional system (yesterday) versus a smart grid (tomorrow). Key components of smart grids are discussed here.

Advanced metering infrastructure enables real-time monitoring and communication between utilities and consumers. This allows for accurate measurement of energy consumption, detection of anomalies, and implementation of demand response programs. Smart grids rely on robust communication networks to exchange data between various components, including meters, sensors, and control systems. These networks ensure seamless information flow, enabling real-time decision-making and grid optimization.

Smart grids are designed to accommodate the integration of microgrids. The latter consists of distributed energy resources such as solar panels, wind turbines, and battery storage systems in a specific area like a university campus, military base, hospital, or a small community. A key feature of a microgrid is its ability to disconnect ("island") from the main grid and operate autonomously during outages or emergencies. Microgrids improve energy resilience and reliability by ensuring local power generation and consumption, reducing reliance on centralized power plants. EMS centralize data collection and analysis, enabling utilities to optimize energy flow, manage demand response programs, and ensure grid stability.

The benefits of smart grids

Smart grids, equipped with advanced communication and control technologies, offer a range of benefits that contribute to a more sustainable and resilient energy future. By enhancing grid resilience, enabling demand response, and empowering consumers, smart grids play a crucial role in integrating renewable energy sources, reducing greenhouse gas emissions, and improving grid efficiency. These are discussed here.

Grid resilience

Smart grids enable utilities to monitor grid conditions in real-time, allowing for early detection and mitigation of potential issues. This enhances grid resilience and reduces the frequency and duration of power outages. Smart grids also facilitate the integration of distributed energy resources, such as solar panels and wind turbines, which can help to balance supply and demand, reduce the strain on the central grid, and improve resilience. In case of disruptions or emergencies, we can allow sections of the grid to operate independently by configuring the smart grids to operate in islanding mode. This enhances grid resilience and reduces the impact of outages.

Demand response

Smart grids empower utilities to implement demand response programs, encouraging consumers to reduce their energy consumption during peak demand periods. This helps to alleviate strain on the grid, reduce the need for additional power plants, and stabilize energy prices. This feature is illustrated in Fig. 10.5.

Smart grids can enable time-of-use pricing, where consumers are charged different rates for electricity depending on the time of day or week. This

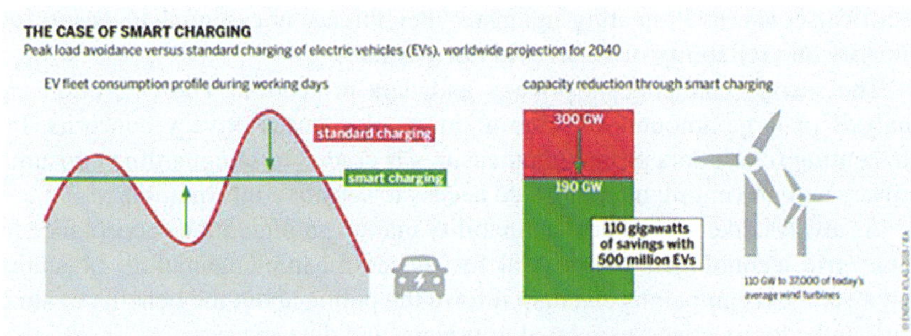

FIGURE 10.5 Peak load avoidance by smart charging of electric vehicles [12].

incentivizes consumers to shift their energy consumption to off-peak hours, reducing peak demand and improving grid efficiency.

In extreme cases of peak demand, smart grids can facilitate controlled load shedding, temporarily reducing power to specific customers to prevent grid collapse. This is a last resort measure, but it can be crucial during extreme events.

Consumer engagement

Smart grids empower consumers with real-time information about their energy consumption, enabling them to make informed decisions and reduce their energy bills. This promotes energy conservation and empowers consumers to play an active role in managing their energy use.

Smart grids also enable two-way communication between utilities and consumers, allowing for personalized messaging, feedback, and program enrollment. This enhances customer engagement and facilitates the implementation of demand response programs. By enabling the participation of consumers in energy management initiatives, community engagement is enhanced. This can include community energy storage projects, demand response programs, and virtual power plants.

Additional benefits

Smart grids can contribute to reducing greenhouse gas emissions by optimizing energy flow and integrating renewable energy sources. Smart grids can also create new economic opportunities, such as jobs in the technology sector and energy management consulting.

As an example, today's photovoltaic inverters not only feed the solar energy into the electric grid, but they also integrate battery storage, optimize EV charging, and take control over flexible loads based on the local weather forecast.

Challenges of smart grids

Cybersecurity is a key challenge for smart grids. As smart grids become increasingly interconnected and reliant on digital technologies, cybersecurity is a critical concern. Protecting against cyber threats is essential to ensure the integrity and reliability of smart grid operations.

The second challenge is privacy and data protection. The collection and analysis of large amounts of data in smart grids raises privacy concerns. Implementing robust data protection measures is crucial to safeguarding consumer privacy and preventing unauthorized access to sensitive information.

As always, like any other sustainability matter, gaining public acceptance for smart grid technologies is essential for successful implementation. Education and awareness campaigns can help inform the public about the benefits of smart grids and address concerns related to privacy and data security.

Finally, international cooperation is needed between countries to facilitate the sharing of knowledge, best practices, and standards for smart grid development. This can accelerate the deployment of smart grid technologies and promote global energy efficiency.

Summary

Increasing industrial energy efficiency could provide more energy at the public's disposal. However, that by itself will not result in a more sustainable future unless energy consumption is reduced. This aspect will be discussed further in Chapter 12. There are many challenges, such as cybersecurity, privacy, and public acceptance, facing this path. Through collaborative efforts and technological advancements, we should be able to harness the full potential of smart grids to build a sustainable and resilient energy future.

Key takeaways

- Implementing and enforcing stringent energy efficiency standards for buildings, appliances, and vehicles can significantly reduce energy consumption. A 10% improvement in efficiency is equivalent to saving 5 Gt CO_2 per year.
- Smart grids, equipped with advanced communication and control technologies, offer a range of benefits that contribute to a more sustainable and resilient energy future.
- International cooperation is needed between countries to facilitate the sharing of knowledge, best practices, and standards for smart grid development.

References

[1] Data source: U.S. Energy Information Administration (EIA). Commercial buildings energy consumption survey (CBECS). (Accessed January 20, 2026) https://www.eia.gov/consumption/commercial/.

[2] Data Source: International Energy Agency (IEA)Global EV Outlook 2024, IEA, Paris, 2024 https://www.iea.org/reports/global-ev-outlook-2024 (Accessed 20 Jan 2026).

[3] Project Drawdown. How the North American grid must change for an electric future. (Accessed January 20, 2026) https://drawdown.org/insights/how-the-north-american-grid-must-change-for-an-electric-future.

[4] U.S. Environmental Protection Agency (EPA). Local energy efficiency benefits and opportunities. (Accessed January 20, 2026) https://www.epa.gov/statelocalenergy/local-energy-efficiency-benefits-and-opportunities.

[5] M.P. Mills, N. Atkinson, Energy delusions: a critique of oil "scenarios" in the IEA World Energy Outlook 2024, National Center for Energy Analytics (NCEA), 2025 https://energyanalytics.org/wp-content/uploads/2025/01/2025-01-NCEA-Energy-Delusions-MillsAtkinson.pdf (Accessed 20 Jan 2026).

[6] The Powers Company. Maximizing profitability: the impact of unmitigated high energy costs on manufacturers. (Accessed January 20, 2026) https://www.thepowerscompany.com/resources/high-energy-costs/.

[7] International Energy Agency (IEA). Multiple benefits of energy efficiency: economic benefits. (Accessed January 20, 2026) https://www.iea.org/reports/multiple-benefits-of-energy-efficiency/economic-benefits-2.

[8] U.S. Environmental Protection Agency (EPA). What you can do about climate change: energy. (Accessed January 20, 2026) https://www.epa.gov/climate-change/what-you-can-do-about-climate-change-energy.

[9] International Energy Agency (IEA). Multiple benefits of energy efficiency: air quality. (Accessed January 20, 2026) https://www.iea.org/reports/multiple-benefits-of-energy-efficiency/air-quality .

[10] U.S. Department of Energy (DOE), Office of energy efficiency and renewable energy (EERE). Energy efficiency: buildings and industry. (Accessed 20 Jan 2026) https://www.energy.gov/eere/energy-efficiency-buildings-and-industry.

[11] Source: Wikipedia Contributors. Staying big or getting smaller (image file) In *Wikipedia, The Free Encyclopedia* Wikimedia Foundation. https://en.wikipedia.org/wiki/Smart_grid#/media/File:Staying_big_or_getting_smaller.jpg, https://energytransition.org/2018/04/europe-must-choose-a-green-future/, (Accessed January 20, 2026).

[12] Source: Wikipedia Contributors. The case of smart charging (image file). In *Wikipedia, The Free Encyclopedia*; Wikimedia Foundation. https://en.wikipedia.org/wiki/Smart_grid#/media/File:The_case_of_smart_charging.jpg, https://energytransition.org/2018/04/europe-must-choose-a-green-future/ (Accessed January 20, 2026).

Further readings

Ipieca, American Petroleum Institute (API), & International Association of Oil & Gas Producers (IOGP), (2020). Sustainability reporting guidance for the oil and gas industry (4th ed.). Ipieca. https://www.ipieca.org/resources/sustainability-reporting-guidance (accessed 4/14/2026).

Sierra magazine fall, (2024). vol. 109, No. 3, https://digital.sierramagazine.org/publication/?i=829286&p=0&view=issueViewer.

Smith, Cedric and Sarah Winstanley (2022). Net-Zero Skills: What will Canada need for the coming energy transition? The Pembina Institute https://www.pembina.org/reports/net-zero-skills.pdf.

M. Wolfe, 2022, The rise and fall of America's environmentalist underground, https://www.nytimes.com/2022/05/26/magazine/earth-liberation-front-joseph-mahmoud-dibee.html. (Accessed April 14, 2026).

Chapter 11

Financial catalysts

Synopsis

In January 2020, BlackRock, the largest asset manager in the world, declared that it intended to use the trillions of dollars under its management to address global warming. Soon, this caused almost every major financial institution to pledge to reduce emissions. However, although the U.S. investment in clean energy reached USD 279 billion in 2025, many large corporations have gone silent on climate change. Major banks and financial institutions have pulled out of the net-zero Asset Managers initiative. This was due to policy changes under the new administration and conservative policies that were more aligned with climate change denial. This shows the critical role of governments in combating climate change.

Renewable energy initiatives and carbon sequestration need to be affordable to the consumers and economical while adhering to strict environmental regulations. Affordability is a remarkably elastic concept. It has to do with the cost of living on one hand and income on the other. But, the inconvenience and insecurity appear when these two concepts are not in balance. The price of goods will go down when the strict environmental regulations are not imposed on manufacturers. This might make some products cheaper (if the savings are passed along to the consumers). However, laxed environmental regulations would cause additional health risks which would make the public to pay very dearly for those savings. In this chapter, some of the financial catalysts, such as government incentives and the carbon economy, are discussed.

F.A.C.T.S. Flexibility, Affordability, Conservation, Transition, Sustainability: Our Path to the Future of Energy and the Environment. DOI: https://doi.org/10.1016/B978-0-443-52574-2.00013-6

Government policies and financial incentives

According to the U.S. Department of the Treasury, [1]. Americans claimed more than USD 8 billion in climate-friendly tax credits under the Inflation Reduction Act in 2023. The bulk of the money, more than USD 6 billion, helped households install rooftop solar panels, small wind turbines, and other renewable energy systems. Credits that helped Americans improve the energy efficiency of their homes by installing an electric heat pump or boiler, adding insulation, replacing windows, and making other upgrades were most popular in the Northeast and Midwest.

Nearly 35% of the world's electricity and 24% of the USA electricity came from renewable energy (in 2025) [2]. Many governments have intervened by setting new regulations that provide incentives for reducing emissions and using renewable energies. For example, several federal government agencies in the USA provide tax credits, grants, and loan programs for qualifying renewable energy technologies and projects. Some states have specific requirements, and some have voluntary goals within a specified time frame for the share of electricity generation or sales in a state that comes from renewable energy. Financial products are available for sale, purchase, or trade that allow a purchaser to pay for renewable energy production without directly producing or purchasing the renewable energy. The most widely available products are renewable energy certificates or credits (RECs) [3]. These products may also be called *green tags, green energy certificates*, or *tradable renewable certificates*, depending on the entity that markets them. Electric utilities can use RECs to comply with state renewable energy portfolio standards. Many companies use RECs or similar products to meet their voluntary targets or goals to reduce greenhouse gas (GHG) emissions from their operations. Several federal and state requirements and incentives are in effect for producing, selling, and using biofuels and other alternative vehicle fuels. The Sustainable Finance Disclosure Regulation in the European Union mandates that financial firms disclose the environmental, social, and governance (ESG) risks of their investments. In addition, to spur interest in carbon capture projects and improve their economic viability, the USA Inflation Reduction Act of 2022 increased the maximum amount of credit for capturing and sequestering carbon dioxide from USD 50 per metric ton to USD 85 per metric ton. Similarly, the European Commission's carbon management strategy has proposed the sequestration of up to 450 million tons of CO_2 annually by 2050 and has kicked off a debate about the future of the EU's emissions trading scheme.

The financial markets have also devised new systems that provide investors with a means of participating in carbon trading. Project financing for clean energy projects has started moving forward with creative solutions. In February 2024, Japan issued the world's first sovereign transition bonds–Japan Climate Transition Bonds. Two offerings of JPY 800 billion (USD 5 billion) each were issued, with tenors of 5 and 10 years. The issuances are certificated by the Climate Bonds Initiative and based on Japan's national transition strategy.

The successful deployment of GHG remediation practices hinges on robust regulatory and legal frameworks. These frameworks must ensure both the safe implementation of these technologies and the security of potential project stakeholders. To this end, federal and state governments have been proactive in creating regulatory environments that incentivize the deployment of GHG reduction projects or safeguard the public against their unsafe implementation.

In many locations, policymakers have incentivized the production and purchase of electric vehicles (EVs) and support wind energy infrastructure by providing tax breaks, subsidies, and investments in battery technology and grid modernization. Governments can also encourage the financial sector to align its financial portfolios with national climate commitments. National development banks, for example, can establish ambitious climate-related lending targets, particularly in the offshore wind sector.

Supportive policies and financial instruments are essential for promoting the widespread adoption of less common electricity production methods, such as small hydropower, in rural communities. This can include providing tax breaks or feed-in tariffs for electricity generated from small hydropower plants, facilitating access to microloans or grants for project development, and streamlining permitting processes.

Supportive government policies and regulatory frameworks, including carbon pricing initiatives, have led to European hydrocarbon-producing nations, notably Norway and the United Kingdom, becoming well-positioned for large-scale carbon capture, utilization, and storage (CCUS) deployment in the coming decade.

In the United States, a robust regulatory and legal framework has been established to ensure the safe and secure transportation and storage of CO_2. Agencies such as the Environmental Protection Agency (EPA), the Department of the Interior (DOI), and the Pipeline and Hazardous Materials Safety Administration (PHMSA) have developed regulations, guidance, and orders that underpin federal CCUS policy. For example, the EPA has established specific regulatory frameworks under the Safe Drinking Water Act to protect underground sources of drinking water and maintain accounting protocols under the Clean Air Act GHG Reporting Program for CO_2 injection into geologic storage. PHMSA sets and regulates standards for the design, construction, and operation of CO_2 pipelines.

However, achieving net-zero emissions will require scaling up GHG remediation efforts. Governments must establish adequate financial incentives through government policies underpinned by a durable regulatory and legal framework. The implementation of these policies should occur in phases, prioritizing projects based on deployment economics and ease of implementation.

Under the USA laws, a project is eligible to claim carbon credits only if it delivers real, additional, measurable, verified, durable, and uniquely claimed emissions reductions or removals under an approved framework (Box 11.1).

BOX 11.1 Tech note—Project eligibility for carbon credits

For any project to be eligible to claim a carbon credit, it needs to meet the following criteria ([4]):

- The project should be in accordance with national policies on sustainability
- The project should avoid negative environmental, social, and cultural impacts
- Credits should be 'additional' to the business-as-usual scenario

As a first step to generate carbon credits, the project should meet the following requirements to be registered with the Clean Development Mechanism (CDM) Executive Board (of the Kyoto Protocol and the Voluntary Carbon Standard):

- Additionality demonstrates the project activity would not be implemented in the 'business-as-usual' scenario due to the existence of a barrier (e.g., investment, technical, and institutional, etc.) or to low financial returns.
- The baseline establishes that, in the absence of geothermal power generation, the equivalent power would have been supplied from a mix of generation sources connected to the power grid, emitting more GHG.
- Eligibility indicates the project meets CDM requirements, such as:
  - Start date of the project
  - Meeting methodology requirements
  - Prior CDM revenue consideration in investment decisions proven by documented evidence
  - Requirements related to host country approval
- Stakeholder Consultation involves meeting with local stakeholders to obtain public input on the environmental and social impacts of the project(s). Mitigation measures should be included in the project implementation plan.

The EPA established requirements for the injection of fluids into the subsurface through the underground injection control (UIC) program. The UIC program's statutory mandate was to protect deep water by ensuring the safe, long-term containment of injected CO_2 streams and displaced formation fluid. Class II wells (specific to oil and gas operations) and Class VI wells (specific to CO_2 injection into saline formations) must meet specific requirements before permits are granted for such operations.

As illustrated in Fig. 11.1, the full permitting and review process typically would take about 6 years. A shorter approval process could significantly benefit such programs.

When developing CO_2 storage projects, project developers must also secure rights to the applicable contiguous pore space. In many countries, subsurface pore space is owned by the federal government or a sovereign entity. In the United States, mineral rights and water rights belong to landowners or those who purchase them from landowners. Under common law, oil and natural gas

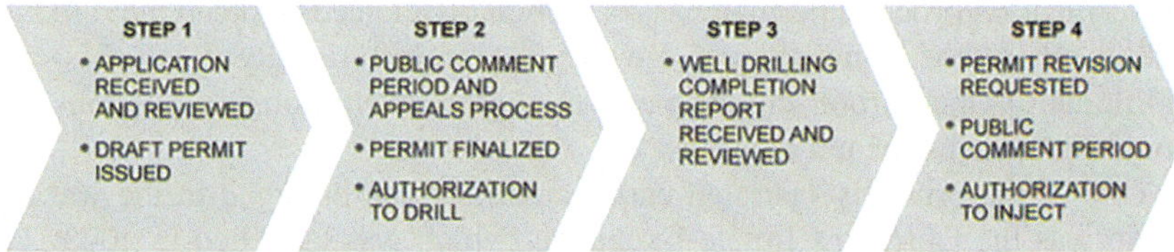

FIGURE 11.1 Class VI well-permitting process [5].

operators have the right to use a surface owner's pore space as reasonably necessary to produce minerals on the property. Therefore, the pore space owner's rights are not violated when CO_2 remains in the pore space.

A key distinction between enhanced oil recovery (EOR) and CO_2 injection is the regulatory obligations imposed during the postinjection period. For instance, Class VI permits for saline storage often require monitoring for 50 years. Shortening this extended liability period could enhance the program. Some states have begun transferring liability to the state after a shorter duration, such as 10 years.

The renewable energy industry offers a precedent for how policy can incentivize large-scale technology deployment. Between 2005 and 2015, the US federal government provided USD 51.2 billion in financial incentives to support solar and wind power development, with tax incentives accounting for 90% of this amount. These financial incentives, combined with renewable energy standards and other supportive policies at the federal and state levels, helped establish the renewable energy industry. Today, wind and solar energy supply over 7% of US electricity.

In the near term, incentives are likely the most effective way to drive deployment. However, in the long-term, a market-based approach is likely a more economically efficient way to reduce CO_2 emissions compared to standards, mandates, or financial incentives. Numerous articles have explored the benefits and drawbacks of incentive-driven programs versus market-based approaches. Most economists agree that a market-based approach is a more effective and efficient way to reduce emissions and benefit the overall economy.

Carbon markets

Carbon markets first came to global attention during talks leading up to the 1997 Kyoto Protocol. The international treaty to lower GHG emissions allowed countries to invest in projects in developing countries to avert or sequester carbon emissions, gaining offsets in return. Kyoto did not specify what the carbon market would look like, and trading never happened under the treaty. But in its wake, multiple private and public carbon markets appeared. In some of these, a private organization served as the intermediary, much like a stock exchange. The credits bought and sold by Kyoto participants were verified by one of a handful of third-party organizations. Two requirements were particularly important to the verification process: permanence (a carbon offset needs to keep the removed carbon out of the atmosphere for a specific period, usually 100 years) and additionality (the carbon credits need to represent emissions reductions that would not have happened otherwise).

One of the privately operated carbon markets that emerged in the wake of Kyoto was the Chicago Climate Exchange, which was launched in 2003. The Exchange folded in 2010 after it became clear that most investors were not interested in a voluntary market and were waiting for the federal government to step in. When US climate legislation failed to pass through the Senate, the market folded. But that has not stopped a new crop of carbon brokers from emerging in recent years. As carbon markets grow in size, resources are being allocated across the financial value chain to enhance the accessibility and liquidity of these markets. Participants in carbon markets are not only companies that need to control emissions but also commercial banks, investment banks, carbon funds, and private equity funds. These participants make the market more liquid and promote innovation in carbon financial products and services. The investors will not only properly price carbon risks in their investment portfolios but also enhance their risk-adjusted returns.

Accelerating annual renewable energy deployment since 2021 has provided a cost-effective solution to the energy crisis's economic challenges. Long-term contracts secured through policy mechanisms and regulations offer stable prices for most wind and solar photovoltaic (PV) power generators in Europe, limiting their exposure to volatile electricity prices. They can also help shelter consumers from rising electricity prices. In 2021 and 2022, the European Union added nearly 90 GW of PV and wind capacity. This capacity has displaced almost 10% of hard coal and natural gas generation, pushing the most expensive power plants out of the market and effectively reducing the price for all consumers. Lawmakers for the group of 27 European countries have adopted rules for taxing imports based on the amount of carbon dioxide that companies emit when making those goods. Experts say it is the first time a major economy has married climate and trade policy and that it could lead other countries to do the same.

The rapid expansion of renewable energy technology use in buildings has eased EU natural gas demand and has contributed to the bloc's energy security in the short term. Projected cumulative new developments in the direct use of renewable heat and expansion of renewable electricity post 2022 displaced almost 8 bcm of EU buildings-related gas consumption annually in 2023 and more than 17 bcm in 2024. This is equivalent to avoided emissions of more than 50 Mt CO_2 in 2023–2024 [6].

Carbon economy

A carbon economy is not exactly a single economic system but rather a financial system that relies on putting a price on carbon emissions. This can be done

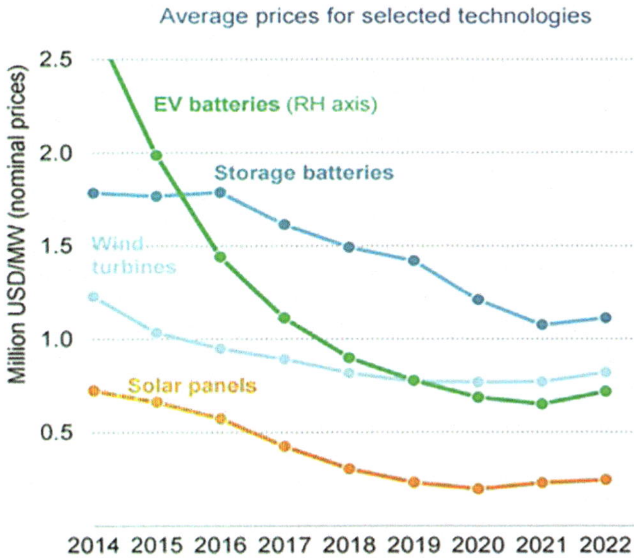

FIGURE 11.2 Average prices for selected technologies [7].

through carbon taxes or emissions trading systems (ETS). A carbon tax involves a government setting a fixed price per ton of CO_2 emitted, with companies paying based on their total emissions. Switzerland holds the highest carbon tax rates, exceeding USD 140/ton as of 2024. ETS describes when governments set a cap on total emissions and issue permits (allowances). The core benefit is that carbon pricing incentivizes polluters to cut back on emissions. When emitting carbon becomes more expensive, businesses and consumers find ways to reduce their carbon footprint. This can involve switching to cleaner energy sources via system design, using more efficient technologies, or simply consuming less energy. Carbon pricing also allows the market to find the most cost-effective ways to reduce emissions. Companies can choose the most economical approach to meet their emission reduction targets, fostering innovation in clean technologies.

Funds generated from carbon taxes or permit sales can be directed toward renewable energy development, clean infrastructure projects, or even offsetting the impact of unavoidable emissions through carbon capture projects. Notice that as the cost of carbon rises, businesses and consumers have a stronger reason to invest in energy-saving measures. This can lead to long-term economic benefits through lower energy bills.

Clean energy costs edged higher in 2022. This could be global inflation following the pandemic. But the trend suggests a lowering of prices across all renewable energy sectors, as shown in Fig. 11.2.

In the CCUS value chain, capturing CO_2 often constitutes the biggest cost component for CCUS and is where significant cost reductions, efficiency gains, and further technological innovations could be achieved. A key factor in the cost of CO_2 capture is the concentration and overall volumes of CO_2 in the source gas, with costs typically decreasing with increased concentration and volume of CO_2 in the flue gas stream. Revenues for capturing CO_2 could be generated in some industrial applications where CO_2 could be utilized (examples include cement, food, and beverages). However, these opportunities remain very limited in scope, and the vast proportion of the captured CO_2 will have no intrinsic economic value. The only well-established application of CO_2 utilization is via CO_2 EOR in the petroleum industry. CCUS project developers can sell carbon credits in the Voluntary Carbon Market based on certified emissions that have been reduced or removed through CCUS.

In designing frameworks to support investment and scaling up of CCUS, the following key elements are essential to mitigate some of the risks and generate a stream of revenues to make projects attractive for private sector investment: [8]

- Stable and supportive legal and regulatory frameworks
- Mechanisms that allow the stacking of revenues for operators in the supply chain and
- Varying degrees of government participation in the CCUS supply chain enable risk-sharing and risk mitigation.

No one CCUS business model fits all projects or sectors. Instead, various business models have emerged with different degrees of integration in the supply chain, including varying degrees of government support and ownership structures (Oxford Energy Group) [8].

Carbon trading

Global investment in the energy sector exceeded USD 3 trillion in 2024, of which USD 2 trillion is associated with clean energy projects (Fig. 11.3) [9]. Electrified transport was the main driver of this spending on the rollout of clean technologies, leapfrogging renewable energy and accounting for more than a third of the investment total. The following figure shows the recent trend in the global energy transition investment volume. The world invested almost twice as much in clean energy as it did in fossil fuels.

Carbon pricing

The nationally determined contributions (NDCs) are commitments that countries make to reduce their GHG emissions as part of climate change mitigation. In 2025, over 100 countries submitted new NDCs for 2035, representing a significant portion of global emissions. However, on the whole, countries' plans are insufficient to meet current NDCs. Projections based on the full implementation of these new NDCs indicate a potential warming of 2.3–2.5°C by the end of

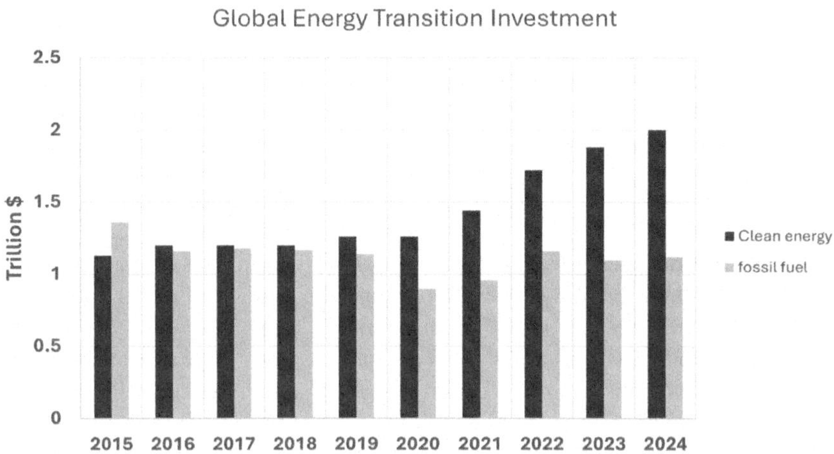

FIGURE 11.3 Global energy transition investment [10]. (Note: Clean power = fossil fuel power with CCUS, hydrogen, ammonia, and large-scale heat pumps. Low-emissions fuels = modern bioenergy, low-emissions H_2-based fuels, and CCUS associated with fossil fuels, which also includes direct air capture. 2024 = estimated values for 2024). *CCUS*; Carbon capture, utilization, and storage.

the century, still well above the Paris Agreement's goals. An immediate and sustained focus on the implementation of new policies is vital.

The carbon market is an important legislative tool for governments to bridge the gap between climate ambitions and policy actions. This includes strengthening emission reduction mandates and expanding to harder-to-abate sectors.

Global initiatives, including the Global Carbon Pricing Challenge (initiated by Canada), the International Carbon Action Partnership, the Coalition of Finance Ministers for Climate Action, the World Trade Organization-led task force on carbon pricing, and the World Bank's own Partnership for Market Implementation, continue work to support international cooperation on the development and implementation of carbon pricing around the world. The average carbon trading price can vary depending on the specific markets. According to the BloombergNEF forecast, [11] prices are low in the early years, reaching just USD 20/ton in 2030. As the milestone years for net-zero goals approach, the price can go as high as USD 238/ton in 2050. Voluntary markets, where companies offset their emissions, prices reach just USD 13/ton in 2030 and a mere USD 14/ton in 2050.

There are two main ways that direct carbon pricing is implemented:

ETS: These set a cap on the total amount of pollution that can be emitted. Companies are given permits to emit a certain amount, and they can trade these permits with each other. This creates a market for pollution, and the price of permits reflects the cost of emitting carbon.

Carbon taxes: A carbon price is a fee on each unit of carbon dioxide (CO_2) or other GHG emissions released into the atmosphere. There are two primary methods of pricing: Carbon taxes and cap-and-trade programs. Carbon taxes would directly establish a price on carbon in dollars per ton of emissions. A carbon tax can be either an emission tax, which is based on the emitted CO_2 level, or a tax on goods or services that are GHG-intensive, such as gasoline. Businesses and consumers will take steps, such as switching fuels or adopting new technologies, to reduce their emissions and avoid paying the tax. As of 2024, 37 carbon tax programs have been implemented across the world [12].

In a cap-and-trade system, the government sets an emissions cap and issues a quantity of emission allowances consistent with that cap. Emitters must hold allowances for every ton of GHG they emit. Companies may buy and sell allowances, and this market establishes an emissions price. Companies that can reduce their emissions at a lower cost may sell any excess allowances to companies facing higher costs to buy. Cap-and-trade differs from a tax in that it provides a high level of certainty about future emissions but not about the price of those emissions. For an industrial entity, a cap may be the preferred policy when it has a specified emission target. Several countries have started participating in this program [13].

Indirect carbon pricing includes a reduction in fuel subsidies, fuel taxes, and other measures. The goal of these approaches is to make it more expensive to pollute, which incentivizes businesses and consumers to reduce their emissions. Carbon taxes promote innovation by encouraging companies to invest in clean energy technologies and develop more efficient processes in order to cut their costs.

Climate economists and major energy firms believe that the price of carbon should be raised to incentivize greater investment in carbon capture technology and renewable energy innovation. Without setting a minimum carbon pricing scheme at the international level or introducing carbon taxes, companies are not being incentivized enough to pledge or meet net-zero carbon goals by 2050.

Carbon credits: Carbon credits can be thought of as certificates representing one ton of carbon dioxide (or its equivalent in other GHG) that is reduced, removed, or avoided from the atmosphere. They are used in what is called carbon offsetting, a way to combat climate change.

Organizations like governments or businesses can create carbon credits by undertaking projects that reduce GHG emissions. Examples include planting trees, capturing methane from landfills, or investing in renewable energy sources. Third-party organizations verify these projects to ensure their legitimacy.

The credits are then sold on carbon markets. Companies that exceed their emission limits can buy credits to offset their emissions. This creates a financial incentive for companies to reduce their carbon footprint.

As demand for credits goes up (due to stricter emission caps), the price goes up as well, encouraging even more emission reduction projects.

The price for carbon credits in the United States fluctuates, but USD 15 to USD 20 per ton (in 2021) of carbon sequestered is common. Firms such as the world's first exchange-traded investment products for high-integrity carbon credits, XTCC, are engaged in carbon trading on a global scale. Tesla has been selling carbon credits to other automakers that did not meet emissions standards. Tesla's carbon credit sales reached USD 1.78 billion in 2022. In 2023, global carbon pricing revenue exceeded USD 100 billion for the first time [14].

Carbon offsetting

Carbon offsets and carbon credits are often used interchangeably, but there is a slight difference. Here is how carbon offsets work:

You cannot target and remove the exact amount of carbon dioxide you emit from your car or plane ride. Carbon offsets provide a way to compensate for your emissions by funding projects that reduce GHG or remove them from the atmosphere elsewhere.

Carbon credit quality determines the price. For carbon offsetting to be effective, the carbon dioxide should not return to the atmosphere after a short period of time. Carbon reduction or removal projects that ensure the carbon dioxide is not returned to the atmosphere for a long time are considered to be higher quality projects, and their associated carbon credits achieve higher prices. This is often referred to as the "permanence premium". There are projects around the world that reduce or capture carbon dioxide, like planting trees, capturing methane from landfills, or building renewable energy sources. These projects are verified to ensure they are legitimate and measurable. One metric ton of carbon dioxide reduction (or equivalent amount of other GHG) is represented by a carbon offset. Individuals or companies can buy these offsets to balance out their own emissions. The effectiveness of carbon offset projects can vary. It is important to choose reputable providers with verified offsets.

A common practice in the carbon offset business is for investors to pay farmers to store carbon and then sell it as offsets to individuals and companies that want to brand themselves as sustainable. Some of these offset sellers are run by majors in the agricultural world: Bayer, Corteva, Farmers Business Network, and Truterra (the last is an offshoot of Land O'Lakes). The carbon offset market tends to draw large farms rather than small ones, since there's more of a payoff for the upfront paperwork and meticulous crop tracking when you have thousands of acres as opposed to a couple of hundred.

Demand for carbon offsets is expected to reach USD 150 billion by 2050—almost 35 times higher than today [11].

Carbon market controversies

Carbon pricing is not without its critics. Some argue that it can be a burden on businesses and consumers, and that the revenue generated from carbon taxes is not always used effectively for GHG reduction activities. Also, some environmentalists have criticized carbon offsetting because it allows carbon dioxide emitters to continue emitting while purchasing carbon credits. Some emitters may be inclined to purchase low-quality carbon credits because of their lower prices, which may not be effective in negating their emissions.

Voluntary carbon credit markets have come under particular scrutiny in 2023, with some projects being accused of over-crediting, lacking additionality, or enabling human rights abuses. This impacted the issuance of energy-related carbon credits in 2023. Credit retirements remained high in 2023, after peaking in 2022. This keeps the costs of most energy-related carbon credits much lower than the real cost of mitigation. To address questions about the integrity of certain legacy credits, cut-off dates for eligible credits have been introduced [9].

To address these concerns, carbon emitters are encouraged to reduce emissions wherever possible and to offset only the emissions that cannot be reduced. Overall, many economists and policymakers believe that carbon pricing is one of the most effective tools we have to address climate change.

Summary

In summary, three carbon markets will help in transforming carbon economics: (1) voluntary carbon markets where some corporations are active in meeting their announced net-zero commitments, (2) compliance carbon markets, where rising carbon prices are boosting CCUS as both mitigation and credit source, and (3) Low-carbon fuel standards are designed to decarbonize the transportation sector.

Unfortunately, the fossil fuel industry has not upped its game in major emission reduction projects. Investment by the oil and gas industry in low-emissions sources of energy in 2023 was less than 5% of its upstream investment. This indicator differed widely by company, with double-digit shares common among the large European companies. Investment by the industry in clean fuels, such as bioenergy, hydrogen, and CCUS, is picking up in response to more supportive policies but remains short of where it needs to be in climate-driven scenarios.

Carbon removal projects take years to materialize, making it even more essential to kickstart the regulatory and technological processes early. It is short-sighted to wait for 2030 to implement all decarbonization options—Countries have to start planning now to scale these technologies by 2050. With respect to the US government's policies and regulations, it remains to be seen how the new administration will impact the incentives for clean energy projects.

Key takeaways

- Products across asset classes that constitute climate solutions are likely to merit a green premium relative to carbon-intensive or grey alternatives.

- The emergence of a green pricing premium is significant for two reasons. First, it tells companies that there is a volume and price benefit for decarbonizing. Second, a premium for green alternatives today is likely to translate into a grey discount over time—A signal that should act as a warning bell for investors not yet integrating sustainability factors into their investment life cycle.
- Carbon pricing has received mixed reviews. Some argue that it can be a burden on businesses and consumers, and that the revenue generated from carbon taxes is not always used effectively for GHG reduction activities. Others have criticized carbon offsetting because it allows carbon dioxide emitters to continue emitting while purchasing carbon credits.
- Projections based on the full implementation of these new NDCs indicate a potential warming of 2.3–2.5°C by the end of the century, still well above the Paris Agreement's goals.

References

[1] U.S. Department of the TreasuryU.S. Department of the Treasury Releases New Data on Inflation Reduction Act Energy Tax Credits, Press Release JY2521, 2024 https://home.treasury.gov/news/press-releases/jy2521 (Accessed Jan 20, 2026).

[2] EmberGlobal Electricity Mid-Year Insights 2025, Ember, 2025 https://ember-energy.org/latest-insights/global-electricity-mid-year-insights-2025/ (Accessed Jan 20, 2026).

[3] U.S. Environmental Protection AgencyRenewable Energy Certificates (RECs), U.S. Environmental Protection Agency, 2025 https://www.epa.gov/green-power-markets/renewable-energy-certificates-recs (Accessed Jan 20, 2026).

[4] World Bank. (2012). State and trends of the carbon market . World Bank. http://hdl.handle.net/10986/13336.

[5] Adapted from National Petroleum Council*Meeting the Dual Challenge: A Roadmap to At-Scale Deployment of Carbon Capture, Use, and Storage*; Chapter 3: *Policy, Regulatory, and Legal Enablers*, National Petroleum Council, Washington, DC, 2019 https://dualchallenge.npc.org/documents/CCUS-Chap_3-030521.pdf (Accessed Jan 20, 2026).

[6] International Energy Agency, How much will renewables reduce gas consumption in european buildings?Renewable Energy Market Update—June 2023, International Energy Agency, Paris, 2023 https://www.iea.org/reports/renewable-energy-market-update-june-2023/how-much-will-renewables-reduce-gas-consumption-in-european-buildings (Accessed Jan 20, 2026).

[7] International Energy AgencyWorld energy investment2023: Overview and key findings, International Energy Agency, Paris, 2023 https://www.iea.org/reports/world-energy-investment-2023/overview-and-key-findings (Accessed Jan 20, 2026).

[8] B Fattouh, *Capture Carbon, Capture Value: An Overview of CCS Business Models*; The Oxford Institute for Energy Studies, Oxford, U.K., 2024 https://www.oxfordenergy.org/publications/capture-carbon-capture-value-an-overview-of-ccs-business-models/ (Accessed Jan 20, 2026).

[9] International Energy Agency, (2024). World Energy Investment 2024: Overview and key findings. https://www.iea.org/reports/world-energy-investment-2024/overview-and-key-findings https://www.iea.org/reports/world-energy-investment-2024/overview-and-key-findings.

[10] Data Source: International Energy AgencyWorld Energy Investment 2024: Overview and Key Findings, International Energy Agency, Paris, 2024 https://www.iea.org/reports/world-energy-investment-2024/overview-and-key-findings (Accessed Jan 20, 2026).

[11] BloombergNEF. Carbon credits face biggest test yet, could reach $238/Ton in 2050, according to BloombergNEF Report. (Accessed Jan 20, 2026). 2024. https://about.bnef.com/insights/commodities/carbon-credits-face-biggest-test-yet-could-reach-238-ton-in-2050-according-to-bloombergnef-report/

[12] Center for Climate and Energy Solutions. *Carbon Tax Basics*. Center for Climate and Energy Solutions https://www.c2es.org/content/carbon-tax-basics/ (Accessed Jan 20, 2026).

[13] Center for Climate and Energy Solutions. *Cap and Trade Basics*. Center for Climate and Energy Solutions https://www.c2es.org/content/cap-and-trade-basics/ (Accessed Jan 20, 2026).

[14] World Bank. Carbon Pricing Dashboard. (Accessed Jan 20, 2026). https://carbonpricingdashboard.worldbank.org/

Further readings

Credit Suisse ESG Report, (2022). Treeprint carbon markets, the beginning of the big carbon age., https://www.globaltrading.net/the-beginning-of-the-big-carbon-age.

Dubai Future Foundation, (2023). How carbon could be a $15 trillion industry of the future. https://www.dubaifuture.ae/reports/, Accessed April 14, 2026.

Japan Organization for Metals and Energy Security, (2023). Handbook for CCS carbon credits: Workshop report of #x201C;Global carbon market and CCS: Towards ASEAN decarbonization#x201D;. https://www.jogmec.go.jp.

Bill McKibben, 2023, A Case for Climate Optimism, and Pragmatism, from John Podesta, https://www.newyorker.com/news/q-and-a/a-case-for-climate-optimism-and-pragmatism-from-john-podesta

United Nations Environment ProgrammeNations Must Go Further than Current Paris Pledges or Face Global Warming of 2.5–2.9°C, United Nations Environment Programme, 2023 https://www.unep.org/news-and-stories/press-release/nations-must-gofurther-current-paris-pledges-or-face-global-warming.

United Nations Framework Convention on Climate Change, "Synthesis Report by the Co-facilitators on the Technical Dialogue of the First Global Stocktake," 2023, https://unfccc.int/sites/default/files/resource/sb2023_09E.pdf?download

World Bank GroupState and Trends of Carbon Pricing, World Bank Group, 2024.

Chapter 12

Empowering the general public

Synopsis

As we have discussed throughout this book, direct communication between energy providers and the public regarding the ethical aspects of sourcing their raw materials, safety protocols, emergency response plans, incident reporting, and waste management practices is crucial to establish trust and to foster informed collaborative decision-making. An additional challenge is correcting misinformation, so readily available in the digital age, concerning all aspects of energy generation. The goal should be to find win-win strategies that provide economic benefits to individual businesses, municipalities, or countries while contributing toward meeting climate mitigation, adaptation, and other sustainability goals.

We showed in Chapter 10 that some of the new technologies, such as smart grids, empower consumers with real-time information about their energy consumption, enabling them to make informed decisions and play an active role in managing their energy consumption. Smart grids also enable two-way communication between utilities and consumers, allowing for personalized messaging, feedback, and program enrollment. This enhances customer engagement and facilitates the implementation of programs for responding to demand fluctuations.

In this chapter, we focus on the future of energy transition and how the public can participate in a global approach to reduce emissions.

Current state of energy transition

We wrote this book to develop readers' understanding of different energy sources to empower them to actively contribute to the creation of a healthier planet

F.A.C.T.S. Flexibility, Affordability, Conservation, Transition, Sustainability: Our Path to the Future of Energy and the Environment. DOI: https://doi.org/10.1016/B978-0-443-52574-2.00004-5

for generations to come. To provide a clear and concise overview, we have synthesized this information in a table. We discussed the necessity for inclusion of the United Nations' sustainable development goals (SDGs) [1] in the energy sector. We used the same framework in Table 12.1 to compare both renewable and non-renewable options in reaching SDG goals. Our exploration has revealed that every energy pathway presents its own set of challenges. The onus, therefore, rests upon all of us to thoughtfully consider our choices and to seek strategies that mitigate the potential harm associated with each energy source.

As expected, energy and the environment are highly intertwined with politics and are subject to the policies of the government in charge. In the USA, many environmental protection initiatives from the previous administration were canceled in 2025. As a consequence, multinational oil companies began cutting planned investments in clean energy and redirected spending toward fossil fuels. ESG funds in the United States are losing funding because of low profitability. These realities underscore a fundamental truth about the global energy system: the transition from a reliance on fossil fuels to a predominantly renewable energy paradigm will be difficult and costly in the beginning and will face many obstacles in a global profit-driven economy.

A key assumption underpinning the anticipated pace of this transition, namely reaching a plateau in fossil fuel consumption, appears to have fallen short of real-world dynamics [2]. Global energy consumption has surged by an astounding 70% since 1990, and the world's population is projected to expand by approximately two billion in the coming decades, with a significant portion of this growth concentrated in the Global South. Additionally, the high energy demands of AI will significantly increase global energy use. Consequently, the demand for fossil fuels may not have yet reached a plateau. Adding another layer of complexity: roughly 30% of the global population continues to rely on traditional biomass for cooking; in Africa, nearly 600 million individuals still lack access to electricity. The financial burden associated with transitioning to renewable energy has proven to be far greater than anticipated, placing a significant strain on developing nations that require considerable economic assistance to facilitate this shift. Moreover, a critical oversight in many initial assessments was the failure to fully appreciate that climate objectives do not exist in isolation but are intrinsically linked to other goals, such as Gross Domestic Product (GDP) growth, economic development, energy security, and the reduction of local pollution. The current climate of heightened global tensions, the Ukraine war, for example, has further impeded progress toward a unified consensus on the way forward.

Adding to these complexities, electricity consumption has begun to outpace recent demand forecasts. This increasing electrification will necessitate a significant and reliable supply of critical minerals, such as lithium. Current projections suggest that the mining industry may struggle to meet this escalating demand. As astutely observed by Yergin et al., [2] the transformation of the global energy system will not unfold in a uniform or linear fashion across different

TABLE 12.1 Energy spectrums and their impacts.

SDG no.	Description	Fossil fuels	Nuclear	Solar	Wind	Geothermal	Hydropower	Hydrogen	Biomass
3	Good health and well-being								
6	Clean water and sanitization	6a	6b					6c	
7	Affordable and		7a		7b	7c		7e	7f
	Clean energy	7g							
		7h						7i	
13	Climate action	13a							
14	Life below water	14a		14c	14d	14e	14f	14g	
15	Life on land	15a	15c	15d	15e		15g		

Table key*:*
3a Generally speaking, all energy sources are intended to provide good health and well-being for humanity.
6a The heavy usage of water in the fossil fuels industries, if not treated properly, could impact the potable water supply.
6b The nuclear industry is relying on a significant water supply. This could impact access to water in some communities.
6c Some of the technologies for generating hydrogen require access to a significant supply of clean water unless their methodology could utilize wastewater.
7a Installing nuclear plants is costly, and when this cost is transferred to the customers, it could make energy unaffordable.
7b Cost of electricity from wind is generally higher. Also, installing windmills has a high upfront cost.
7c Similar to Wind, geothermal projects are associated with high upfront costs.
7e Current methods of hydrogen generation are very expensive.
7f biomass is an emerging technology and is not available everywhere.
7g Fossil fuels are considered polluting and not clean.
7h Once CO_2 is captured via the CCUS process, fossil fuels approach net zero energy.
7i Although green hydrogen is considered clean, other forms of hydrogen generation are causing a lot of emissions and, similar to fossil fuels, they need CCUS to be considered green.
13a Climate action is required to increase investment in renewable and nuclear energy while making fossil fuels more carbon neutral.
14a Careful management of groundwater, seas, lakes, and oceans is a key feature of a sustainable future. Some experts have expressed concerns about how fossil fuels have impacted this asset with spills, mining, and fracturing.
14c The Mining industry, in its search for minerals and metals, has accumulated a lot of wastewater in tailings dams. Careful management of these sites is essential to prevent groundwater contamination.
14d Similar to solar, all windmill parts might come from high-emission sources.
14e Geothermal plants might require additional hydraulic fracturing to provide more access to hot rocks.
14f Hydropower dams generally disrupt the river flow and impact fishing life.
14g High water use in hydrogen production could impact the water sources.
15a Many exploration, production, and development activities might impact terrestrial ecosystems, which could lead to biodiversity loss.
15c There are many risks associated with nuclear energy that need close management.
15d Presence of solar panels on land could be disruptive to livestock. However, in a desert environment, the presence of shade under panels could help the soil and potential vegetation.
15e The same is true of solar panels, plus the associated risks to birds and impact on the landscape.
15g The disruption caused by water dams could impact the community's way of living.

Note: *Green, orange and red colors designate good alignment, somewhat challenged and severely challenged respectively in delivering SDG goals.*

regions of the world. Instead, it will proceed at varying speeds, characterized by diverse combinations of fuels and technologies. Ultimately, the energy transition represents nothing less than a profound re-engineering of the entire global economy, a multifaceted endeavor demanding innovation, collaboration, and a steadfast commitment to a sustainable future.

Energy transition impact on jobs

A "just" transition to a low-carbon economy requires economic, environmental, and social policies that aim to create decent work, quality jobs, and an equitable, prosperous future for all workers and communities. This necessitates understanding, anticipating, and responding to the full spectrum of impacts that the energy transition will have on employment and related human rights. On a positive note, the workforce in energy production sectors can transition to emerging clean energy fields. Moreover, the renewable energy sector is more labor-intensive than fossil fuels and is likely to generate a greater number of both low-skilled and skilled jobs.

The energy transition will require a highly skilled workforce with advanced expertise in research, innovation, and the development of pioneering renewable energy technologies. This calls for renewed investments in science, technology, engineering, and mathematics education and training for both men and women. The existing workforce needs to be retooled, and the new generation needs to be educated using new approaches in order to fill in the competency gaps in the application of new technologies. For instance, coal miners, who already have strong skills in handling heavy equipment and working in physically demanding environments, can be retrained as solar panel installers or wind turbine technicians. Some of the changes have already started. Professionals are trained in electrical basics, renewable energy systems, and safety certifications, including climbing and rigging for wind projects. Similarly, coal plant operators, who are accustomed to overseeing complex power generation systems, have found new opportunities, such as battery storage operators or grid technicians, by learning about power electronics and digital monitoring. College courses are being continuously updated to include renewable energy fundamentals.

In the oil and gas sector, offshore rig operators could transition into offshore wind technician roles. They are already familiar with marine safety and challenging offshore conditions, so the additional training would focus on turbine maintenance and electrical systems. Petroleum engineers, with their expertise in fluid dynamics and reservoir modeling, are well-suited to shift toward geothermal energy or hydrogen systems engineering, though they would need renewable-specific training to adapt their knowledge. Pipeline workers, currently transporting oil and gas, could be redeployed to manage hydrogen or CO_2 transport systems, provided they receive training in gas handling, leak detection, and hydrogen safety standards. Refinery workers might find new employment in biofuel or hydrogen plants, where their process knowledge can be extended through learning biochemical processes and updated safety protocols. Such training needs to be initiated in many city colleges.

The automotive industry is another sector experiencing significant disruption. Workers in assembly lines focused on internal combustion engines have been retrained for electric vehicle (EV) assembly, where new skills in drivetrains, battery integration, and high-voltage safety are essential. Mechanics specializing in gasoline engines could also adapt to become EV and hybrid technicians by learning to diagnose batteries, handle high-voltage systems, and manage software-driven maintenance. In heavy industry, steel workers operating blast furnaces may find new roles in green steel production, which uses hydrogen instead of coal. This transition would involve developing knowledge in hydrogen metallurgy and process automation. Chemical plant workers, who have long operated fossil-based production processes, have started redirecting their expertise toward bio-based chemicals or circular economy facilities, which focus on waste-to-product innovation.

Beyond direct industrial roles, logistics and construction are also transforming. Some workers previously employed in fossil fuel transport, whether by rail, trucking, or shipping, have shifted toward renewable energy supply-chain management and EV charging logistics. This involved training in fleet electrification, charging infrastructure, and digital logistics platforms. Some construction workers whose primary experience has been in building fossil fuel plants or pipelines have transitioned into building and retrofitting for energy efficiency. Their new skills include installing insulation, electrifying HVAC systems, and implementing smart building technologies.

Across all these transitions, several training themes are emerging. Electrical and digital skills are becoming increasingly essential in renewables, EVs, and smart grids. Safety and certification remain critical, whether it involves working at heights for wind, managing high-voltage systems for EVs, or handling hydrogen and CO_2 safely. Green process knowledge is also necessary, especially in biofuels, green steel, and carbon capture technologies. Finally, IT and data literacy are vital as the energy system grows more reliant on smart systems, predictive maintenance, and digital twins. Some of the training courses for energy transition and relevant new job skills are included in the Attachment at the end of this chapter.

The good news is that a green energy transition means many more jobs and career opportunities will open up. The International labour organization (ILO) estimates that net employment gains from a sustainable energy scenario could amount to around 2.5 million jobs, more than offsetting the loss of approximately 400,000 jobs in the fossil fuel sector [3]. However, these new opportunities may arise in different locations, require diverse new skill sets, and operate in different sectors than those from which jobs are being lost. Targeted training programs and strategic investments are essential to ensure that workers at risk of losing their

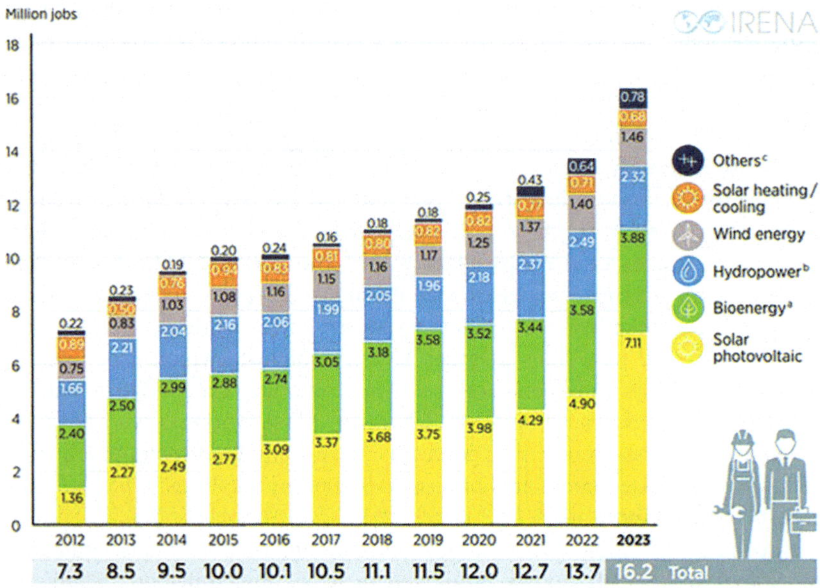

FIGURE 12.1 Evolution of global renewable energy jobs between 2012 and 2023 ('a' includes biofuels, solid biomass, and biogas, 'b' direct jobs only, 'c' includes geothermal, concentrated solar power, heat pumps, municipal and industrial waste, and ocean energy) [5].

jobs due to the energy transition can benefit from it. Robust social protection schemes, including unemployment benefits, can provide some level of safety net [4].

To support the workers' upskilling for a carbon neutral transition, we need to focus on:

- Financial support for worker upskilling and training.
- Aligning federal skills programming with climate change initiatives.
- Providing workers with career pathways, including options for early retirement.

Fig. 12.1 presents the evolution of jobs in the renewable industries in the past few years.

It is important to note that the impact of energy transition and sustainability on employment will vary depending on factors such as government policies, technological advancements, and economic conditions. While some jobs may decline, exciting new opportunities will emerge in related sectors, such as renewable energy, energy storage, and energy efficiency.

The ILO [6] identified skills development as a key policy area for addressing environmental, economic, and social sustainability. Ultimately, the goal is to create an energy sector that is both sustainable and just. This requires a commitment to human rights, environmental protection, and equitable access to energy for all. If we can all work together, we can build a future where energy serves as a force for positive change rather than a source of harm and inequality [7].

A just transition will require integrated government policies and significant adaptive capacity and flexibility from workers and employers. New and evolving skill requirements to serve new technologies may require workers to undergo significant career changes and frequent training measures. For employers, on the other hand, fast-paced change in an environment of labor and skills shortages also means that, across sectors, skills requirements could become more important than degree requirements. That means preparing for the new jobs will be cost-effective and will make labor markets more inclusive for individuals with non-standard education and for older workers.

Countries should take urgent action to anticipate the skills needed for the transition to greener economies and provide new training programs. The transition to more sustainable agricultural systems has started to create jobs in medium and large organic farms and allow smallholders to diversify their sources of income.

Policymakers will need to work globally on minimizing the uncertainties and disruptions inherent in the energy transition. Efforts should be made to close social protection gaps in the transformation process. Measures to develop skills should be integrated into the planning stages of national transition frameworks and new renewable energy projects while encouraging enterprises and industry to invest in upskilling and reskilling their workforce.

An example of how this can be done is the EU Skills Partnership for the renewable energy industrial ecosystem [8]. The European Commission has launched the Pact for Skills, a comprehensive partnership aimed at enhancing skills within the renewable energy ecosystem. This initiative unites over 40 employers' organizations and 35 chambers of commerce across the continent to establish robust skills partnerships. Its objectives include monitoring skill supply and demand and anticipating future needs, fostering a culture of lifelong learning, and promoting gender equality and equal opportunities while combating discrimination. The partnership has moved beyond launch into active implementation through linked projects to develop training programs and create EU-wide certifications tailored to high-demand renewable energy roles. It has already supported millions of workers with reskilling and upskilling opportunities across sectors, indicating that the model is functioning and growing.

If timely solutions are not developed, the consequences could be dire, as was demonstrated by the experience of UK coal miners in the 1980s. Within a decade, most UK coal mines were closed, leading to 90% of the workforce – more than 200,000 people – losing their jobs. A recent analysis of this experience [9] reveals that hourly wages fell by 40% and earnings dropped by 80%–90% one year after the job losses and that these losses persisted for many years. These displacements had profound impacts on health, education, and demographic trends. Similar social stress could arise in fossil-fuel-dependent regions, including oil-producing countries such as Iraq, Nigeria, the Russian Federation, and the United States, if the energy transition fails to adopt a broad, people-centered approach.

A just transition will also require bringing the people who have borne the brunt of both industrial and fossil fuel pollution to the front of the line. Some of the provisions would need to be quite specific, like a low-income allocated credit that would build out rooftop and community solar in low-income communities and tribal communities. There could be a bonus in production and investment tax credits to site facilities for clean energy in traditional-energy communities and disadvantaged communities (Please see Chapter 11 for more details).

People's behavioral changes

A survey of nearly 100,000 residential energy consumers over three years, spanning 21 markets in North and South America, Europe, and the Asia Pacific [10], was conducted in 2024. It revealed that 70% of consumers were unwilling to spend more on energy sustainability. The study indicated that consumer fatigue (due to repeated policy changes, rising costs, etc.) is setting in, eroding confidence and stalling progress. The study also identified a disconnect between government clean energy policies and consumer behavior. For example, while several jurisdictions are increasingly focusing on replacing gas boilers with heat pumps, only 11% of consumers indicated that heat pumps would be their top choice for energy-related investments in the next three years.

Ernst and Young's Energy Consumer Confidence Index [11] is a measure of how confident consumers feel about the energy system and its transition to a cleaner future. It is based on a survey of thousands of residential energy consumers across nearly 20 markets globally, providing insights into consumer attitudes and behaviors related to energy. The most recent data (2024) showed declining confidence scores in almost all markets, with the steepest drops occurring in regions experiencing significant price volatility, such as Australia, the Netherlands, and the UK. One major barrier to clean energy investment, according to the report, is the growing pressure of energy affordability. The survey found that two-thirds of American consumers felt that they would be unable to handle a 10% increase in their energy bills. Access to clean energy options is another critical obstacle, with only 33% of respondents confident that they could access these solutions. Renters and low-income individuals were two to three times less likely to plan investments in new energy technologies, highlighting a significant gap in the adoption of clean energy. Another aspect of people's behavioral change is their high usage of digital media and their lack of awareness of its carbon footprint. This is discussed in the following section.

Use of social media and digital activities

Many people spend hours a day using their phones and digital media. Also, video streaming is a common daily activity for many individuals. The carbon footprint of digital activities and social media use is often underestimated but has a significant environmental impact due to the electricity required to power data centers, servers, and devices. Only a few companies have started switching from fossil fuels to alternative energy for powering their data centers.

Streaming video content, particularly in high definition (HD), is one of the most carbon-intensive digital activities [12]. Major platforms like Netflix, YouTube, and Twitch generate considerable emissions due to the large volumes of data transferred through the internet. During the COVID-19 pandemic, for example, Netflix reported a 16% spike in traffic, and video streaming became a major contributor to the increased demand for electricity in data centers, driving up CO_2 emissions globally. The Yale study estimates that if remote working and high streaming rates were sustained, they could contribute an additional 34.3 million tons of CO_2 emissions [13,14].

To mitigate this, users can reduce video quality to standard definition when HD is not necessary by simply adjusting the video quality on their TV sets. This small change could collectively save millions of tons of CO_2 annually.

Social media platforms such as Facebook, Instagram, TikTok, and Twitter (X) consume significant energy because they rely on vast server farms to store and process large volumes of data (photos, videos, and messages). Facebook's data centers, for instance, are known to be large consumers of electricity, but efforts have been made to power them with renewable energy. Instagram, which hosts massive amounts of photo and video content, also generates a notable carbon footprint due to its extensive use of cloud storage.

While exact footprints vary by platform, estimates suggest that the average social media user generates around 6 gm of CO_2 per minute of browsing, depending on the platform's data intensity. Activities such as watching videos on TikTok or Instagram tend to be more energy-demanding than simply scrolling through a Facebook feed.

Emails might seem harmless, but every email sent, received, and stored also adds to your carbon footprint. A standard email (without attachments) can result in emissions of around 4g of CO_2, while an email with a large attachment can generate up to 50 g of CO_2. With billions of emails sent daily, the cumulative impact is significant. Reducing unnecessary emails or cleaning out unused email accounts can make a difference.

Cloud services like Google Drive, Dropbox, and Microsoft OneDrive contribute to carbon emissions due to the constant storage and retrieval of data in cloud data centers. While major providers are working toward carbon neutrality, the energy required to maintain cloud infrastructures is still considerable.

Platforms like Zoom, Microsoft Teams, and Google Meet saw enormous growth during the pandemic. A one-hour video call can emit between 150 and 1000 gm of CO_2, depending on video resolution and the platform used. Reducing video quality or opting for voice-only calls, when possible, can significantly reduce emissions.

Online gaming and downloading software or updates also have a measurable carbon footprint. The gaming industry, especially cloud-based gaming, consumes substantial energy to process real-time data and manage graphics-intensive tasks. For example, downloading a video game or large files multiple times (instead of storing them locally) increases the energy needed to keep servers running.

Consumers and companies alike can take steps to reduce the carbon footprint of online activities. By using alternative energy sources to run their data centers, major companies could impact their carbon footprint significantly. On the consumer side, reducing video streaming quality, using energy-efficient devices, and turning off idle devices can help with reducing the digital media impact on the environment.

Conscious consumerism

When the movie Blood Diamond was released in 2006, it reignited the discussion about the "right" sourcing of consumer products. The film explores themes of conflict diamonds (blood diamonds), war profiteering, child soldiers, and the exploitation of African nations in the global diamond trade. It highlights the brutal reality of how diamonds have been used to finance civil wars, leading to widespread human suffering. There is a growing consumer movement focused on identifying the ethical sources of products and avoiding those with dubious histories, particularly when it comes to human rights violations, environmental destruction, and unethical labor practices. This movement spans various industries, including fashion, food, electronics, diamonds, and other consumer goods. Consumers actively seek products labeled as Fair Trade Certified, Rainforest Alliance Certified, or Ethically Sourced, ensuring that workers receive fair wages and that environmental sustainability is considered. Even social media and online activism have fueled boycotts of companies that engage in child labor, unfair wages, or environmental destruction [15].

Online activism could also be applied to identifying where each source of energy is coming from. Raising awareness can apply to all levels of energy production. Some consumer advocacy groups are asking to trace each barrel of produced oil to the conditions under which it was produced. The same applies to each kWh of electricity and how it was generated. Many consumers and advocacy groups push for conflict-free minerals in electronics, ensuring that materials like titanium, tin, tungsten, and gold (3TG minerals) do not fund warlords. Fortunately, some ethical companies are starting to use blockchain technology to allow consumers to track where their products come from, ensuring no forced labor or unethical sourcing is involved. Several trusted organizations assess companies' supply-chain ethics:

KnowTheChain – ranks companies on forced labor and supply-chain transparency.

Corporate Human Rights Benchmark evaluates human rights due diligence.
Carbon Disclosure Project covers supply-chain environmental data.
MSCI ESG, Sustainalytics provide ESG risk scores (subscription-based).
Fair labor association membership indicates independent monitoring.

Companies that track ethical practices often align with recognized standards:

ISO 20400 (sustainable procurement)
SA8000 (social accountability)
Fairtrade, Rainforest Alliance, FSC (sector-specific)
OECD due diligence guidance compliance (especially for minerals and energy)

Plastic recycling

Plastics are the most common items in landfills. Petrochemicals, like plastics, are custom-designed hydrocarbon chains tailored to highly specific applications. About 14% of global oil and 8% of natural gas are consumed to produce petrochemicals and pharmaceuticals. Some of the most common plastic materials are polyvinyl chloride, polyethylene, low-density polyethylene, polyester, polypropylene (PP), Styrofoam, Plexiglas, Mylar, Teflon, and polyethylene terephthalate (commonly known as PET). The annual global production of plastic currently runs to more than eight hundred billion pounds [16]. What was a problem of scarcity is now a problem of superabundance. Plastic waste, including water bottles, shopping bags, and snack packages, turns up pretty much everywhere today. It has been found at the bottom of the ocean, thirty-six thousand feet below sea level, on beaches and shores around the world. The Great Pacific Garbage Patch [17], a collection of floating debris that stretches across 600,000 square miles between California and Hawaii, is thought to contain some 1.8 trillion plastic shards. Environmentalists are especially concerned about microplastics, plastic bits smaller than 5 mm across. They can contaminate our surroundings, including food products. In addition to their main ingredients, plastics may contain any number of additives. Many of these, for example, polyfluoroalkyl substances (PFASs), are suspected carcinogens [18]. Many of the others have never been adequately tested.

Americans produce more plastic waste each year than the residents of any other country—almost 500 pounds per person, nearly twice as much as the average European and sixteen times as much as the average Indian [19]. Thus, recycling plastic is a must for sustainability.

The problem with plastic recycling is that a polymer degrades each time it is heated. Thus, even under ideal circumstances, plastic can be reused only a couple of times, and in the waste management business, very little is ideal.

It is not easy to find alternative materials to replace plastics. Precisely because plastic is now ubiquitous, it is difficult to imagine how to replace all of it, or even much of it. Even in cases where substitutes are available, it is not always clear that they are preferable. The Danish Environmental Protection Agency [20] analyzed how different kinds of shopping bags compare in terms of lifecycle impacts. The study found that, to have a lower environmental impact than a plastic bag, a paper bag would have to be used forty-three times, and a cotton tote would have to be used an astonishing seventy-one hundred times. Generally, it is understood that exchanging plastic for other materials may involve tradeoffs, including energy and water use and carbon emissions. However, if not recycled properly, plastics are by far more damaging to the environment than paper (Box 12.1). The readers who are interested in learning more about recycling can access publicly accessible references [21,22].

BOX 12.1 Tech note: Recyclable plastics

The most commonly recycled plastics are those with the recycling symbols 1, 2, 4, and 5 [23].

1. PET: Water bottles, soda bottles, food packaging.
2. High-density polyethylene: Milk jugs, detergent bottles, shampoo bottles.
4. Low-density polyethylene: Grocery bags, plastic wrap, bubble wrap.
5. PP: One Yogurt containers, medicine bottles, bottle caps.

There have been many reports about inadequate plastic recycling in different cities, or a lack of funds to do so. Perhaps, providing incentives for consumers, such as the ones used for glass bottles or aluminum can recycling, might promote plastic recycling. However, we need to be proactive in reducing our plastic consumption. When faced with the question of "Paper or Plastic" in a store, a better approach would be to use our own bags.

Food composting

According to the UN Environment Program [24], over a billion tons of food, roughly one-fifth to a third of all food produced, is wasted each year globally. Aside from methods such as better ways of demand forecasting for retailers and restaurants (using AI) or smart packaging that tracks freshness, composting is one way of reducing the unavoidable food waste. Composting is the controlled decomposition of organic materials (food scraps, yard waste) into a soil additive called compost, which is rich in nutrients and microbes. The outcome of composting is healthier soils, lower emissions, less waste for landfills, less methane emissions, and a more circular and resilient food system. Open piles, tumbler bins, wood bins, and wire mesh bins can be used for this purpose. The compost can be ready within 3–6 months [25].

The end-of-life and the right to repair advocacy

End-of-life (EoL) advocacy for all consumer products promotes responsible and sustainable product design, use, and disposal, placing the responsibility for product disposal on manufacturers. It seeks to minimize environmental impact and maximize resource efficiency by extending product lifespans, recovering valuable materials, and reducing waste. This approach aligns with the principles of a circular economy in which materials and products are kept in circulation for as long as possible. A circular economy reduces material use, redesigns materials and products to be less resource-intensive, and recaptures "waste" as a resource to manufacture new materials and products. This is in contrast with the traditional linear "take-make-dispose" model.

To implement EoL advocacy effectively, products should be designed with their entire lifecycle in mind, considering factors such as material selection, energy efficiency, and recyclability. Manufacturers should be held accountable for the entire lifecycle of their products, including EoL management and disposal. Companies can implement take-back programs to collect and recycle used products, such as electronics, batteries, and appliances. Designing products for easy disassembly and recycling allows for the recovery and reuse of components, while strategies like reuse, repair, and sharing help reduce waste generation.

When properly implemented, EoL advocacy can reduce pollution, conserve resources, and mitigate climate change. It can also create new jobs and industries related to recycling, remanufacturing, and repair while driving innovation in product design and manufacturing processes. Closely related to this concept is the "Right to Repair" movement, which grants consumers the legal right to repair their own products without voiding the manufacturer's warranty. This movement also pushes manufacturers to make repair information publicly available, extending the longevity of consumer products and reducing waste. Organizations like iFixit provide repair manuals, teardowns, and right to repair explainers. They also provide step-by-step instructions and repairability scores.

Unfortunately, the right to repair provisions were stripped from the compromise version of the 2026 defense policy bill after industry pushback [26]. The public needs to be proactive in reinstating such provisions. The impact of consumerism on the environment is thoroughly examined in the 2024 Netflix special *Buy Now: The Shopping Conspiracy.*

While EoL advocacy is gaining traction, significant challenges remain. Public awareness of sustainable consumption and responsible disposal needs to increase. Infrastructure for recycling, remanufacturing, and waste management must be developed, and policies and regulations should incentivize these practices. Effective implementation will require collaboration between manufacturers, retailers, and consumers to promote circular economy principles.

Case study: Empowering rural communities with hybrid renewable energy

This case study serves as an innovative model for rural communities seeking to transition to clean energy while promoting economic development and environmental sustainability. It presents the initiatives taken by a local co-op in bringing renewable energy to its customers.

New Mexico (NM) has an outstanding resource base for natural gas, wind, and solar. Formed in 1944, Kit carson electric cooperative (KCEC) is a member-owned electric distribution cooperative in Taos, NM, and the second-largest cooperative in the state. Driven by state mandates and a commitment to sustainability from its members, KCEC's journey toward renewable energy is fundamentally driven by its members. Dissatisfied with escalating wholesale power rates and a reliance on coal power plants, the co-op's members sought a more localized and sustainable energy solution. This democratic approach, inherent to the co-op model, empowered the community to dictate its energy future (personal interview 2025).

Luis Reyes, CEO and General Manager of KCEC, explained, "We have to complement our economy, our outdoor recreation, and the tourist economy. So, keeping the air and water clean is paramount, and it is just not just an environmental issue. I mean, we have hunters and fishermen. I mean, we have everything for the outdoors."

He explained that co-op owners pushed for these changes: "Our investors dictated what kind of power supply was right and then did it locally. A lot of challenges and criticism were received because we're breaking that status quo model."

Started around 2005, KCEC leveraged NM's abundant solar resources to provide 42 MW of distributed solar resources to its 30,000 members. This is a win-win model mixing agriculture with solar farms. Solar photovoltaics (PV) arrays are well-accepted in Taos County, and they are being deployed locally. Fig. 12.2 shows the Taos Mesa solar and battery facility. The project spans roughly 170 acres and has approximately 43,680 bifacial solar panels.

The co-op's commitment to renewable energy extended beyond solar power. Battery storage is crucial for achieving round-the-clock renewable energy and bolstering grid resilience (see Fig. 12.3). They operate 16.25 MW of battery storage across their service territory. The battery energy storage system, composed of 10 Tesla Megapacks are located on the east side of the Taos Mesa solar and battery facility.

To mitigate the community impact of catastrophic events such as wildfires and safeguard the region's tourism-based economy, KCEC has added three new microgrid local electrical power resources and distributed generation, powered by batteries. KCEC is also actively exploring alternative technologies like lithium iron phosphate and iron rust batteries, prioritizing safety, longevity, and

FIGURE 12.2 KCEC's Taos Mesa solar array (Courtesy KCEC). *KCEC*, Kit carson electric cooperative.

FIGURE 12.3 KCEC's battery energy storage system (Courtesy KCEC). *KCEC*, Kit carson electric cooperative.

cost-effectiveness. The objective is to increase storage capacity and duration to cover nighttime energy demands and provide essential grid support services.

The community's electricity supply is already 100% solar during daylight hours. However, KCEC is mandated by state law to deliver 100% renewable electricity 24-7 by 2050, and its membership is interested in transitioning sooner

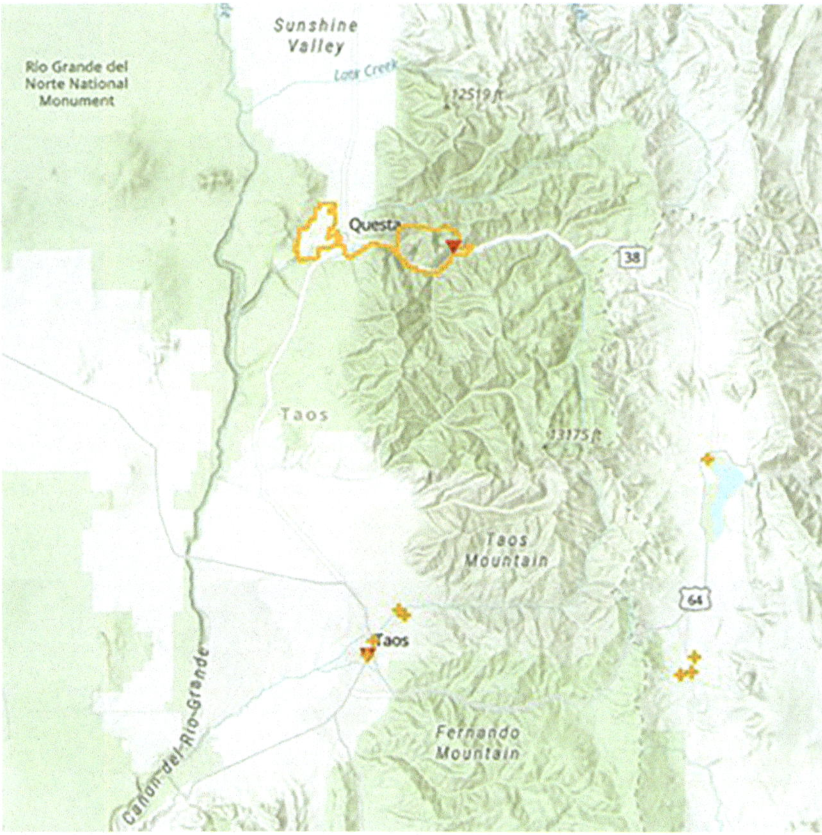

FIGURE 12.4 Map of former molybdenum mine site and tailings facility located in Questa, New Mexico. (Map by EPA) [30]. *EPA*, Environmental Protection Agency. (*Map lines delineate study areas and do not necessarily depict accepted national boundaries*).

if possible. In order to fulfill this target, KCEC has initiated a project to generate clean hydrogen using solar power for water electrolysis. Since potable water is scarce, they are focusing on using the abundant wastewater resource from a nearby molybdenum mine tailings dam (see Fig. 12.4 for location map). It is part of the legacy of the Questa Mine, formerly known as the Molycorp Mine (Fig. 12.5). Mining operations began in 1920. Open pit mining took place from 1965 to 1983. Mining operations placed over 328 million tons of acid-generating waste rock into nine piles surrounding the open pit and disposed of over 100 million tons of tailings in the tailings ponds. The mine was decommissioned in 2014 and is currently under reclamation activities, including wastewater treatment [27]. This wastewater requires treatment to remove contaminants such as the heavy metals arsenic, cadmium, and molybdenum, and suspended solids. The US Department of Energy, as part of the community's local energy action

FIGURE 12.5 Questa mine open pit [31].

program (LEAP) [28], granted $231 million to KCEC to carry out feasibility studies and project planning [29]. LEAP was established to facilitate sustained economic empowerment through clean energy, improve local environmental conditions, and open the way for other benefits primarily through the Department of Energy (DOE) clean energy deployment work. This opportunity was specifically open to low-income, energy-burdened communities that are also experiencing either direct environmental justice impacts or direct economic impacts from a shift away from historical reliance on fossil fuels. The technical work was authored by the National Renewable Energy Laboratory for the DOE.

The project's design seeks to generate 460 tons/year of green hydrogen. This hydrogen will then be used for power generation or as fuel in hydrogen cars. The selection of the optimal electrolysis technology is contingent on thorough due diligence, with water quality and cost being paramount considerations. Collaboration with universities and national labs is helping KCEC in its endeavor. The project is designed to create a replicable model for rural communities, fostering local economic opportunities through job creation and the development of a hydrogen economy.

Project details

Following an initial feasibility study in 2022 [32], the Questa project is going forward with a 2028 startup milestone [33]. The conceptual design calls for withdrawing 5.34 acre-ft/year of wastewater and 25.9 GWh/yr of solar electricity.[1] They have allocated up to 105 acres of land for the 34 MW of solar PV and H_2 facility. The hydrogen storage is designed to hold 29 tons of compressed hydrogen gas [34].

This project is expected to generate 460 tons/year of H_2, which in turn can generate 9.0 GWh/year of electricity. As a byproduct of this operation, 0.27 acre-ft/year of condensed water will be produced. Levelized cost of hydrogen is estimated at USD 4.77/kg of 700 bar H_2 (or USD 3.82/gallon of diesel equivalent with incentives). The estimated carbon credit for this project is 456 metric tons of avoided CO_2e/year.

The inherent uncertainties surrounding the total cost of the hydrogen project are primarily due to fluctuating market conditions, inflation, tariffs on essential materials like steel, and the evolving requirements for infrastructure development. KCEC will need supplementary funding, including private investors, while balancing it with the implications of such partnerships on the co-op's governance.

The Questa hydrogen project was originally planned to use treated effluent from the wastewater treatment plant for hydrogen production through electrolysis. Shortly after the project received its USDA grant, the operator changed the proposed water source to a well located at the old tailings site, a portion of which KCEC has leased from the operator for the project. Due to drought conditions, some residents have objected to such use. Additional studies are underway to rectify the issues surroung the water source [35].

The outcome of the Questa project is powering 25,000 homes while reducing pollution by 98,000 tons each year, equivalent to removing 20,700 gasoline-powered vehicles annually. Another positive aspect of this project is generating 350 local jobs during the construction phase. Use of compressed hydrogen for fuel cell usage in heavy machinery is also planned.

Summary

By embracing collective responsibility, fostering innovation, and adopting sustainable practices, we can move toward a future powered by clean, secure, and equitable energy sources. The choices we make now will shape the future of our planet and the well-being of future generations. The journey toward a sustainable energy future demands a collective effort. Taos KCEC co-op members pushed

1. 1 acre foot = 1 year of indoor water use for 4 NM households or 325,851 gallons. 1 metric ton H_2 = 1,250 gallons of diesel fuel, KCEC approximates 6-7 acres per 1 MW of solar PV power generation capacity

for clean sources of energy for their community and also advocated for better conservation practices at home.

Key takeaways

As individuals, we can contribute by [36]:

- Being aware of the pros and cons associated with all sources of energy and opt for the one that will result in fewer emissions to the environment.
- Supporting policies that promote renewable energy development, energy efficiency, and a just transition to cleaner energy sources is crucial for long-term sustainability.
- Supporting research and the development of technologies that enhance clean energy options will accelerate our progress toward a sustainable energy future.
- Reducing global waste via recycling, composting, and mindful consumption. Even small daily changes can contribute to a cleaner planet.
- Choosing active transportation such as walking, cycling, or using public transport. If adopted collectively, this will significantly reduce carbon emissions while improving personal health. For longer distances, trains and buses are more eco-friendly alternatives to cars and planes.
- Using more efficient appliances for heating and cooling. Choosing heat pumps, electric tankless water heaters, electric stoves, or induction burners can lower your energy consumption significantly and reduce carbon emissions.
- Shifting toward more sustainable consumption patterns. While changing lifestyles means confronting ingrained social habits and the private interests that sustain current consumption patterns, these changes are essential to achieving 'regulation' and 'early action' policies aimed at the fossil fuel and energy sectors.
- Simple changes can have a significant impact and are the best place to start. Turning off lights, using lids on pots, lowering thermostats, washing clothes in cold water, taking shorter showers, and unplugging unused devices can all save energy. Replacing traditional bulbs with energy-efficient LEDs can further reduce environmental harm.
- The fashion industry has enormous environmental and social costs. It generates over 92 million tons of waste annually, much of which ends up incinerated in landfills or exported to developing countries. Choosing fewer fast-fashion items and opting for higher-quality, longer-lasting clothing can help reduce this waste.

Attachments

1. ***Academic courses for energy transition and sustainability***

Here are some academic courses that can provide a strong foundation for students interested in energy transition and sustainability:

a. Core science courses

Physics: Understanding fundamental concepts of energy, mechanics, and thermodynamics is crucial for energy analysis and technology development.

Chemistry: Knowledge of chemical reactions, energy transformations, and materials science is essential for understanding energy sources, storage, and conversion.

Biology: A grasp of biological systems, ecosystems, and environmental processes is important for understanding the impacts of energy production and consumption on the natural world.

b. Engineering courses

Mechanical engineering: Courses in thermodynamics, fluid mechanics, and heat transfer are particularly relevant for energy systems design and analysis.

Electrical engineering: Understanding electrical circuits, power systems, and renewable energy technologies is crucial for developing sustainable energy infrastructure.

Chemical engineering: Knowledge of process design, materials science, and energy systems is valuable for optimizing energy production and efficiency.

c. Environmental science and sustainability courses

Environmental science: Courses on environmental impacts, pollution control, and sustainable practices provide a broad understanding of the environmental challenges and opportunities related to energy.

Sustainability studies: Courses exploring sustainable development, environmental ethics, and policy analysis help students develop a holistic perspective on energy transition.

d. Economics and policy courses

Energy economics: Understanding energy markets, pricing, and policy analysis is essential for making informed decisions about energy investments and transitions.

Environmental economics: Courses on environmental valuation, cost-benefit analysis, and policy instruments provide a framework for evaluating the economic impacts of energy policies.

Energy policy: Courses on energy governance, regulation, and international cooperation explore the role of policy in shaping energy transitions.

e. Specialized courses (depending on interest)

Renewable energy technologies: Courses on solar, wind, hydro, geothermal, and biomass energy systems provide in-depth knowledge of specific renewable energy technologies.

Energy storage: Courses on battery technology, energy storage systems, and grid integration help students understand the challenges and opportunities for energy storage.

Energy efficiency: Courses on energy efficiency measures, building design, and industrial processes explore strategies for reducing energy consumption.

By taking a combination of these courses, students can gain a strong foundation in the technical, environmental, economic, and policy aspects of energy transition and sustainability.

2. Required jobs for energy transition and sustainability [37]

The energy transition and sustainability sectors are experiencing rapid growth, creating a wide range of job opportunities. Here are some examples:

a. Renewable energy

Renewable energy engineers: Design, develop, and maintain renewable energy systems like solar, wind, hydro, and geothermal.

Renewable energy technicians: Install, operate, and maintain renewable energy equipment. There were more than 760,000 electricians in the United States in 2022, according to the Bureau of Labor statistics. The number of electrician jobs is expected to grow by 6 percent between 2022 and 2032, double the average growth rate for all occupations.

Renewable energy project managers: Oversee the planning, development, and implementation of renewable energy projects.

b. Energy efficiency and sustainability

Energy efficiency consultants: Assess energy consumption and recommend efficiency improvements for businesses and households.

Sustainability managers: Develop and implement sustainability strategies for organizations.

Environmental scientists and engineers: Conduct research and develop solutions to environmental challenges related to energy production and consumption.

c. Energy storage

Battery engineers: Research, develop, and test battery technologies for energy storage.

Grid storage specialists: Manage energy storage systems connected to the electrical grid.

Energy storage project managers: Oversee the development and implementation of energy storage projects.

d. Energy policy and regulation

Energy policy analysts: Conduct research and provide analysis on energy policies and regulations.

Energy lawyers: Specialize in energy law, advising clients on regulatory compliance and legal issues.

Energy lobbyists: Advocate for energy policies and initiatives on behalf of various stakeholders.

e. Energy technology development

Energy researchers: Conduct research on new energy technologies and materials.

Energy technology developers: Develop and commercialize new energy technologies.

Energy entrepreneurs: Start and manage businesses focused on energy innovation.

f. Energy finance and investment

Energy analysts: Evaluate investment opportunities in the energy sector.

Energy bankers: Provide financial services to the energy industry.

Energy fund managers: Manage investment funds focused on energy transition and sustainability.

g. Climate change mitigation and adaptation

Climate change scientists: Conduct research on climate change impacts and mitigation strategies.

Climate change policy analysts: Develop and implement climate change policies and regulations.

Climate change adaptation specialists: Develop strategies to help communities and businesses adapt to climate change.

This is just a sampling of the many job opportunities available in the energy transition and sustainability sectors. As these fields continue to grow, new and innovative job roles will emerge.

References

[1] United Nations. Sustainable development. (Accessed January 19, 2026). https://www.un.org/sustainabledevelopment/.

[2] Yergin, D., Orszag, P., Arya, A. The troubled energy transition, How to Find a Pragmatic Path Forward. Foreign Affairs. 2025. (Accessed January 19, 2026) https://www.foreignaffairs.com/united-states/troubled-energy-transition-yergin-orszag-arya.

[3] International Labour Organization (ILO), in: World Employment and Social Outlook 2018, ILO, Geneva, 2018, p. 5. https://www.ilo.org/publications/world-employment-and-social-outlook-2018-greening-jobs. 42.

[4] United Nations Department of Economic and Social Affairs (UN DESA), in: Report2020, United Nations, New York, 2020, p. 21. https://sdgs.un.org/publications/report-2020-sdg7-tag-policy-briefs-accelerating-sdg7-achievement-time-covid-33178.

[5] International Renewable Energy Agency (IRENA); International Labour Organization (ILO)Renewable Energy and Jobs: Annual Review 2024, IRENA, Abu Dhabi; ILO: Geneva, 2024 Accessed January 19, 2026 https://www.ilo.org/publications/renewable-energy-and-jobs-annual-review-2024 .

[6] International Labour Organization (ILO)Guidelines for a Just Transition toward Environmentally Sustainable Economies and Societies for All, ILO, Geneva, 2015 https://www.ilo.org/publications/guidelines-just-transition-toward-environmentally-sustainable-economies Accessed January 19, 2026.

[7] K. Huhta, Conceptualizing energy justice in the context of human rights law, Nord. J. Hum. Rights 41 (4) (2023) 378–392. https://doi.org/10.1080/18918131.2023.2210443.

[8] International Labour Organization (ILO)Renewable Energy and Jobs – Annual Review 2024, ILO, Geneva, 2024 https://www.ilo.org/publications/renewable-energy-and-jobs-annual-review-2024 Accessed January 19, 2026.

[9] J.-P. Rud, M. Simmons, G. Toews, F. Aragon, Job Displacement costs of phasing out coal, J. Public Econ. 236 (2024) 105167. https://doi.org/10.1016/j.jpubeco.2024.105167.

[10] ESG Today. 70% of consumers unwilling to spend more on energy sustainability: EY survey. https://www.esgtoday.com/70-of-consumers-unwilling-to-spend-more-on-energy-sustainability-ey-survey/ (Accessed January 19, 2026).

[11] EY. EY research shows declining consumer confidence in the U.S. Energy transition. (Accessed January 19, 2026) https://www.ey.com/en_us/newsroom/2024/02/ey-research-shows-declining-consumer-confidence-in-the-us-energy.

[12] Everything Energy. Learning center hub. (Accessed January 19, 2026) https://everythingenergy.com/learning-center-hub/page/2/.

[13] WebFX. Carbon footprint of the internet. (Accessed January 19, 2026) https://www.webfx.com/blog/marketing/carbon-footprint-internet/.

[14] YaleNews. Surge in digital activity has hidden environmental costs. 2021, (Accessed January 19, 2026) https://news.yale.edu/2021/01/27/surge-digital-activity-has-hidden-environmental-costs.

[15] Ethical Consumer. What is ethical consumerism? (Accessed January 19, 2026) https://www.ethicalconsumer.org/what-ethical-consumerism.

[16] Kolbert, E. How plastics are poisoning us. The New Yorker 2023, (Accessed January 19, 2026) https://www.newyorker.com/magazine/2023/07/03/book-reviews-plastic-waste.

[17] The New Yorker. A grand plan to clean the great pacific garbage patch. 2019, (Accessed January 19, 2026) https://www.newyorker.com/magazine/2019/02/04/a-grand-plan-to-clean-the-great-pacific-garbage-patch.

[18] National Cancer Institute, Division of cancer epidemiology and genetics (DCEG). PFAS exposure and risk of cancer. (Accessed January 19, 2026) https://dceg.cancer.gov/research/what-we-study/pfas.

[19] U.S. Environmental Protection Agency (EPA). National strategy to prevent plastic pollution. (Accessed January 19, 2026) https://www.epa.gov/circulareconomy/national-strategy-prevent-plastic-pollution.

[20] Danish Environmental Protection Agency. Life cycle assessment of grocery carrier bags; 2018. (Accessed January 19, 2026) https://www2.mst.dk/udgiv/publications/2018/02/978-87-93614-73-4.pdf.

[21] K. Wells, *12 Ways to Break Up With Single-Use Plastics; Wirecutter*, *The New York Times*, New York, NY, 2024 https://www.nytimes.com/wirecutter/reviews/stop-using-single-use-plastics/ Accessed 20 Jan 2026.

[22] U.S. Environmental Protection AgencyWhat You Can Do to Reduce Plastic Waste, U.S. EPA, Washington, DC, 2025 https://www.epa.gov/plastics/what-you-can-do-reduce-plastic-waste Accessed January 20, 2026.

[23] Walker-Franklin, I., Jambeck, J. Plastics; Penguin Random House: New York, 2023. (Accessed January 19, 2026) https://www.penguinrandomhouse.com/books/730885/plastics-by-imari-walker-franklin-and-jenna-jambeck/.

[24] UN Environment Program. The facts; stop food loss and waste: global, 2026 (Accessed January 20, 2026) https://www.stopfoodlosswaste.org/about/facts.

[25] U.S. Environmental Protection AgencyCompost Guide, U.S. EPA, Washington, DC, 2009 Accessed January 20, 2026 https://www3.epa.gov/recyclecity/pdf/compost-guide.pdf .

[26] Obis, A. Congress quietly strips right to repair provisions from 2026 NDAA despite wide support; *Federal News Network*: Washington, DC, 2025. (Accessed January 20, 2026) https://federalnewsnetwork.com/congress/2025/12/congress-quietly-strips-right-to-repair-provisions-from-2026-ndaa-despite-wide-support/.

[27] U.S. Environmental Protection Agency (EPA). Case summary: $143 million settlement, cleanup work to prevent future contamination (Red River). (Accessed January 19, 2026) https://www.epa.gov/enforcement/case-summary-143-million-settlement-cleanup-work-prevent-future-contamination-red-river
[28] U.S. Department of Energy. Communities LEAP: Questa, New Mexico. (Accessed January 19, 2026) https://www.energy.gov/communitiesLEAP/questa-new-mexico.
[29] Santa Fe New Mexican. Questa green hydrogen project awarded $231 million. (Accessed January 19, 2026) https://www.santafenewmexican.com/news/local_news/questa-green-hydrogen-project-awarded-231-million/article_4439f366-d390-11ef-87d0-6b2b10f40321.html.
[30] U.S. Environmental Protection Agency (EPA). Superfund site profile (Site ID 0600806). (Accessed January 19, 2026) https://cumulis.epa.gov/supercpad/SiteProfiles/index.cfm?fuseaction=second.photovideoaudio&id=0600806.
[31] U.S. Environmental Protection Agency. Superfund site profile: EPA ID 0600806; U.S. EPA: Washington, DC. (Accessed January 20, 2026) https://cumulis.epa.gov/supercpad/SiteProfiles/index.cfm?fuseaction=second.photovideoaudio&id=0600806.
[32] National Renewable Energy Laboratory (NREL)Is clean hydrogen production a good fit for questa? Intermediate feasibility study results, NREL, Golden, CO, 2022 https://www.nrel.gov/docs/fy23osti/84348.pdf Accessed January 19, 2026.
[33] National Renewable Energy Laboratory (NREL). Pathways to commercial liftoff: clean hydrogen; NREL: Golden, CO, 2023. (Accessed January 19, 2026)https://docs.nrel.gov/docs/fy23osti/86665.pdf
[34] Kit Carson Electric Cooperative (KCEC). Press release /newsroom. (Accessed January 19, 2026) https://kitcarson.com/.
[35] Taos News, Questa hydrogen water study underway, Emily Sender, May 28, 2026, https://www.taosnews.com/news/environment/questa-hydrogen-water-study-underway/article_048dc0d8-1571-549d-b0a9-622386354001.html, accessed on June 1, 2026.
[36] J. Williams, H. Brock, How to make your life greener in 2024, The Conversation, Melbourne, Australia, 2023 https://theconversation.com/how-to-make-your-life-greener-in-2024-219365.
[37] International Renewable Energy Agency, IRENA, Renewable energy and jobs annual review 2024. https://www.irena.org/-/media/Files/IRENA/Agency/Publication/2024/Oct/IRENA_Renewable_energy_and_jobs_2024.pdf. (Accessed: April, 10,2026).

Glossary

Abatement Reduction in the degree or intensity of emissions or other pollutants.

Absorption Chemical or physical take-up of molecules into the bulk of a solid or liquid, forming either a solution or compound.

Acid gas Any gas mixture that turns to an acid when dissolved in water (normally refers to $H_2S + CO_2$ from sour gas).

ACTL Alberta carbon trunk line

Adsorption The uptake of molecules on the surface of a solid or a liquid.

Advanced bioenergy Sustainable fuels produced from nonfood crop feedstocks, which are capable of delivering significant lifecycle greenhouse gas emissions savings compared with fossil fuel alternatives, and which do not directly compete with food and feed crops for agricultural land or cause adverse sustainability impacts.

Afforestation Planting of new forests on lands that historically have not contained forests.

AGR Advanced gas-cooled reactor

AGS Advanced geothermal systems

AI Artificial intelligence

AMI Advanced Metering Infrastructure

Amine Organic chemical compound containing one or more nitrogen in $-NH_2$, -NH or -N groups.

Ammonia (NH_3) This is a compound of nitrogen and hydrogen. It can be used directly as a fuel in a direct combustion process, in fuel cells, or as a hydrogen carrier.

Anthropogenic source A source that is man-made as opposed to natural.

ANWR Arctic National Wildlife Refuge

APS Announced Pledges Scenario

Aquifer Geological structure containing water and with significant permeability to allow flow; it is bound by seals.

Auto thermal reforming (ATR) A process in which the heat for the reaction of CH_4 with steam is generated by partial oxidation of CH_4.

Basalt A type of basic igneous rock that typically erupts from a volcano.

Basin A geological region with strata dipping towards a common axis or centre.

Bathymetric Pertaining to the depth of water.

BE Binding energy, the energy required to overcome this attractive force and split the nucleus

BECCS Bioenergy with carbon capture and storage

BEV Battery electric vehicles

Biodiesel Diesel-equivalent, processed fuel made from the transesterification (a chemical process that converts triglycerides in oils) of vegetable oils and animal fats.

Bioenergy Energy content in solid, liquid, and gaseous products derived from biomass feedstocks and biogas. It includes solid biomass, liquid biofuels, and biogases.

Biogas A mixture of methane, carbon dioxide, and small quantities of other gases produced by anaerobic digestion of organic matter in an oxygen-free environment, such as landfills.

Biomass Matter derived recently from the biosphere.

Biomethane Biomethane is a near-pure source of methane produced either by upgrading biogas (a process that removes any CO^2 and other contaminants present in the biogas) or through the gasification of solid biomass followed by methanation. It is also known as renewable natural gas.

BIPV Building-integrated photovoltaics

Bituminous coal An intermediate rank of coal falling between the extremes of peat and anthracite, and closer to anthracite.

BMS Battery management system

Building sector Includes energy used in residential, commercial, and institutional buildings, and nonspecified other. Building energy use includes space heating and cooling, water heating, lighting, appliances, and cooking equipment.

Buoyancy Tendency of a fluid or solid to rise through a fluid of higher density.

CO_2 equivalent (CO_2e) The universal unit of measurement to indicate the global warming potential (GWP) of each greenhouse gas, expressed in terms of the GWP of one unit of carbon dioxide. It is used to evaluate releasing (or avoiding releasing) different greenhouse gases on a common basis.

CAES Compressed Air Energy Storage

Cap rock Rock of very low permeability that acts as an upper seal to prevent fluid flow out of a reservoir.

Capacity credit Proportion of the capacity that can be reliably expected to generate electricity during times of peak demand in the grid to which it is connected.

Capillary entry pressure Additional pressure needed for a liquid or gas to enter a pore and overcome surface tension.

Capture efficiency The fraction of CO_2 separated from the gas stream of a source

Carbon credit A convertible and transferable instrument that allows an organization to benefit financially from an emission reduction.

Carbon trading A market-based approach that allows those with excess emissions to trade that excess for reduced emissions elsewhere.

Carbonate Natural minerals composed of various anions bonded to a CO_3^{2-} cation (e.g., calcite, dolomite, siderite, limestone).

CBM Coal-bed methane

CCUS Carbon dioxide (CO_2) capture, utilization, and storage is a process consisting of the separation of CO_2 from industrial and energy-related sources, transport to a storage location and long-term isolation from the atmosphere.

CDM Clean development mechanism

CDR Carbon dioxide removal

Certification In the context of carbon trading, certifying that a project achieves a quantified reduction in emissions over a given period.

CIS Commonwealth of Independent States

Clean energy Includes renewables, energy efficiency, low-carbon fuels, nuclear power, battery storage and carbon capture, utilization, and storage.

Coal Includes both primary coal (including lignite, coking, and steam coal) and derived fuels (including patent fuel, brown-coal briquettes, coke-oven coke, gas coke, gas-works gas, coke-oven gas, blast furnace gas and oxygen steel furnace gas). Peat is also included.

Concentrating solar power (CSP) Solar thermal power/electric generation systems that collect and concentrate sunlight to produce high temperature heat to generate electricity.

Conventional liquid biofuels Fuels produced from food crop feedstocks. These liquid biofuels are commonly referred to as first generation and include sugar cane ethanol, starch-based ethanol, fatty acid methyl esther (FAME) and straight vegetable oil (SVO).

CO_2 avoided The difference between CO_2 captured, transmitted, and/or stored, and the amount of CO_2 generated by a system without capture, net of the emissions not captured by a system with CO_2 capture.

Containment Restriction of movement of a fluid to a designated volume (e.g., reservoir).

COP Conference of the Parties refers to countries that initially signed the U.N. climate agreement in 1992.

CRI Carbon Recycling International

Cryogenic Pertaining to low temperatures, usually under about $-100°C$.

CSP Concentrated solar power

CATF Clean Air Task Force

CTL Coal-to-liquids

DAC Direct air capture

DBD Deep borehole disposal

Dense fluid A gas compressed to a density approaching that of the liquid.

DER Distributed energy resources

Diagenesis Processes that cause changes in sediment after it has been deposited and buried under another layer.

DLE Direct lithium extraction

DOI Department of the Interior

Downstream emissions Indirect GHG emissions from sold goods and services. Downstream emissions also include emissions from products that are distributed but not sold (i.e., without receiving payment).

EGS Enhanced geothermal systems

EIA US Energy Information Administration

EIC Energy Infrastructure Council

EMF Electromagnetic fields

Emissions factor A factor that converts activity data into greenhouse gas emissions data (e.g., kg CO_2-e emitted per GJ of fuel consumed, kg CO_2-e emitted per kWh of electricity used).

Emissions credit A commodity giving its holder the right to emit a certain quantity of GHGs (q.v.).

Emissions trading A trading scheme that allows permits for the release of a specified number of tons of a pollutant to be sold and bought.

EMS Energy management systems

Energy content factor The energy content of a fuel is an inherent chemical property that is a function of the number and types of chemical bonds in the fuel.

Energy coal Used as a fuel source in electrical power generation, cement manufacture, and various industrial applications. Energy coal may also be referred to as steaming or thermal coal.

EOR Enhanced oil recovery: the recovery of oil in addition to that produced naturally by fluid injection or other means.

EPA Environmental Protection Agency

EPR European pressurized reactor

ESG Environmental, Social, and Governance

ETC Energy Transition Commission

ETS Emissions trading systems
EV Electric vehicle
Exothermic Concerning a chemical reaction that releases heat, such as combustion.
Fault In geology, a surface at which strata are no longer continuous, but displaced.
Fault reactivation The tendency for a fault to become active, i.e., for movement to occur.
FEED Front end engineering design
Fischer-Tropsch synthesis Catalytic production process to produce synthetic fuels. Natural gas, coal, and biomass feedstocks can be used. Gases includes natural gas, biogases, synthetic methane, and hydrogen.
Flue gas Gases produced by the combustion of a fuel that are normally emitted to the atmosphere.
FNR Fast neutron reactors
Formation A body of rock of considerable extent with distinctive characteristics that allow geologists to map, describe, and name it.
Fuel cell Electrochemical device in which a fuel is oxidized in a controlled manner to produce an electric current and heat directly.
Fugitive emission Any releases of gases or vapours from anthropogenic activities such as the processing or transportation of gas or petroleum.
Gasification Process by which a carbon-containing solid fuel is transformed into a carbon- and hydrogen-containing gaseous fuel by reaction with air or oxygen and steam.
GCRs Gas-cooled reactors
Geothermal Geothermal energy is heat derived from the sub-surface of the earth. Water and/or steam carry the geothermal energy to the surface.
Geochemical trapping The retention of injected CO_2 by geochemical reactions.
Geosphere The earth, its rocks and minerals, and its waters.
GHG (Greenhouse gas) These are the aggregate anthropogenic carbon dioxide equivalent emissions of carbon dioxide (CO_2), methane (CH_4), nitrous oxide (N_2O), hydrofluorocarbons (HFCs), perfluorocarbons (PFCs), and sulphur hexafluoride (SF6).
GHP Geothermal heat pump
GISTM Global industry standard on tailings management
GMST Global mean surface temperature
GPU Graphics processing unit
GTL Gas-to-Liquids
GWP (Global warming potential) A factor describing the radiative forcing impact (degree of harm to the atmosphere) of one unit of a given greenhouse gas relative to one unit of CO_2. (see GWP from the Intergovernmental Panel on Climate Change (IPCC) Assessment Report 4 (AR4) based on a 100-year timeframe).
HAWT Horizontal-axis wind turbines
Hazardous and non-hazardous waste Potentially harmful and nonharmful substances that have been released or discarded into the environment.
HDI Human development index
HDR Hot dry rocks
HEV Hybrid electric vehicles
HF Hydraulic fracturing
HVAC Heating, ventilation, and air conditioning
HWR Heavy water reactors

Hydrate An ice-like compound formed by the reaction of water and CO_2, CH_4, or similar gases.

Hydrodynamic trap A geological structure in which fluids are retained by low levels of porosity in the surrounding rocks.

Hydrofracking (hydroshearing) creating fractures in rocks using liquids

IAEA International Atomic Energy Agency

IEA International Energy Agency

ILO The International Labor Organization

Injectivity A measure of the rate at which a quantity of fluid can be injected into a well.

In-situ mineralization A process where minerals are not mined; carbon dioxide is injected in the silicate formation where it reacts with the minerals, forming carbonates, and silica.

IPCC Intergovernmental Panel on Climate Change.

IRENA International Renewable Energy Agency

Kyoto Protocol Protocol to the United Nations Framework Convention on Climate Change, which was adopted at Kyoto on 11 December 1997

LCA Life cycle assessment

LNG (liquefied natural gas) Consists largely of methane that has been liquefied through chilling and pressurization. One ton of LNG is approximately equivalent to 46,000 cubic feet of natural gas.

L-DAC Liquid Direct Air Capture

Leakage In respect of carbon trading, the change of anthropogenic emissions by sources or removals by sinks that occur outside the project boundary, the escape of injected fluid from storage.

Lignite/sub-bituminous coal Relatively young coal of low rank with a relatively high hydrogen and oxygen content

Lithosphere The outer layer of the Earth, made of solid rock, which includes the crust and uppermost mantle up to 100 km thick.

London Convention On the Prevention of Marine Pollution by Dumping of Wastes and Other Matter, which was adopted at London, Mexico City, Moscow, and Washington on 29 December 1972.

Low-carbon energy carrier Fuel that provides low fuel-cycle-wide emissions of CO_2, such as methanol.

LWR Light water reactors

Madrid Protocol A protocol to the 11th Antarctic Treaty to provide for Antarctica's environmental protection.

Market-based method (for reporting) Scope 2 greenhouse gas emissions based on the generators (and therefore the generation fuel mix from which the reporter contractually purchases electricity and/or is directly provided via a direct line transfer).

Metallurgical coal A broader term than coking coal, which includes all coals used in steelmaking, such as coal used for the pulverized coal injection process.

Mineral trap A geological structure in which fluids are retained by the reaction of the fluid to form a stable mineral.

MIT Massachusetts Institute of Technology

Monitoring The process of measuring the quantity of carbon dioxide stored and its location.

MRV Monitoring, reporting, and verification

NCEA National Center for Energy Analytics

NCG Non-condensable gases

NDC Nationally determined contributions
NET Negative emissions technology
NGL Natural gas liquid
NOAA National Oceanic and Atmospheric Administration
NOx Nitrogen oxides
NPC National Petroleum Council
NPRA National Petroleum Reserves-Alaska
NZE Net zero emission
OCDR Ocean-based carbon dioxide removal
OECD Organization for Economic Co-operation and Development
Operational boundaries The boundaries that determine the direct and indirect emissions associated with operations owned or controlled by the reporting company.
OTEC Ocean Thermal Energy Conversion
OWC Oscillating water column
Oxyfuel combustion Combustion of fuel with pure oxygen or a mixture of oxygen, water, and carbon dioxide.
Paris Agreement The Paris Agreement is an agreement between countries under the United Nations Framework Convention on Climate Change (UNFCC) to strengthen efforts to combat climate change and adapt to its effects, with enhanced support to assist developing countries. Paris Agreement goals were: to hold global average temperature increase to well below 2°C above preindustrial levels and pursue efforts to limit the temperature increase to 1.5°C above pre-industrial levels.
Permeability Ability to flow or transmit fluids through a porous solid, such as rock.
PEMEL Polymer electrolyte membrane electrolyzers
PHEV Plug-in hybrid electric vehicles
PHMSA Pipeline and hazardous materials safety administration
PM Particulate matter
POP Persistent organic pollutants
Porosity Measure for the amount of pore space in a rock.
Post-combustion capture The capture of carbon dioxide after combustion.
Pre-combustion capture The capture of carbon dioxide following the processing of the fuel before combustion.
PSE Passive solar energy
PV Photovoltaic
QCD Quantum chromodynamics
REC Renewable energy certificates or credits
Reforestation Planting of forests on lands that have previously contained forests but that have been converted to some other use.
Safe Drinking Water Act An Act of the US Congress originally passed in 1974. It regulates, among other things, the possible contamination of underground water.
Saline formation Sediment or rock body containing brackish water or brine.
SDG Sustainable development goals
S-DAC Solid direct air capture
Seal An impermeable rock that forms a barrier above and around a reservoir such that fluids are held in the reservoir.
SFR Sodium-cooled fast reactors

SMR Steam methane reforming: a catalytic process in which methane reacts with steam to produce a mixture of H_2, CO, and CO_2.
SMRs Small Modular Reactors
Solubility trapping A process in which fluids are retained by dissolution in liquids naturally present.
SOx Sulphur oxides
Sour gas Natural gas containing significant quantities of acid gases like H_2S and CO_2.
STE Solar thermal energy
Steam reforming A catalytic process in which a hydrocarbon is reacted with steam to produce a mixture of H_2, CO, and CO_2.
STEPS Stated policies scenario
Supercritical At a temperature and pressure above the critical temperature and pressure of the substance concerned. The critical point represents the highest temperature and pressure at which the substance can exist as a vapour and liquid in equilibrium
Synfuel Fuel, typically liquid fuel, produced by processing fossil fuel.
TAPS Trans-Alaska Pipelines
TPU Tensor processing unit
UNEP United Nations Environment Program
UNFCCC United Nations Framework Convention on Climate Change, which was adopted at New York on 9 May 1992.
USD US dollar
UV Ultraviolet light
VAWT Vertical-axis wind turbines
VCUs Verified carbon units
VAWT Vertical axis wind turbines
Verification The proving, to a standard still to be decided, of the results of monitoring (q.v.).
VOC Volatile organic compounds
ZED Zero energy discharge
ZGD Zero gas discharge
ZLD Zero liquid discharge
ZMD Zero mineral discharge

Postface

In this book, we aimed to provide a balanced and unbiased review of various forms of energy, examining both the advantages and disadvantages of each source. Our primary focus was to raise awareness, particularly for future generations, about the importance of energy choices and their far-reaching impacts.

Chapter 1 introduced different energy sources and offered a brief overview of each. We believe that understanding these sources and the challenges they present is essential for making informed decisions. Transitioning to renewable energy is not just beneficial but necessary to reduce environmental harm and ensure the long-term availability of resources. At the same time, we must address broader societal challenges—such as energy security, equitable access, and a fair transition for workers—if we are to create a truly sustainable future of energy.

In Chapter 2, we took a deeper look at fossil fuels, exploring their advantages, the challenges they pose, and what the future may hold for this energy source. As we look ahead, it is clear that two key stakeholders play central roles in the conversation around fossil fuels and emissions reduction. The first is the fossil fuel industry itself, which faces an increasingly uncertain future. The second stakeholder is governments, which carry the responsibility of ensuring that the industry's emissions are aligned with national climate goals. At the same time, the governments must manage a transition toward more sustainable energy systems.

In Chapter 3, we turned our attention to nuclear energy, examining its challenges, future trends, and the evolving landscape of reactor technology. We provided an overview of various types of nuclear reactors, highlighting both established designs and emerging innovations. Among the newer developments, small modular reactors (SMRs) and microreactors stand out for their increased flexibility, scalability, and cost-effectiveness. These advanced designs have the potential to make nuclear energy more accessible and adaptable, especially in remote or decentralized settings.

In Chapter 4, we explored solar energy, reviewing recent technological advances as well as the ongoing challenges associated with its broader adoption. As we consider the potential of solar power to become one of the main energy sources for future generations, it is important to recognize the factors that will influence its success. This includes not only technical aspects like efficiency, storage, and grid integration but also the socioeconomic challenges that may arise. Understanding these dynamics is essential for shaping a solar-powered future that is both sustainable and inclusive.

In Chapter 5, we explored wind power's current role and future challenges. As a key solution in the fight against climate change, wind energy offers significant benefits through various technologies, including onshore, offshore, and vertical-axis turbines. We discussed the need to address concerns such as visual impact, bird mortality, and siting through responsible development and community engagement. The chapter also highlighted wind power's potential to support clean transportation and generate economic benefits like job creation, enhanced energy security, and reduced fuel costs.

Chapter 6 focused on water-based energy sources such as hydropower, geothermal, and ocean energy. These technologies tap into Earth's natural heat and water movement to produce electricity. While geographically dependent and not without environmental concerns, they offer strong potential for clean, reliable energy. With proper collaboration and planning, they can play a key role in reducing fossil fuel reliance and advancing sustainable development.

Chapter 7 explored emerging electrical energy sources, focusing on electric vehicles, hydrogen fuel, and biomass. These technologies represent key pathways for achieving net-zero carbon emissions by 2050. While batteries, hydrogen, and biomass offer promising alternatives for electricity generation, they also pose ethical and environmental challenges. The chapter emphasized that through responsible sourcing, sustainable practices, innovation, and collaboration, these energy sources can evolve into essential components of a clean and resilient energy future.

Chapter 8 dealt with clean air, providing an in-depth examination of the challenges associated with carbon sequestration, methane emissions, plastic pollution, and public perception. It examined the complexities of managing these environmental issues and emphasized the importance of public understanding and support in driving effective solutions. This chapter highlighted that addressing air quality and pollution requires a multifaceted approach combining technology, policy, and community engagement.

Chapter 9 focused on the interconnectedness of clean water, soil, and the path to net-zero emissions. It highlighted the need for responsible water stewardship across clean energy technologies like solar, geothermal, and hydrogen, each of which requires careful water management. Solar energy, despite its low water footprint, still relies on water for panel cleaning and material processing. Geothermal energy must use water for steam, and hydrogen production relies on water for electrolysis. This chapter emphasized that a sustainable future depends on innovations in water management, soil preservation, and responsible mineral extraction, requiring collaboration across sectors to balance clean energy development with environmental health.

In Chapter 10, we addressed increasing energy efficiency, particularly in the industrial sector, as a solution for providing more accessible energy to the public. However, energy efficiency alone will not ensure a sustainable future unless energy consumption is also reduced. This chapter acknowledged challenges such as cybersecurity, privacy concerns, and public acceptance that need to be addressed. Through collaboration and technological advancements, we can leverage the full potential of smart grids to create a more sustainable and resilient energy future.

Chapter 11 focused on the financial aspects of climate solutions, highlighting the growing green premium for decarbonized products over carbon-intensive alternatives. This premium signals benefits for companies adopting low-emission technologies and warns investors not yet considering sustainability in their strategies. However, the fossil fuel industry has lagged in emission reduction investments, with less than 5% of oil and gas upstream investment going into low-emission sources. Although investments in clean fuels are increasing, they remain insufficient. This chapter stressed the need to start planning and scaling decarbonization technologies now, as carbon removal projects take time, and the impact of government policies on clean energy remains uncertain.

Chapter 12 emphasized the role of individuals in creating a sustainable energy future. Key actions include choosing cleaner energy sources, supporting policies for renewable energy, reducing waste, using efficient appliances, and adopting sustainable transportation and consumption habits. This chapter highlighted that collective responsibility and innovation are crucial for building a cleaner, more sustainable future for generations to come.

Final thoughts

In the race toward more efficient use of energy, China has been taking huge steps recently, and to some extent, they are beating the competition. For example, the United States exported about $3 billion worth of batteries in 2024. China exported $65 billion worth of batteries. The United States exported about $69 million worth of solar panels. China exported about $40 billion worth of solar panels. When it comes to electric cars, the United States exported about $12 billion. Most of that was Tesla. China exported $38 billion. And that number is growing incredibly fast.

The World Economic Forum (WEF) has proposed the Energy Transition Index (ETI) as a means of assessing how 118 countries are positioned to navigate the evolving energy landscape.[1] It measures both system performance (security, equity, and sustainability outcomes) and transition readiness (enablers of progress, including regulations and political commitment, finance and investment, education and human capital, infrastructure, and innovation), resulting in

1. World Economic Forum. *Fostering Effective Energy Transition 2025: Digest.* **2025**, June 18. https://www.weforum.org/publications/fostering-effective-energy-transition-2025/digest/ (accessed Jan 20, 2026).

an ETI score. Overall, 65% of countries improved their ETI scores in 2025. Average sustainability scores also improved, highlighting lower energy and emissions intensities and clean energy's increased share of energy consumption. However, energy security saw little improvement, indicating persistent vulnerabilities in energy supply flexibility and diversity for many countries. Overall, only 28% of countries made simultaneous progress across security, equity, and sustainability in 2025, reflecting uneven progress.

A key assumption underpinning the anticipated pace of this transition appears to have fallen short of real-world dynamics.[2] Global energy consumption has surged by an astounding 70% since 1990, and the world's population is projected to expand by approximately two billion in the coming decades, with a significant portion of this growth concentrated in the Global South. Consequently, the demand for fossil fuels has not yet reached a plateau. Adding another layer of complexity, roughly 30 percent of the global population continues to rely on traditional biomass for cooking, while in Africa, nearly 600 million individuals still lack access to electricity. Affordability and reliability have taken the front seat in the energy transition. The financial burden associated with transitioning to renewable energy has proven to be more substantial than anticipated, placing a significant strain on developing nations that require considerable economic assistance to facilitate this shift. Moreover, a critical oversight in many initial assessments was the failure to fully appreciate that climate objectives do not exist in isolation but are intrinsically linked to other vital goals, such as GDP growth, economic development, energy security, and the reduction of local pollution. The current climate of heightened global tensions has further impeded progress toward a unified consensus on the way forward.

Adding to these complexities, electricity consumption has begun to outpace recent demand forecasts. This increasing electrification will necessitate a significant and reliable supply of critical minerals, such as lithium. Current projections suggest that the mining industry may struggle to meet this escalating demand. As astutely observed by Yergin et al., the transformation of the global energy system will not unfold in a uniform or linear fashion across different regions of the world. Instead, it will proceed at varying speeds, characterized by diverse combinations of fuels and technologies. Ultimately, the energy transition represents nothing less than a profound re-engineering of the entire global economy, a multifaceted endeavor demanding innovation, collaboration, and a steadfast commitment to a sustainable future.

Today, humanity consumes resources equivalent to 1.8 Earths, meaning that our ecological footprint, the demand placed on biologically productive areas for food, timber, fiber, carbon sequestration, and infrastructure, exceeds the planet's

---

2. Yergin, D., Orszag, P., Arya, A. *The Troubled Energy Transition. Foreign Affairs* **2025**, February 25. https://www.foreignaffairs.com/united-states/troubled-energy-transition-yergin-orszag-arya (accessed Jan 20, 2026).

regenerative capacity by about 80%. This growing strain on natural resources has severe consequences for biodiversity and ecosystems, reflected in a 73% decline in the average size of monitored wildlife populations worldwide between 1970 and 2020.[3] Sustaining economic growth and long-term prosperity will therefore require operating more effectively within the ecological limits that nature provides. While the path to decarbonization is still long, the message for all of us is clear: even if the future cannot be predicted, we should prepare for it by supporting actions that will build flexibility, affordability, conservation, transition, and sustainability in the energy sector.

3. World Economic Forum. *Nature Positive: Role of the Technology Sector*. **2025**, December 4. https://www.weforum.org/publications/nature-positive-role-of-the-technology-sector/ (accessed Jan 20, 2026).

Index

Page numbers followed by "*f*" and "*t*" indicate, figures and tables respectively.

M

N

O

P

R

S